Preface

These proceedings contain the texts of 37 contributions presented at the International Conference on Engineering Optimization in an Industrial Environment, which took place on 3 - 4 September 1990 at the Karlsruhe Nuclear Research Center, FR Germany.

The presentations consisted of oral and poster contributions arranged in five sessions:

- Shape and layout optimization
- Structural optimization with advanced materials
- Optimal designs with special structural and material behaviour
- Sensitivity analysis - Programme systems
- Optimization with stability constraints - Special problems

The editors wish to express their appreciation to all authors and invited speakers for their interesting contributions.

The proceedings cover a wide range of topics in structural optimization representing the present state of the art in the fields of research and in the industrial environment as well. The editors hope that this book will also contribute towards new ideas and concepts in a world of ever decreasing natural resources and ever increasing demands for lighter and yet stronger and safer technical components.

Finally, the editors wish to thank all colleagues who helped in the organisation of the conference, especially Mrs. E. Schröder and Dr. K. Bethge, as well as Mr. A. von Hagen and Mrs. E. Raufelder, Springer Publishing Company, Heidelberg for the good cooperation and help in the publication of these proceedings.

| H. Eschenauer | C. Mattheck | N. Olhoff |
| Siegen | Karlsruhe | Aalborg |

December 1990

Programme of the International Conference on Engineering Optimization in Design Processes

Sept. 3 - 4, 1990, Karlsruhe Nuclear Research Center

Sept. 3, 1990

9.00 - 9.10 h	Welcome Addresses
9.10 - 9.30 h	H. Eschenauer (FRG) Structural Optimization - a Need in Design Processes?
9.30 - 10.10 h	C. Mattheck, S. Burkhardt, D. Erb (FRG) Shape Optimization of Engineering Components by Adaptive Biological Growth

Session 1 **Chairpersons:**	**Shape and Layout Optimization** **H. Eschenauer, N. Olhoff**
10.30 - 10.50 h	M.P. Bendsøe (DK), H.C. Rodrigues (P) Integrated Topology and Boundary Shape Optimization of 2D-Solids
10.50 - 11.10 h	H.P. Mlejnek, U. Jehle, R. Schirrmacher (FRG) Some Approaches to Shape Finding in Optimal Structural Design
11.10 - 11.30 h	G. Krzesinski (PL) Shape Optimization and Identification of 2D Elastic Structures by the BEM
11.30 - 11.50 h	G. Rozvany, M. Zhou (FRG) Applications of the COC - Algorithm in Layout Optimization
11.50 - 12.10 h	M. Weck, W. Sprangers (FRG) Shape Optimization in Machine Tool Design
12.10 - 12.30 h	H. Baier, B. Specht (FRG) Shape Optimization and Design Element Selection for Problems with Constrained Static and Dynamic Response

Session 2 **Chairpersons:**	**Structural Optimization with Advanced Materials** **M. Bendsøe, H. Eschenauer**
14.00 - 14.20 h	P. Pedersen (DK) Optimal Design with Advanced Material
14.20 - 14.40 h	P.U. Post (FRG) Optimization of the Long-Term Behaviour of Composite Structures under Hygrothermal Loads
14.40 - 15.00 h	J. Bühlmeier (FRG) Design of Laminated Composites under Time Dependent Loads and Material Behaviour

15.00 - 15.20 h	M. Rovati, A. Taliercio (I) Optimal Orientation of the Symmetry Axes of Orthotropic 3D-Materials
15.20 - 15.40 h	Ch. Seeßelberg, G. Helwig, H. Baier (FRG) Strategies for Interactive Structural Optimization and (Composites) Material Selection

Session 3
Chairpersons:

Optimal Designs with Special Structural and Material Behaviour
P. Pedersen, G. Cheng

16.00 - 16.20 h	H. Eschenauer, Th. Vietor (FRG) Some Aspects on Structural Optimization of Ceramic Structures
16.20 - 16.40 h	J. Blachut (GB) Shape Optimization of FRP Dome Closures under Buckling Constraints
16.40 - 17.00 h	V.V. Kobelev (USSR) Numerical Method for Shape Optimization using BEM
17.00 - 17.20 h	L. Harzheim, C. Mattheck 3D-Shape Optimization: Different Ways Leading to an Optimized Design
17.20 - 17.40 h	D. Radaj, Sh. Zhang (FRG) Multiparameter Design Optimization in Respect of Stress Concentrations

Sept. 4, 1990

Session 4a
Chairpersons:

Sensitivity Analysis - Programme Systems
A. Osyczka, G. Rozvany

9.00 - 9.20 h	N. Olhoff (DK) On Elimination of Inaccuracy in Semi-Analytical Sensitivity Analysis
9.20 - 9.40 h	V. Braibant, P. Morelle, C. Fleury (B) Recent Advances in the SAMCEF Optimization System
9.40 - 10.00 h	G. Müller, P. Tiefenthaler (FRG) Design Optimization with the Commercial FEM Program ANSYS
10.30 - 12.30 h	Poster Session and Exhibition

Session 4b
Chairpersons:

Sensitivity Analysis - Programme Systems
H. Baier, A. Gajewski

14.00 - 14.20 h	G. Cheng, Y. Gu, W. Xicheng (China) Improvement of Semi-Analytical Sensitivity Analysis and MCADS
14.20 - 14.40 h	V.P. Malkov, V.V. Toropov (USSR) Simulation Approach to Structural Optimization using Design Sensitivity Analysis
14.40 - 15.00 h	S. Kibsgaard (DK) MODULEF - A FEM Tool for Optimization

15.00 - 15.20 h	A. Osyczka (PL) Computer Aided Multicriterion Optimization System in Use
15.20 - 15.40 h	W. Heylen, S. Lammens, P. Sas (B) Dynamic Structural Optimization by Means of the U.A. Model Updating Technique

Session 5 **Optimization with Stability Constraints -**

Special Problems

Chairpersons: **C. Mattheck, N. Olhoff**

16.00 - 16.20 h	B. Bochenek, M. Życzkowski (PL) Optimal I-Section of an Elastic Arch under Stability Constraints
16.20 - 16.40 h	H.H. Müller-Slany (FRG) Application of Optimization Procedures to Substructure Synthesis
16.40 - 17.00 h	A. Gajewski (PL) Multimodal Optimization of Uniformly Compressed Cylindrical Shells
17.00 - 17.20 h	B.W. Kooi (NL) The "Cut-and-Try" Method in the Design of the Bow
17.20 - 17.40 h	H. Huber-Betzer, C. Mattheck (FRG) Computer-Simulated Self-Optimization of Bony Structures
17.40 - 17.50 h	Closure Address

Poster presentations

A. Borkowski, S. Jozwiak, M. Danicka (PL)
Truss Optimization Using Knowledge Base

J. Holnicki-Szulc, J.T. Gierlinski (PL)
Optimal Redesign Process Simulated by Virtual Distortions

P. Knödel (FRG)
On Optimal Structures in Engineering

H. Moldenhauer, K. Bethge (FRG)
Application of the CAO-Method to Axisymmetrical Structures under
Non-Axisymmetrical Loading

T.J. Reiter, F.G. Rammerstorfer, H.J. Böhm (A)
Numerical Simulation of Internal and Surface Bone Remodeling

Y. Gu, G. Cheng (China)
Engineering Oriented Approaches of Structural Shape Optimization

List of Participants

Austria
A.K. Maschinek, Th. Reiter

Belgium
C. Fleury, S. Lammens

China
G. Cheng

Denmark
M.P. Bendsøe, S. Kibsgaard, N. Olhoff, P. Pedersen

Finland
J. Koski, R. Silvennoien

FR Germany
U. Adriany, J. Aktaa, R. Allrutz, F. Ansorge, H. Baier, C.U. Bauer, M. Beller, K. Bethge, J. Bühlmeier, S. Burkhardt, H. Bolz, M. Deligeorges, K. Depta, W. Eggert, D. Erb, T. Ertl, H. Eschenauer, M. Fanni, D. Freitag, H. Friedmann, E. Gaber, F. Geyer, D. Glötzel, B. Harter, L. Harzheim, H. Hempel, K. Hornberger, H. Huber-Betzer, O. Iancu, U. Jehle, R. Kahn, K. Kasper, H. Kim, D. Klein, P. Knödel, D. Koch, R. Kussmaul, M. Lawo, C. Mattheck, M. Miller, H.P. Mlejnek, H. Moldenhauer, J. Müller, H.H. Müller-Slany, H. Pfeifer, A. Pompetzki, U. Post, M. Prinz, D. Radaj, N. Rashwan, A. Rebetzky, G. Rozvany, J. Sauter, J. Schäfer, R. Schirrmacher, U. Schneider, G. Schuhmacher, Ch. Seeßelsberg, B. Specht, W. Sprangers, M. Teschner, Th. Thiemeier, P. Tiefenthaler, Th. Victor, U. Vorberg, F. Walther, H.Ch. Wille, B. Yuan, S. Zhang, M. Zhou, J.E. Zimmermann

Great Britain
J. Blachut

Italy
C. Cinquini, M. Rovati, A. Taliercio

The Netherlands
B.W. Kooi

Poland
B. Bochenek, A. Borkowski, A. Gajewski, J. Holnicki-Szulc, G. Krzesinski, A. Osyczka

Switzerland
P. Kim, H. Meier

USA
A.R. Diaz

USSR
V.V. Kobelev, V.V. Toropov

CONTENTS

H. Eschenauer (FRG)
Structural Optimization - a Need in Design Processes? 1

C. Mattheck, S. Burkhardt, D. Erb (FRG)
Shape Optimization of Engineering Components by Adaptive
Biological Growth 15

Shape and Layout Optimization

M.P. Bendsøe (DK), H.C. Rodrigues (P), J. Rasmussen (DK)
Topology and Boundary Shape Optimization as an Integrated
Tool for Computer-Aided Design 27

H.P. Mlejnek, U. Jehle, R. Schirrmacher (FRG)
Some Approaches to Shape Finding in Optimal Structural Design 35

G. Krzesinski (PL)
Shape Optimization and Identification of 2D-Elastic Structures
by the BEM 51

G. Rozvany, M. Zhou (FRG)
Applications of the COC - Algorithm in Layout Optimization 59

M. Weck, W. Sprangers (FRG)
Shape Optimization in Machine Tool Design 71

B. Specht, H. Baier, (FRG)
Shape Optimization and Design Element Selection for
Problems with Constrained Static and Dynamic Response 79

Structural Optimization with Advanced Materials

P. Pedersen (DK)
Optimal Design with Advanced Material 91

P.U. Post (FRG)
Optimization of the Long-Term Behaviour of Composite
Structures under Hygrothermal Loads 99

J. Bühlmeier (FRG)
Design of Laminated Composites under Time-Dependent
Loads and Material Behaviour 107

M. Rovati, A. Taliercio (I)
Optimal Orientation of the Symmetry Axes of Orthotropic 3D-Materials 127

Ch. Seeßelberg, G. Helwig, H. Baier (FRG)
Strategies for Interactive Structural Optimization and Composites
Material Selection 135

Optimal Designs with Special Structural and Material Behaviour

H. Eschenauer, Th. Vietor (FRG)
Some Aspects on Structural Optimization of Ceramic Structures 145

J. Blachut (GB)
Shape Optimization of FRP Dome Closures under Buckling Constraints 155

V.V. Kobelev (USSR)
Shape Optimization Using Boundary Elements 165

L. Harzheim, C. Mattheck (FRG)
3D-Shape Optimization: Different Ways Leading to an Optimized Design 173

D. Radaj, Sh. Zhang (FRG)
Multiparameter Design Optimization in Respect of Stress Concentrations 181

Sensitivity Analysis - Programm Systems

N. Olhoff, J. Rasmussen (DK)
Method of Error Elimination for a Class of Semi-Analytical
Sensitivity Analysis Problems 193

G. Müller, P. Tiefenthaler (FRG)
Design Optimization with a Commercial Finite Element Program 201

G. Cheng, Y. Gu, W. Xicheng (China)
Improvement of Semi-Analytical Sensitivity Analysis and MCADS 211

V.P. Malkov, V.V. Toropov (USSR)
Simulation Approach to Structural Optimization using Design
Sensitivity Analysis 225

S. Kibsgaard (DK)
MODULEF - An FEM Tool for Optimization 233

A. Osyczka (PL)
Computer-Aided Multicriterion Optimization System in Use 241

W. Heylen, S. Lammens, P. Sas (B)
Dynamic Structural Optimization by Means of the U.A. Model
Updating Technique 249

Optimization with Stability Constraints - Special Problems

B. Bochenek, M. Zyczkowski (PL)
Optimal I-Section of an Elastic Arch under Stability Constraints 259

H.H. Müller-Slany (FRG)
Application of Optimization Procedures to Substructure Synthesis 267

A. Gajewski (PL)
Multimodal Optimization of Uniformly Compressed Cylindrical Shells 275

B.W. Kooi (NL)
The "Cut-and-Try" Method in the Design of the Bow 283

X

H. Huber-Betzer, C. Mattheck (FRG)
Computer-Simulated Self-Optimization of Bony Structures 293

A. Borkowski, S. Jozwiak, M. Danicka (PL)
Truss Optimization Using Knowledge Base 301

J. Holnicki-Szulc, J.T. Gierlinski (PL)
Optimal Redesign Process Simulated by Virtual Distortions 309

P. Knödel (FRG)
On Optimal Structures in Engineering 317

H. Moldenhauer, K. Bethge (FRG)
Application of the CAO-Method to Axisymmetrical Structures under
Non-Axisymmetrical Loading 325

T.J. Reiter, F.G. Rammerstorfer, H.J. Böhm (A)
Numerical Simulation of Internal and Surface Bone Remodeling 333

Y. Gu, G. Cheng (China)
Engineering-Oriented Approaches of Structural Shape Optimization 341

Author Index 355

List of Contributors

Baier, H.
Dornier GmbH
Postfach 1420, 7990 Friedrichshafen, FR Germany

Bendsøe, M.P.
The Technical University of Denmark, Mathematical Institute, Bldg. 303
2800 Lyngby, Denmark

Bethge, K.
Kernforschungszentrum Karlsruhe, IMF IV
Postfach 3640, 7500 Karlsruhe, FR Germany

Blachut, J.
Liverpool University, Department of Mechanical Engineering
P.O. Box 147, Liverpool, L69 3BX, Great Britain

Bochenek, B.
Technical University of Cracow, Institute of Mechanics and Machine Design
Warszawska 24, 31-155 Cracow, Poland

Böhm, H.J.
TU Wien, Institut für Leicht- und Flugzeugbau
Gußhausstr. 25 - 29/317, 1040 Wien, Austria

Borkowski, A.
Polish Academy of Sciences, Institute of Fundamental Technological Research
Swietokrzyska 21, 00-049 Warsaw, Poland

Braibant, V.
Samtech S.A.
Bd. Frère Orban 25, 4000 Liège, Belgium

Bühlmeier, J.
Institute for Computer Applications, Universität Stuttgart
Pfaffenwaldring 27, 7000 Stuttgart 80, FR Germany

Burkhardt, S.
Kernforschungszentrum Karlsruhe, IMF IV
Postfach 3640, 7500 Karlsruhe, FR Germany

Cheng, G.
Dalian University of Technology, Research Institute of Engineering Mechanics
116024 Dalian, China

Danicka, M.
Polish Academy of Sciences, Adaptive Systems Laboratory, Institute of Fundamental
Technological Research
ut. Swietokrzyska 21, 00-049 Warsaw, Poland

Erb, D.
Kernforschungszentrum Karlsruhe, IMF IV
Postfach 3640, 7500 Karlsruhe, FR Germany

Eschenauer, H.
Universität-GH Siegen, Fachbereich 11, Inst. für Mechanik und Regelungstechnik
Paul-Bonatz-Str. 9 - 11, 5900 Siegen, FR Germany

Gajewski, A.
Technical University of Cracow, Institute of Physics
ul. Podchorazych 1, 30-084 Cracow, Poland

Gierlinski, J.T.
WS Atkins Engineering Sciences
Epsom, Great Britain

Gu, Y.
Dalian University of Technology, Research Institute of Engineering Mechanics
116024 Dalian, China

Harzheim, L.
Kernforschungszentrum Karlsruhe, IMF IV
Postfach 3640, 7500 Karlsruhe, FR Germany

Helwig, G.
Dornier GmbH
Postfach 1420, 7990 Friedrichshafen, FR Germany

Heylen, W.
Kath. Universiteit Leuven, Mechanische Konstruktie en Produktie
Celestijnenlaan 300 B, 3030 Heverlee, Belgium

Holnicki-Szulc, J.
Polish Academy of Sciences, Institute of Fundamental Technological Research
Swietokrzyska 21, 00-049 Warsaw, Poland

Huber-Betzer, H.
Kernforschungszentrum Karlsruhe, IMF IV
Postfach 3640, 7500 Karlsruhe, FR Germany

Jehle, U.
Institute for Computer Applications, Universität Stuttgart
Pfaffenwaldring 27, 7000 Stuttgart 80, FR Germany

Jozwiak, S.
Polish Academy of Sciences, Adaptive Systems Laboratory, Institute of Fundamental
Technological Research
Swietokrzyska 21, 00-049 Warsaw, Poland

Kibsgaard, S.
The University of Aalborg, Institute of Mechanical Engineering
Pontoppidanstraede 101, 9220 Aalborg East, Denmark

Knödel, P.
Universität Karlsruhe, Versuchsanstalt für Stahl, Holz und Steine
Kaiserstr. 12, 7500 Karlsruhe 1, FR Germany

Kobelev, V.V.
USSR Academy of Sciences, Institute for Problems in Mechanics
Vernadskogo, 101, Moscow, 117526, USSR

Kooi, B.W.
Free University, Department of Biology
P.O. Box 7161, 1007 MC Amsterdam, The Netherlands

Krzesinski, G.
Technical University of Warsaw, Institute of Aircraft Technology and Applied Mechanics
ul. Nowowiejska 22/24, 00-665 Warsaw, Poland

Lammens, S.
Kath. Universiteit Leuven, Mechanische Konstruktie en Produktie
Celestijnenlaan 300 B, 3030 Heverlee, Belgium

Malkov, V.P.
Gorky University, Department of Solid Mechanics
Gorky, 603600, USSR

Mattheck, C.
Kernforschungszentrum Karlsruhe, IMF IV
Postfach 3640, 7500 Karlsruhe, FR Germany

Mlejnek, H.P.
Institute for Computer Applications, Universität Stuttgart
Pfaffenwaldring 27, 7000 Stuttgart 80, FR Germany

Moldenhauer, H.
Im Brückengarten 9a, 6074 Rödermark, FR Germany

Morelle, P.
Samtech S.A.
Bd. Frère-Orban, 25, 4000 Liège, Belgium

Müller, G.
CAD-FEM GmbH
8017 Ebersberg/München, FR Germany

Müller-Slany, H.H.
Universität-GH Siegen Fachbereich 11, Inst. für Mechanik und Regelungstechnik
Paul-Bonatz-Str. 9 - 11, 5900 Siegen, FR Germany

Olhoff, N.
Aalborg University, Institute of Mechanical Engineering
9220 Aalborg, Denmark

Osyczka, A.
Technical University of Cracow, Institute of Machine Technology
Warszawska 24, 31-155 Cracow, Poland

Pedersen, P.
The Technical University of Denmark, Department of Solid Mechanics, Bldg. 404
2800 Lyngby, Denmark

Post, P.U.
Festo KG
Postfach, 7300 Esslingen, FR Germany

Radaj, D.
Daimler Benz AG, Forschung Grundlagen
Postfach 60 02 02, 7000 Stuttgart 60, FR Germany

Rammerstorfer, F.G.
TU Wien, Institut für Leicht- und Flugzeugbau
Gußhausstr. 25 - 29/317, 1040 Wien, Austria

Rasmussen, J.
Aalborg University, Institute of Mechanical Engineering
Pontoppidanstraede 101, 9220 Aalborg, Denmark

Reiter, Th.
TU Wien, Institut für Leicht- und Flugzeugbau
Gußhausstr. 25 - 29/317, 1040 Wien, Austria

Rodrigues, H.C.
Technical University of Lisbon, Mechanical Engineering Department, Instituto Superior Tecnico
1096 Lisbon Codex, Portugal

Rovati, M.
University of Trento, Department of Structural Mechanics and Automation in Design
Via Mesiano 77, 38050 Trento, Italy

Rozvany, G.
Universität-GH Essen, Fachbereich 10, Bauwesen
Postfach 10 37 64, 4300 Essen 1, FR Germany

Sas, P.
Kath. Universiteit Leuven, Mechanische Konstruktie en Produktie
Celestijnenlaan 300 B, 3030 Heverlee, Belgium

Schirrmacher, R.
Institute for Computer Applications, Universität Stuttgart
Pfaffenwaldring 27, 7000 Stuttgart 80, FR Germany

Seeßelberg, Ch.
Dornier GmbH
Postfach 1420, 7990 Friedrichshafen, FR Germany

Specht, B.
Dornier GmbH
Postfach 1420, 7990 Friedrichshafen, FR Germany

Sprangers, W.
RWTH Aachen, Institut für Werkzeugmaschinen und Betriebslehre
Steinbachstr. 53/54, 5100 Aachen, FR Germany

Taliercio, A.
Politecnico di Milano, Dept. of Structural Engineering
Piazza Leonardo da Vinci, 32, 20 133 Milano, Italy

Tiefenthaler, P.
CAD-FEM GmbH
8017 Ebersberg/München, FR Germany

Toropov, V.V.
Gorky University, Department of Solid Mechanics
Gorky, 603600, USSR

Vietor, Th.
Universität-GH Siegen, Fachbereich 11, Inst. für Mechanik und Regelungstechnik
Paul-Bonatz-Str. 9 - 11, 5900 Siegen, FR Germany

Weck, M.
RWTH Aachen, Lehrstuhl für Werkzeugmaschinen
Steinbachstr. 53/54, 5100 Aachen, FR Germany

Xicheng, W.
Dalian University of Technology, Research Institute of Engineering Mechanics
116024 Dalian, China

Zhang, Sh.
Daimler Benz AG, Forschung Grundlagen
Postfach 60 02 02, 7000 Stuttgart 60, FR Germany

Zhou, M.
Universität-GH Essen, Fachbereich 10, Bauwesen
Postfach 10 37 64, 4300 Essen 1, FR Germany

Zyczkowski, M.
Technical University of Cracow, Institute of Mechanics and Machine Design
Warszawska 24, 31-155 Cracow, Poland

Structural Optimization –
a Need in Design Processes?

H.A. Eschenauer
Research Laboratory for Applied Structural Optimization
at the
Institute of Mechanics and Control Engineering
University of Siegen
FRG

Abstract: After being able to determine the structural behaviour by means of finite methods, an important goal of engineering activities is to improve and to optimize technical designs, structural components and structural assemblies. The task of structural optimization is to support the engineer in searching for the best possible design alternatives of specific structures. The "best possible" or "optimal" structure is the structure which is highly corresponding to the designer's desired concept whilst at the same time meeting the functional, manufacturing and application, e.g. all multidisciplinary requirements. In comparison to the "trial and error"- method generally used in the engineering environment and based on intuitive heuristic approach, the determination of optimal solutions by applying mathematical optimization procedures is more reliable and efficient if correctly applied. These procedures will be a need in the design process and they are already increasingly entering the industrial practice.

1. INTRODUCTION

The recent rapid development of the so-called "new" or "High Technologies" (Computer Technology, Material Sciences, Robotics, etc.) suggests to pause and to contemplate the question whether the worldwide activities of scientists do correspond to what great scholars in the 16th and 17th century thought about already. On the GAMM-Conference 1990 in Hannover, an exhibition was devoted to the philosopher, mathematician, physicist, technician and author GOTTFRIED WILHELM LEIBNIZ (1646 - 1716) as one of the last universal scholars of Modern Times (Fig. 1 [1]). His merits especially in the fields of mathematics, natural and engineering sciences can be viewed as the foundation of multidisciplinary, analytical or, in other words, coherent thinking. We should remember that Leibniz built the first calculator with a stepped drum; achievements which nowadays make optimization computations possible on a larger scale. Apart from that, Leibniz gave impetus to many areas of engineering sciences. Of special importance in this respect are his connection of "theoria cum praxii" and his inventions like the drive units for wind and water art-works.

The principles of mechanics allow the formulation of classical problems of natural and engineering sciences by means of the calculus of variation. LEIBNIZ and L. EULER (1707-1783) established the required mathematical tool for finding the extreme values of given functions by introducing the infinitesimal calculus. Herewith, it is possible to carry out an integrated and modern treatment of energy principles in all fields of mechanics with application to dynamics of rigid bodies, general elasticity theory, analysis of supporting structures (frames, trusses, plates, shells), the theory of buckling, the theory of vibrations, etc. Some very interesting examples, among others the "curve of the shortest falling time" ("brachistochrone") and the isoperimetric problem, were investigated by JACOB BERNOULLI (1655-1705) and DANIEL BERNOULLI (1700-1782). A further task was solved by I. NEWTON

Fig. 1:
Gottfried Wilhelm
Leibniz (1646 - 1716)

(1643-1727), namely the determination of the smallest resistance of a body of revolution. With the principle of least action and the integral principle, J.L. LAGRANGE (1736-1813) and W.R. HAMILTON (1805-1865) contributed to the perfection of the calculus of variation. Useful approximation methods basing on variational principles of mechanics were devised by Lord Rayleigh (1842-1919), W. RITZ (1878-1909), B.G. GALERKIN (1871-1945) and others. In a first application on optimum structural design variational methods have been treated by J.L LAGRANGE, T. CLAUSEN and B. DE SAINT-VENANT. The investigations on finding the optimal design of one-dimensional structures under various loadings should be mentioned here. Typical examples are the bar subjected to buckling loads or the cantilever beam under a single load or dead weight, respectively, for which optimal cross-sectional shapes were found by means of the calculus of variation. For this purpose, optimality criteria are derived in terms of necessary conditions, e.g. Euler's equations in the case of unconstrained problems. If constraints are considered additionally, the Lagrangian multiplier method is employed; it correspond to the solution of an isoperimetric problem. During the last decades of this century it was especially WILLIAM PRAGER (1903-1980) who gained merit in the development of structural optimization [20].

It is especially we who, in our theory and application of optimization problems in technology, should try to come close to the way of thinking of such important scholars. In order to increasingly refute the opinion of so-called "practical men" that optimization is "the playground of mathematicians", it is important to re-model reality precisely with multidisciplinary models to gain sufficiently precise results from optimization calculations. This is especially essential for the field of structural optimization which begins to develop slowly from the phase of trial-and-error procedures and thereby begins to enter into the design process.

2. NOTES ON STRUCTURAL OPTIMIZATION AND DESIGN PROCESSES

As the term "structural optimization" is nowadays interpreted in many different ways, we should try to first of all precisely define the term "structure" or establish it from the point of view of structural mechanics. GORDON [2] defined "structure" as a material arrangement which serves to resist mechanical loads. The keyword "serves" shall be used in the sense of serviceable. All kinds of constructions, buildings, components and also skeletons of life-forms and plants are understood by that. This division leads to a distinction between "artificial" and "living" structures. As the term "material arrangement" says, a distinction between materials on the one hand and structures on the other hand will not be possible any more in future.

The information exchange between special branches is beginning to improve. Terms like biomechanics, mechatronics, structural optimization prove this. A further example is given by the aircraft and space technology. Here, weight used to be and still is a "luxury", and material failure can have disastrous or even fatal consequences. In this field, special attention has for that reason always been paid to the investigation into materials **and** structures. Because of the increased demands, this nowadays also applies to other branches as for example the car industry. In order to develop optimal solutions here, it is simply necessary to make efficient optimization procedures more and more accessible in future. For the initial design of new developments, nature is taken as an example.

Nowadays, the obligation of multidisciplinary cooperation regards all branches of engineering sciences. The mode of cooperation for structural optimization tasks can be seen in a VENN-Diagramme in Fig. 2. It shows that the structural optimization forms a joint quantity from the engineering disciplines, structural mechanics and the mathematical optimization algorithm. The rapid development of the computer technology and, connected to that, the efficiency increase of algorithms enables us to fast and exactly find the "optimal design" of components and constructions. Furthermore, the VENN-Diagramme gives the basic idea, too, of

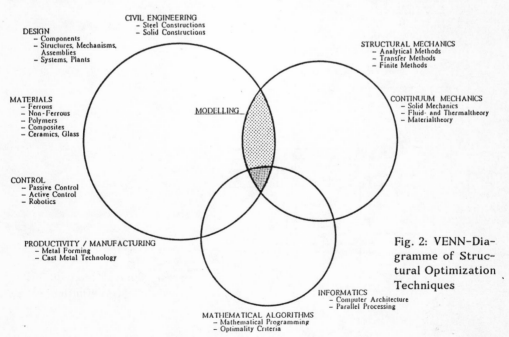

Fig. 2: VENN-Diagramme of Structural Optimization Techniques

the fundamental structure of an optimization procedure, i.e. the so-called "Three-Columns-Concept". consisting of a structural model, optimization algorithms and an optimization model [3, 4].

Depending on the complexity, the development of components and systems for mass-production is carried out in several phases:

1) Feasibility study, } Concept phase
2) Project definition phase, } Concept phase
3) Development phase
   - main system
   - subsystem,
4) Design phase
   - assemblies
   - components,
5) Checking and testing of a prototype,
6) Production, manufacturing,
7) Erection phase,
8) Acceptance phase.

Each phase is a control process under a systems engineering point of view at the end of which decision criteria have to be established for the successive phase. Fig. 3 shows such a control loop for the project definition phase.

The most important steps are:

1) Definition of the problem formulation, preliminary specification,
2) Preparation and design of concept variants,
3) Search for solution principles and calculation methods,
4) Technical-economical evaluations,
5) Formulation of the final specification-list,
6) Optimization of the construction and of the components.

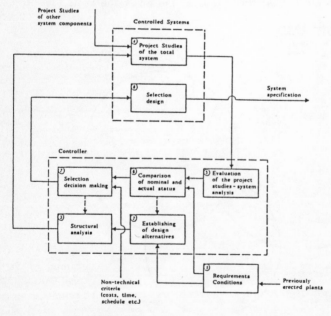

Fig. 3:
Control loop of a
project definition phase

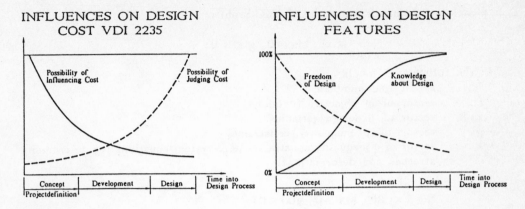

Fig. 4: Influences on design in the single phases of a design process [5,6]

The process within the systems engineering control loop can be made considerably more efficient by inserting optimization procedures between the single stages. An optimization cycle could be inserted between the structural analysis (box 3) and the decision making (box 7) by means of which certain relevant objectives for components could be laid-out optimally.

The utility of introducing specific optimization strategies into the design process is represented in Fig. 4. Fig. 4a shows the freedom of design and the knowledge about design depending on the different phases. The two curves with their quantitative tendencies demonstrate how the knowledge about design [5] increases in the course of the design process while the freedom of design decreases. Almost the same characteristics for the possibility of influencing the costs and the specific estimation of costs in the different phases are given in Fig. 4b [6]. Here, the present problem for the design engineer to make a decision for a special design version is demonstrated as it is impossible to give any complete statements about the stresses or the final shape of a component or the structure.

This proves the necessity to use optimization strategies as early as possible in the design process. The obligation of economic production and the tendency to reduce weight, save energy, increase accuracy and reliability which lead to modern and modified constructions do suggest at the same time the use of advanced materials.

3. OPTIMIZATION MODELLING - SENSITIVITY ANALYSIS

When dealing with an optimization problem, it is recommendable to proceed according to the already mentioned "Three-Colums-Concept" [3, 4]. The first step is the theoretical formulation of the optimization problem, taking all of the relevant requirements into account. In many cases the structural optimization task can be considered as a Multicriteria-Optimization-Problem (MC-Problem), i.e. a design variable vector $\mathbf{x}$ has to be found which makes the m components of the objective function vector $\mathbf{f}$ as small as possible while fulfilling all constraints. Thus, MC-Problems can mathematically be defined by the following model formulations.

a) **Model 1**: Continuous, deterministic MC-Problem

$$\text{"Min"} \{ f(x) : h(x) = 0 , g(x) \le 0 \}$$
$$x \in \mathbb{R}^n$$

with the following symbols

$\mathbb{R}^n$	set of real numbers,
f	vector of m objective functions,
$x \in \mathbb{R}^n$	vector of n design variables,
g	vector of p inequality constraints ,
h	vector of q inequality constraints (e.g. system equations for determining stresses and deformations),

and

$$X := \{ x \in \mathbb{R}^n : h(x) = 0 , g(x) \le 0 \}$$

"feasible domain" where $\le$ is to be interpreted for each single component.

b) **Model 2**: Discrete, deterministic MC-Problem

$$\text{"Min"} \{ f(x) \}$$
$$x \in X_d$$

with the discrete design space

$$X_d := \{ x \in \mathbb{R} \mid x_j \in X_j ; \; j = 1, \ldots, N ; \; g(x) \le 0 , h(x) = 0 \}$$

and the N sets of discrete values

$$X_j = \{ x_j^{(1)} , x_j^{(2)} , x_j^{(3)} , \ldots , x_j^{(nj)} \}, \; X_j \subset \mathbb{R} \; \forall \; j = 1, \ldots, N$$

with n_j number of discrete values of the j-th design variable.

c) **Model 3**: Stochastic MC-Problem

$$\text{"Min"} \{ f(y) \mid P[f(y)] = r^f . \; P[g(y) \le 0] = r^g \}$$
$$y \in Y$$

with

y	vector of N random variables (loads, dimensions, characteristic values of the material) including design variables,
Y	vector of the expected values of the N random variables,
$P[\;]$	probability,
r^f, r^g	vectors of the m or q reliabilities concerning the objective and inequality constraints.

These three models represent the corresponding strategies which are adapted to given cases of application.

Fig. 5 shows the structure of an optimization loop consisting of the columns "Optimization Algorithms", "Structural Model", "Optimization model". The optimi-

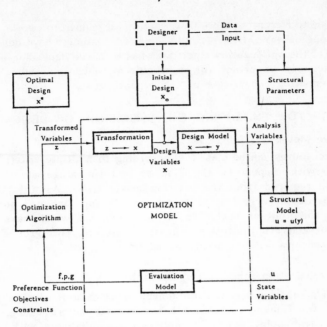

Fig. 5: The structure of an optimization loop

zation procedure consists in selecting and combining an appropriate optimization algorithm with the structural model and the optimization model. An essential requirement of structural optimization is the formulation of the structural behaviour in mathematical terms (Structural Modelling). In the case of mechanical systems this refers to the typical structural response of static and dynamic loads, such as deformations, stresses, eigenvalues, buckling loads etc. Here, all of the state variables required for formulating the objective function and constraints are provided. The structural calculation is carried out using efficient analysis procedures such as the finite element method or transfer matrices procedures. In order to ensure the largest possible field of application, it should be possible to use several structural analysis methods.

From an engineer's point of view, the optimization modelling is the relevant column of the optimization procedure. First of all, the quantities which are to be changed during the optimization process, i.e. the analysis variables, are selected from the structural parameters. The design model (variable linking, variable fixing, approach functions etc.) provides a mathematical link between analysis variables and the design variables. In order to increase efficiency and to improve the convergence of the optimization computation, the optimization problem is adapted to meet the special requirements of the optimization algorithm by transforming the design variables into transformation variables. When formulating the optimization model the engineer must take the demands from the fields of design, production, assembly and operation into account.
The use of numerous optimization algorithms, gradients of objective functions and constraints must be determined by a sensitivity analysis with regard to the design variables. These gradients also provide the design engineer with information about the sensitivity of the structure [5].

In recent years different authors have published numerous works on sensitivity methods (among others [7-11]). Thereby, numerous studies have demonstrated that the selection of the optimization algorithms has to ensue depending on the problem. This is particularly important for a reliable optimization and a high level of efficiency (computing times, rate of convergence etc.).

4. CONCEPTS FOR ENHANCING THE EFFICIENCY OF DESIGN PROCESSES

4.1 Interactive Methods [12-14]

Interactive procedures can be classified according to multiple criteria. A distinction can be made with respect to the kind of and the stage at which preference information is required from the decision maker (DM). As a further distinctive feature the particular organization scheme of an approach can be considered as it prescribes the kind of scalar substitute problems which have to be solved during the interactive optimization process. Therefore, two groups of interactive organization schemes will be distinguished here:

a) Superior organization schemes (Fig. 6/1 a, b)

Methods with a superior organization scheme use scalar substitute problems with the preference functions described in [3]. The free parameters leading to a particular functional-efficient point can be established in a dialogue with the DM. Each substitute problem can be solved by means of any problem-adjusted mathematical programming procedure and thus leads to a functional-efficient point so that functional-efficient alternative solutions are offered to the DM.

b) Extended organization schemes (Fig. 6/1 a, c)

Extended organization schemes are based on the extension of particular mathematical programming algorithms so that the preference function can be changed according to the preferences of the DM before the scalar substitute problem is finally solved. This procedure provides optimization steps within the criterion space in the direction of the functional-efficient boundary whereby the DM influences the step direction. The method of the mathematical programming cannot be adjusted according to the present optimization problem.

For the real design process interactive procedures have been mentioned or applied in only a few cases [3, 14]. This, first of all, results from the nonlinearity of the problems, and furthermore, it is due to the fact that the structural analyses such as finite element methods must be carried out numerically which is a time-consuming process. In particular, such problems are regarded by DIAZ who presents an effective sensitivity analysis to variations in the DM's preferences based on sequential quadratic programming [13].

4.2 Integrated Topology and Shape Optimization

In [15-17], a new concept is introduced which incorporates the determination of the general design or topology of a component into the optimization process. Thus, methods of formal structural optimization are already used in the project stage. A methodology, consisting of three phases, which uses the so-called homogenization method as a tool for finding the topology is described here. The methodology couples the homogenization method with a versatile structural optimization procedure via image processing method and geometrical modelling techniques.

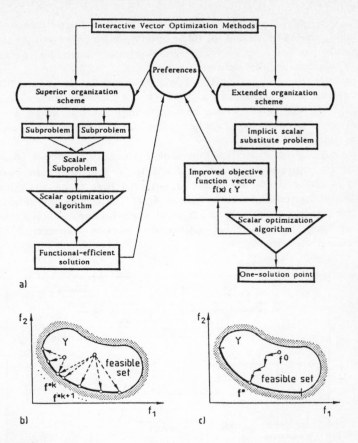

a)

b)   c)

Fig. 6 : Organization schemes of interactive vector optimization methods [3,14]
 a) Optimization schemes of both structures
 b) Iteration of superior organization schemes in the criterion space
 c) Iteration path of the extended organization schemes in the criterion
 space

The integration of a topology generating step into the optimization process follows the following phases:

Phase I: Generating information about the optimal topology for a structure,

Phase II: Processing and interpreting the information about the topology,

Phase III: Construction of a detailed structural model and optimization model based upon the above-mentioned topology and optimization of the design with conventional structural optimization techniques

Phase I is carried out by means of the homogenization method by BENDSØE and KIKUCHI, which has proved to be very promising for finding the topology. This method requires the definition of a design space (i.e. a space in which the designer wants the material to be optimally distributed) and also of material data, loads and boundary conditions.

10

Because of the very accurate discretization of the design space into variable standard cells, the optimization problem which is to be solved by means of the homogenization method can increase immensely in dimension. Several thousand design variables are usually defined.

In order to keep the computing times within a reasonable range, the problem is solved by means of a specialized optimality criteria procedure. The optimization model should for the same reason be kept simple at this stage, i.e. the number of different parts of objective functions and the constraint types should be kept as small as possible.

Detailed, locally defined specifications which are only expected to have a minor influence on the optimal topology need not to be considered in Phase I but can be included in Phase II and III. What is obtained here is an analytically deduced, approximate optimal configuration, a first draft for the design problem based on fulfilling the primary structural mechanical specifications. This model forms the basis for an integrated structural optimization system. The run of the system is roughly demonstrated in Fig. 7.

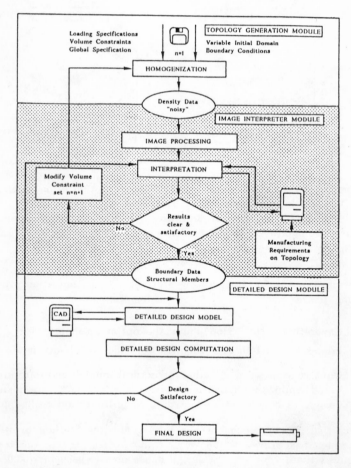

Fig. 7: Principle flowchart of the Three-Columns-Concept [15-17]

4.3 Decomposition Strategy for Treating Large Scale Systems

It is possible to subdivide complex structural optimization problems (Large Scale Systems) into small and clear parts by means of decomposition strategies. Details about this are to be found in [18-19].

The general optimization problem is described as follows:

$$\text{Min } \{f(x, u(x)) \mid h(u(x)) = 0 \ , \ g(u(x)) \leq 0 \ , \ g_G(x, u(x)) \leq 0, x_L \leq x \leq x_U \} , \qquad (1)$$

where f is an objective function, x a design variable vector, h state equations, $u(x)$ state variable vector, g local constraints, g_G global constraints, x_L, x_U side constraints (bounds). The optimization carried out by means of a decomposition technique refers to the following steps:

Step 1: Establishing a main system model,
Step 2: Establishing subsystem models,
Step 3: Determination of coupling information,
Step 4: Optimization of substructures,
Step 5: Iteration loop.

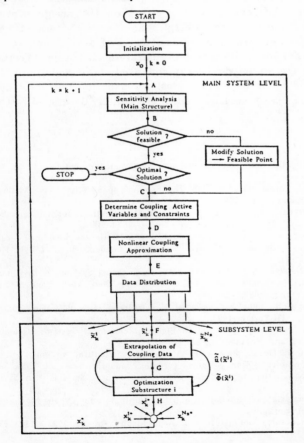

Fig. 8:
Flowchart of decomposition steps [18,19]

Fig. 8 gives a survey of the complete decomposition strategy and is divided into main system modules (approximation and control) and substructure modules (optimization of the substructures). The decomposition strategy starts with an initialization phase providing all data for structural and optimization models of main structures and substructures including the starting point x_0 of the design variables as well as further data for defining the interfaces of the subsystem. The data exchanges are carried out with the help of a data-management system.

CONCLUSION

The necessity of introducing optimization techniques into the design process is beyond question because of the following reasons:

1) Increasing the quality and quantity of products and plants and at the same time reducing costs and thereby being competitive.

2) Fulfilling the permanently increasing specification demands as well as considering reliability and safety, observing severe pollution regulations and saving energy and raw materials.

3) Introducing inevitable rationalization measures in development and design offices (CAD, CAE) in order to save more time for the staff to work creatively.

Starting with a definition of a structure, the design process and the structural optimization it is shown which influences have an effect on the design process, and that decisions in form of optimization strategies are necessary. An efficient sensitivity analysis of the different structural parameters is of special importance. The sensitivity analysis is a tool for a number of optimization algorithms. Furthermore, interactive procedures are to be developed and used more and more. The topology optimization for finding optimal initial designs, and the multilevel optimization for treating Large-Scale-Systems have to be preferentially developed further.

References:

[1]   Stein, E.: Gottfried Wilhelm Leibniz. Hannover Schlütersche Verlagsanstalt und Druckerei. 1990. 151 p.

[2]   Gordon, J.E. : Strukturen unter Stress (Originaltitel: The Science of Structure and Material). Heidelberg, Spektrum der Wissenschaft. 1989. 205p.

[3]   Eschenauer, H.A.; Koski, J.; Osyczka, A.: Multicriteria Design Optimization. Berlin, Heidelberg, New York, London, Paris, Tokyo, Hong Kong. Springer-Verlag. 1990. 481 p.

[4]   Eschenauer, H.A.: The "Three Columns" for Treating Problems in Optimum Structural Design. In: H.W. Bergmann: Optimization. Berlin, Heidelberg, New York, London, Paris, Tokyo, Hong Kong. Springer-Verlag, 1989, 1 - 21.

[5]   Sobieszcanski-Sobieski, I.: Multidisciplinary Optimization for Engineering Systems. Achievements and Potential. In: H.W. Bergmann: Optimization. Berlin, Heidelberg, New York, London, Paris, Tokyo, Hong Kong. Springer-Verlag, 1989, 42 - 62.

[6]   VDI-Richtlinie 2235: Wirtschaftliche Entscheidung beim Konstruieren (Economic Decisions in Design Processes). Düsseldorf. VDI-Verlag, 1987.

[7]   Haug, E.J.; Arora, J.S.: Design Sensitivity Analysis of Elastic Mechanical Systems. Computer Methods in Applied Mechanics and Engineering 15. (1978) 35-62.

[8]   Haug, E.I.; Choi, K.K.; Komkov V.: Design Sensitivity Analysis of Structural Systems. Orlando, San Diego, New York, London, Montreal, Sydney, Tokyo, Toronto: Academic Press, Inc. 1986, 381 p.

[9]   Adelman, H.M.; Haftka, R.T.: Sensitivity Analysis for Discrete Structural Systems - A Survey. NASA TM 85333, 1984.

[10] Pedersen, P.: On Sensitivity Analysis and Optimal Design for Laminates. In: Green: Mechanical Behaviour of Composites and Laminates.

[11] Olhoff, N.; Taylor, J.E.: On Structural Optimization. J. Appl. Mech., Vol. 50, 1983, 1139-1151.

[12] Arora, I.S.: Interactive Design Optimization of Structural Systems. In: Eschenauer, H.; Thierauf, G.: Discretization Methods and Structural Optimization - Procedures and Applications. Springer-Verlag 1989, 10-16.

[13] Diaz, A.: Interactive Solution to Multiobjective Optimization Problems. Inst. f. Num. Methods in Eng. 24, (1987), 1865-1877.

[14] Schäfer, E.: Interaktive Strategien zur Bauteiloptimierung bei mehrfacher Zielsetzung und Diskretheitsforderungen. Dissertation, Universität-GH Siegen, 1990.

[15] Bremicker, M.: Ein Konzept zur integrierten Topologie- und Gestaltsoptimierung von Bauteilen. In: Müller-Slany, H.H. (ed).: Beiträge zur Maschinentechnik, Festschrift Prof. Dr. H. Eschenauer, 1990, 13-39.

[16] Bendsøe, M.; Kikuchi, N.: Generating Optimal Topologies in Structural Design Using a Homogenization Method. Computer Methods in Applied Mechanics and Engineering (1988), 197-224.

[17] Papalambros, P.Y.; Chreshdust, M.: An Integrated Environment for Structural Configuration Design. J. Eng. Design, vol. 1, No 1 (1990), 73-96.

[18] Bremicker, M.: Dekompositionsstrategien in Anwendung auf Probleme der Strukturoptimierung. Dissertation, Universität-GH Siegen 1989. Düsseldorf: VDI-Verlag 1989, 170 p.

[19] Bremicker, M.; Eschenauer, H.: Application of a Decomposition Technique for Treating a Shape Optimization Problem. In: Ravani, B. (ed.): Advances in Design Automation - 1989, Vol. 2 Design Optimization. New York, ASME, 1989, 1 - 6.

[20] Rozvany, G.I.N.: Structural Design via Optimality Criteria. Dordrecht, Boston, London, Kluwer Academic Publishers, 1989, 463 p.

SHAPE OPTIMIZATION OF ENGINEERING COMPONENTS
BY ADAPTIVE BIOLOGICAL GROWTH

C. Mattheck, S. Burkhardt, D. Erb
Nuclear Research Center Karlsruhe GmbH
Institute for Material and Solid State Research IV
Postfach 3640
7500 Karlsruhe, West Germany

Abstract

Biological load carriers always grow into a shape whereby a constant stress can be found everywhere along the surface of the biological component for the most significant natural loading applied. This avoids local stress peaks and therefore pre-defined failure points in the design. This mechanism of adaptive growth is copied by the so called CAO-method (Computer Aided Optimization). The method is briefly described and the shape optimization of a tree fork illustrates the adaptive growth. Furthermore a rubber bearing, a bending bar with rectangular window as well as a joint of metal sheets are shape-optimized as engineering examples. In cases where the design proposal which the CAO-method starts from cannot be guessed easily, an oversized rough proposal can be analysed by FEM. After cutting off unloaded parts, the remaining structure can then be used as a starting design for CAO.

1. Introduction

According to the authors knowledge, the first publication concerning with the constant stress state at the surface of biological load carriers was written by Metzger [1] in 1893. He had found a relation of $h \simeq D^3$ between height h and diameter D of spruce tree stems. The same relation is valid for a cantilever beam loaded by a lateral end load if a constant bending stress along its length is required. The lateral load is of course the effective wind load acting at a point within the crown of the spruce tree. In further studies [2,3,4] Mattheck et al. have shown that this constant stress hypothesis holds true also for all other parts of the tree such as root-stem joints, branch-stem joints etc.. There are indeed no notch stresses at these joints although the force flow has to be redirected from the branch into the tree stem and from the stem into the roots, respectively. The ingenious tree obviously knows how to design its unavoidable notches in order to avoid any notch stress. Even if the tree is injured (notched) mechanically, it will hurry to reduce the related notch stresses by adaptive growth, restoring the constant stress state. The principle of adaptive growth is as easy as nature itself: More material will be attached at overloaded places where the stresses are higher than a reasonably defined surface stress. Either no material will be attached (as in trees) or can even be removed (as in bones) at underloaded places characterized by stresses lower than the reference stress. Exactly this mechanism is computer-simulated by the CAO-method [5,6].

In a previous paper [7] the method is described in an earlier state where volumetric swelling is used to simulate growth. In the next chapter it will be shown how the CAO-method works at present state, e.g. by stress-controlled thermal expansion simulating adaptive growth. In this way not only biological growth may be simulated, but also engineering components may be optimized starting from a reasonable design proposal which may grow into an optimized design with a homogeneous stress state and thus avoiding any notch stresses.

2. CAO: Computer Aided Optimization

The CAO-method in the present state consists of the following steps:

1. A reasonable design proposal is made which growth can start from.

2. A FEM-mesh has to be generated. The outermost layer of the Finite Elements has to be of equal thickness at least at these parts of the surface where growth will later be allowed.

3. Now an elastic FEM-run is performed with the load and boundary conditions expected in later service.

4. The Mises equivalent stresses of step 3 are now set formally equal to a fictitious temperature field.

5. This temperature field is the only loading in the next FEM-run where only the surface layer of Finite Elements has a non-zero heat expansion coefficient. Furthermore the elastic modulus E of this thin surface layer is reduced $E \rightarrow E/400$.

6. The thermal displacements calculated in step 5 are now scaled up and added to the nodal point coordinates of the original structure. This is now already an improved design. After mesh correction and resetting the elastic modulus equal to its former value, the procedure starts again at step 3.

7. The procedure will be stopped if the Mises stress is completely constant along the surface of the component or if further growth is restricted by design limitations.

The procedure is sketched schematically in Fig.1.
If biological growth is simulated in a tree, bone etc. another result can be gained by CAO. One will only get agreement with the biological design in reality if the loading and boundary conditions assumed for CAO-application agree well with those from real nature. In this way the natural loading may be determined by comparison of the natural growth product with the shape of its CAO-simulation. This is of extreme importance, because in many cases the load situation in plants and animals is unknown and subject to intense research and heated discussion.

3. Application of CAO

The method described in the previous section will now be demonstrated by practical examples.

3.1 Tree forks

At first view a tree fork looks very dangerous and pre-defined to failure. Especially if one imagines the loading case of bending two individual stems away from each other. Indeed this loading case would be risky if the inner contour line of the fork would be a semi-circle as the non-optimized design proposal in Fig.2 shows. However, adaptive growth simulated by CAO will completely reduce the notch stresses in the tree fork. The optimized design fits exactly the contour of a really grown fork of the walnut tree shown. It is interesting that trees of completely different outer contour may have the same inner contour of the fork which therefore seems to be most important for the degree of shape optimization.
The example of the tree fork illustrates that biological notches do not cause notch stresses, because they are adapted to the force flow.
In the following, examples from engineering will be considered.

3.2 A bending bar with rectangular window

The non-optimized design in Fig.3 shows high localized notch stresses. It is one of the advantages of the CAO-method that it works for 3D-problems just as easy as for 2D-cases. The only extra effort is the generation of a 3D-mesh. However, this has to be done in any FEM-analysis, too. The optimization leads straight forward to the optimized design which is free of any notch stress. Fatigue tests have proved a more than 40 times longer fatigue life of the optimized structure without breakage or even visible crack initiation compared to the non-optimized design.
The next examples also of 3D-quality, however are more difficult because of some restrictions.

3.3 Welding of sheet metals

Two components of sheet metal are welded together. The problem was to optimize the contour of the welding in order to avoid localized stress peaks. As a further restriction, the thickness of each individual metal had to be constant. Therefore one side of the metal was allowed to 'grow' whilst the other side has been moved the same local amount in the same direction during the step of mesh correction. Beyond this, no problem-specific modifications have been necessary. Although design changes are limited here to changes of the weld curvature, a stress reduction of 14% was reached (Fig.4). Up to now the design proposal could easily be guessed. This will be frequently the case if a trouble component always breaking in service is the subject of the optimization. In this case just the trouble component itself may be used as a starting design and the CAO-method will do the rest.
However, under complex loading and boundary conditions the design proposal cannot be guessed. For these probably rare cases the following sample procedure has been used successfully.

4. The KILL OPTION - a straight way to pre-optimize design proposals

The procedure consists of only three steps:

1. A rough design proposal of simplest geometrical shape (cylinder, brick, sphere etc.) within the design limitations is generated as a FEM-mesh.

2. A FEM run is performed with the load and boundary conditions expected in later service. This will give the distribution of Mises equivalent stress within the design.

3. Now the non-load bearing parts of the structure having small Mises stresses are 'killed', e.g. cut off along the threshold-isoline of the Mises stress.

The remaining design will be used now as a pre-optimized design proposal for a following CAO-application.

Because it may cause trouble to generate a mesh bordered by isolines of the Mises stresses, a sophisticated FEM-mesh generator 'POLYPHEM' [8] is used in the Nuclear Research Center which allows to take a picture of the cut-off contour line by a video camera followed by automatic generation of a FEM-mesh within this contour by POLYPHEM. Furthermore this comfortable program takes care that the border of the mesh has the equal thickness layer of finite elements which is inadmissable for CAO-use.

5. Conclusions

The major results of the paper may be summarized in the following statements.

- CAO is a powerful method for the simulation of biological growth as well as for engineering shape optimization leading to light weight and fatigue resistant components.

- The KILL OPTION is an easy and straight forward method providing the user with pre-optimized design proposals for further optimization with CAO.

- POLYPHEM helps to generate easily FEM-meshes even of structures with complex boundaries as it may result from the use of the KILL OPTION.

- The combination of KILL OPTION and CAO is a complete lay-out theory.

- CAO works well for 2D- and 3D-problems and the user needs only a FEM-program and nothing else.

- Because no problem-specific modifications beyond the generation of the individual FEM-mesh is necessary, the CAO-method is a very effective tool especially with practical problems in industrial enviroments.

6. References

[1] K. Metzger
Der Wind als maßgeblicher Faktor für das Wachstum der Bäume
Mündener Forstliche Hefte, Springerverlag 1893 (in German)

[2] C. Mattheck
Why they grow how they grow - the mechanics of trees
Arboricultural Journal 14 (1990) 1-17

[3] C. Mattheck, G.Korseska
Woundhealing in a plane (Platanus Acerifolia (Ait.) Willd.)
an experimental proof of its mechanical stimulation
Arboricultural Journal 13 (1989) 211-218

[4] C. Mattheck, H. Huber-Betzer, K.Keilen
Die Kerbspannungen am Astloch als Stimulanz der Wundheilung bei Bäumen
Allg. Forst- und Jagdzeitung 161 (1990) 47-53 (in German)

[5] C. Mattheck
Engineering components grow like trees
Materialwissenschaft und Werkstofftechnik 21 (1990) 143-168

[6] C. Mattheck, S. Burkhardt
A new method of structural shape optimization based on biological growth
Int. J. of Fatigue 12 No 3 (1990) 185-190

[7] C. Mattheck, H. Moldenhauer
An intelligent CAD-method based on biological growth
Fatigue Fract. Engng. Mater. Struct. 13 (1990) 41-51

[8] POLYPHEM
Information brochure by Science&Computing GmbH at the Institute for
Theoretical Astrophysics, University of Tübingen, FRG

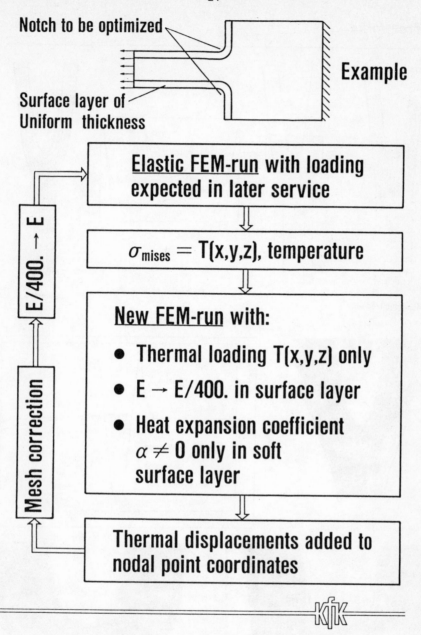

Fig.1: Procedure of CAO

Tree forks

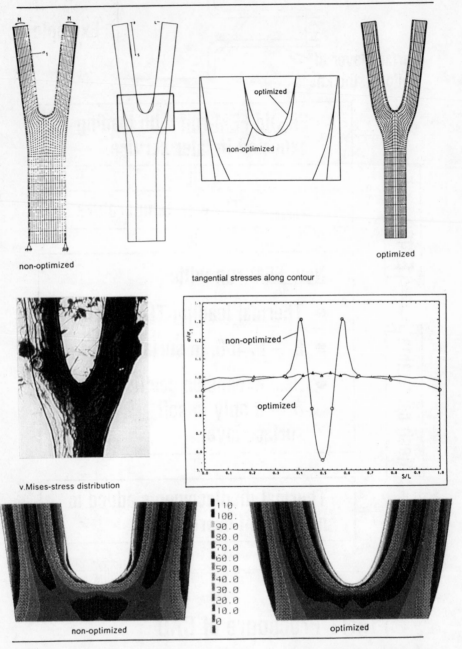

non-optimized

optimized

tangential stresses along contour

v.Mises-stress distribution

non-optimized

optimized

FEM: Uwe Vorberg

Fig.2: Tree fork under bending load

SHAPE AND LAYOUT
OPTIMIZATION

TOPOLOGY AND BOUNDARY SHAPE OPTIMIZATION
AS AN INTEGRATED TOOL FOR COMPUTER AIDED DESIGN

Martin Philip Bendsøe
Mathematical Institute
The Technical University of Denmark
Building 303
DK–2800 Lyngby, Denmark

John Rasmussen
Institute of Mechanical Engineering
Aalborg University
Pontoppidanstræde 101
DK–9220 Aalborg Ø, Denmark

Helder Carrico Rodrigues
Mechanical Engineering Department
Technical University of Lisbon
Instituto Superior Tecnico
1096 Lisbon Codex, Portugal

Abstract: The optimal topology of a two dimensional linear elastic body can be computed by regarding the body as a domain of the plane with a high density of material. Such an optimal topology can then be used as the basis for a shape optimization method that computes the optimal form of the boundary curves of the body. This results in an efficient and reliable design tool, which can be implemented via common FEM mesh generator and CAD input–output facilities.

1. Introduction

Traditionally, in shape design of mechanical bodies, a shape is defined by the orientated boundary curves of the body and in shape optimization the optimal form of these boundary curves is computed. This approach is very well established (cf. review paper by Haftka, [1]) and commercial software using this method is available. The boundary variations methods predicts the optimal form of boundaries of a fixed, a priori chosen topology. However, it is well known that the topology is a very important element of the final performance of a mechanical body. As an alternative to the boundary parametrization of shape, a mechanical body can be considered as a domain in space with a high density of material, that is, the body is described by a global density function that assigns material to points that are part of the body. By introducing composites with microvoids, such shape design problems appear as sizing problems for fixed reference domains, and a prediction of topology and boundary shape is possible ([2]–[6]).

$$\text{Volume} = \int_\Omega (\mu + \gamma - \mu\gamma)d\Omega \tag{5}$$

Alternative microstructures such as square or rectangular holes in square cells can also be used, the important feature being the possibility of having density values covering the full interval [0, 1].

The optimization problem can now be solved either by optimality criteria methods ([3]) or by duality methods, where advantages is taken of the fact that the problem has just one constraint. The angle θ of layer rotation is controlled via the results on optimal rotation of orthotropic materials as presented in Ref. [7].

It turns out that this method allows for the prediction of the shape of the body and it is possible to predict placement and shape of holes in the structure.

3. Integration

In order to finalize a design obtained by the material density approach, it is reuired to optimize the final shape of the boundaries of the optimal topology. The choice of initial proposed form for the boundary optimization methods is usually left entirely to the designer but the material distribution optimization gives the designer a rational basis for the choice of initial form.

Interfacing the topology optimization method with the boundary variations method is a problem of generating outlines of objects from grey level pictures. A procedure for an automatic computation of the proposed initial form for the boundary variations technique could thus be based on ideas and techniques from image analysis and pattern recognition. For the examples presented in this paper, the outlines for the initial proposed form were generated manually thus mimicking a design situation where the ingenuity of the designer is utilized to generate a 'good' initial form from the topology optimization results. The term 'good' in this context covers considerations such as ease of production, aesthetics etc. that may not have a quantified form. A reduction of the number of holes proposed by the topology optimization by ignoring relatively small holes exemplifies design decisions that could be taken before proceeding with the boundary variations technique.

The material density approach should be seen as a preprocessor for boundary optimization and by integrating the two methods a very efficient design tool can be developed. In an integrated system, common CAD input—output facilities can be used as well as a common mesh generator for the FEM analysis. Interfacing the two methods by a CAD based module added to the input facility for the boundary variations method, allows the designer to actively control the information used and such interactive possibilities have been found to be very important.

2. Topology optimization

For the topology optimization we minimize compliance for a fixed, given volume of material, and use a density of material as the design variable. The density of material and the effective material properties related to the density is controlled via geometric variables which govern the material with microstructure that is constructed in order to relate correctly material density with effective material property.

The problem is thus formulated as

$$\min \quad L(w)$$
$$\text{so:} \quad a_D(w, v) = L(v) \quad \text{for all} \quad v \in H \qquad (1)$$
$$\text{Volume} \leq V$$

where

$$L(v) = \int_\Omega f \, v \, d\Omega + \int_\Gamma t \, v \, d\Gamma \qquad (2)$$

$$a_D(w, v) = \int_\Omega E_{ijkl}(D) \, \epsilon_{ij}(w) \, \epsilon_{kl}(v) d\Omega \ . \qquad (3)$$

Here, f, t are the body load and surface traction, respectively, and ϵ_{ij} denotes linearized strains. H is the set of kinematically admissable deformations. The problem is defined on a fixed reference domain Ω and the rigidity E_{ijkl} depend on the design variables used. For a so—called second rank layering constructed as in Fig. 1, we have a relation

$$E_{ijkl} \equiv E_{ijkl}(\mu, \gamma, \theta) \qquad (4)$$

where μ, γ denote the densities of the layering and θ is the rotation angle of the layering. The relation (4) can be computed analytically ([3]) and for the volume we have

4. Boundary optimization

Once the optimal topology and initial boundary shape is defined, the objective is to refine this initial shape, such that the final structure will be the optimum structure of a more refined design model, where global as well as local displacement and stress constraint (objectives) are taken into consideration. For example, we could seek a structure where the von–Mises equivalent stress in the body is minimized, subject to a resource and compliance constraint:

$$\begin{array}{ll} \min & \max \quad \overline{\sigma}_{eq} \\ \Omega \leq D & x \in \Omega \end{array} \qquad (6)$$

$$\text{so:} \quad \text{Equilibrium}$$

$$\int d\Omega \leq V$$

$$\text{Compliance} \leq \overline{\phi}$$

Here D denotes the set of admissable boundary shapes, defined through local geometric constraints, and with boundaries defined through boundary nodes or spline control points.

Alternatively, we could seek to minimize weight for a given set of stress and displacement constraints:

$$\min \int d\Omega$$

$$\text{so:} \quad \text{Equilibrium}$$

$$\overline{\sigma}_{eq}(x) \leq \overline{\sigma}_{max} \qquad \text{for all } x \qquad (7)$$

$$\left| u(x) \right|_{\infty} \leq \overline{u} \qquad \text{for all } x$$

The integration has been carried out for two different boundary shape optimization systems ([8], [9]). In the case of (6), this problem is solved with a gradient technique, with shape sensitivities obtained via the speed method ([5]). For the sensitivity analysis, very precise estimates of stress is required and for this reason, the equilibrium is defined via the stationarity condition for the Hu–Washizu variational principle ([8]). Also, a boundary fitted elliptic mesh generator is used to generate the FEM–mesh used for the numerical solving procedure for the mixed analysis problem. This mesh generator is employed at each iteration step of the boundary optimization, thus maintaining good mesh properties throughout the shape modification process [5]. In order to cater for the non–simply connected domains predicted by the topology optimization system, the mesh generator is based on a subdivision of the domain by blocks. The remeshing is a crucial element in the boundary optimization procedure and together with the use of a mixed FEM method, allows for the boundary movements to be parametrized by movement of the FEM nodes along the design boundaries ([8]).

For the problem defined by Eqs. (7), the CAD–based system CAOS was employed [9]. This shape design system is fundamentally of similar type to the one described above, i.e. it is based on a boundary description of shape. However, the implementation is somewhat different in detail. One important common procedure is to use an efficient meshing technique, and to employ this throughout the iterative design process. The meshing is based on the division of the structural domain into design elements (quadrilaterals) and boundaries are parametrizised through control nodes for a spline representation of boundaries. Optimization techniques employed cover SLP and CONLIN (see [9]). The integration with CAOS was greatly simplified by the geometric design model and showed the ease of integration of the topology optimization method into such CAD–based shape optimization systems.

5. Examples

Figures 2 through 4 show examples of 2–dimensional structures optimized through the material distribution method followed by a boundary variations technique, as described above.

As can be seen, the topology optimization results in very good initial forms obtained for the boundary variations technique. Generally, only small and localized design changes occur during the boundary optimization, if the problem formulation (6) is used. Typically, the minimization of the stress level during the boundary optimization also results in some decrease in the compliance, but this is not unexpected as the drawing of the initial form from the topology data constitutes a not insignificant pertubation of the minimum compliance design. For the problem formulation (7), the topology optimization results in so good initial proposed forms that the boundary shape optimization gives rise to some quite large design changes, albeit mostly in the sense of scaling; however, the gains obtained through the use of topology optimization are quite significant.

Acknowledgements

This work has been supported in part by the European Communities through the SCIENCE programme, contract no. SCI–0083 (IICR). Support from the Danish Technical Research Council, programme for Computer Aided Mechanical Design (MPB, JR) and from AGARD, project P86 (SMP/ASP 31) (MPB, HCR) is also gratefully acknowledged.

References

[1] R.T. Haftka and R.V, Gandhi: Structural shape optimization — A survey. Comp. Meths. Appl. Mech. Engrg. 57 (1986) 91—106.

[2] M.P. Bendsøe and N. Kikuchi: Generating optimal topologies in structural design using a homogenization method. Comput. Meths. Appl. Mechs. Engrg. 71 (1988), 197—224.

[3] M.P. Bendsøe: Shape design as a material distribution problem. Struct. Optim., 1 (1989), 193—202.

[4] Suzuki, N., Kikuchi, N.: Shape and Topology Optimization for Generalized Layout Problems Using the Homogenization Method. Comp. Meths. Appl. Mechs. Engrg. (to appear).

[5] M.P. Bendsøe, and H.C. Rodrigues: Integrated topology and boundary shape optimization of 2—D solids. Comp. Meths. Appl. Mechs. Engrg. (to appear).

[6] Bremicker, M.; Chirehdast, M.; Kikuchi, N.; Papalambros, P.: Integrated Topology and Shape Optimization in Structural Design. Dept. of Mech. Engng., Univ. of Mich., USA, 1990.

[7] P. Pedersen: On optimal orientation of orthotropic materials. Struct. Optim., 1(1989), 101—106.

[8] II. Rodrigues: Shape optimal design of elastic bodies using a mixed variational formulation. Comp. Meths. Appl. Mechs. Engrg. 69 (1988) 29—44.

[9] J. Rasmussen: The Structural Optimization System CAOS. Struct. Optim., 2(1990), 109—115.

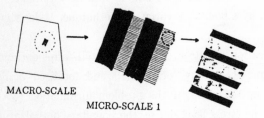

MACRO-SCALE

MICRO-SCALE 1

MICRO-SCALE 2

Fig. 1. Construction of a layering of second rank.

a

Fig. 2. Optimal design of a beam. A: Optimal topology with outline showing reference domain. B and C: Initial and final design using the boundary variations method. Two blocks are used for the elliptic mesh generator. Only the boundaries of block 1 can move. The maximum stress is reduced by 55.7% and the compliance by 7.3%. Block divisions are shown as hatched, bold lines: design boundaries as bold solid lines.

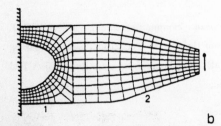

b

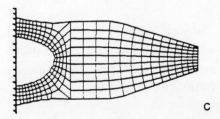

c

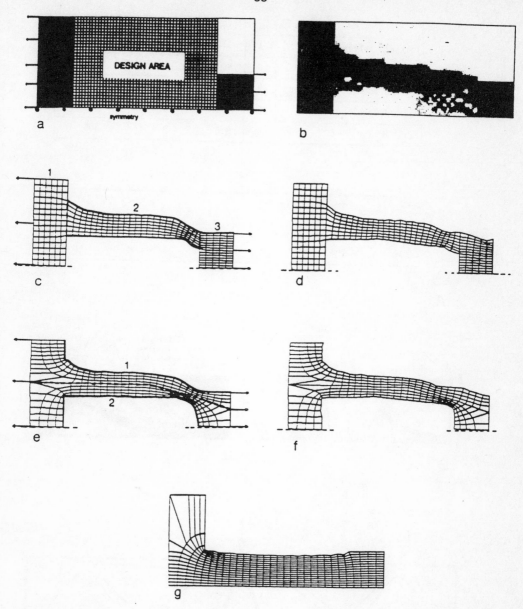

Fig. 3. Optimal design of a fillet. A: The reference domain showing loads etc. and the reduced design area. B: The optimal topology. C and D: Initial design and optimal design with the boundary variations method. Mesh generated using three blocks. E and F: Initial and final optimal design using different meshes (two blocks). G: Optimal design with no hole (minimum compliance design).

	Fig.	Volume	Deflection	Max. stress
Initial, infeasible design	A	1.07	10.1	292
Optimal Circular Holes		1.10	9.4	248
Optimal Boundary of Holes	B	1.02	9.4	372
Optimal Topology	C	1.10	6.0	227
Final Design	D	0.62	9.4	305

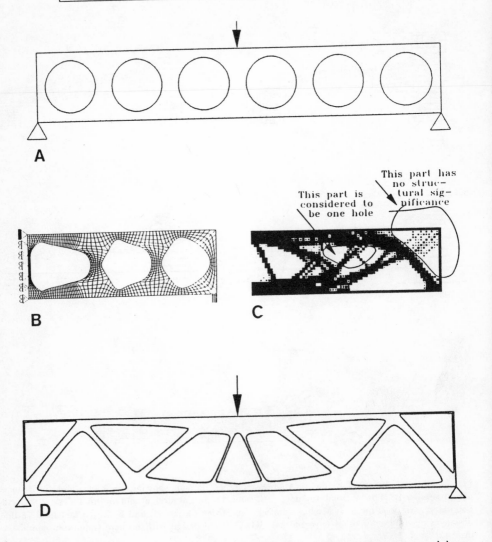

Fig. 4. Optimal design of an aircraft support beam. Design requirements are: upper and lower surfaces must be planar and the beam of constant depth: there must be a number of holes in the structure to allow for running wires, pipes etc. through the beam. The optimal design problem is to minimize weight so that the maximum deflection does not exceed 9.4 (mm), and the maximum von Mises stress should not exceed 385 (N/mm²). Only half of the beam needs to be analyzed. The final design is 64% better than the design with optimally shaped boundaries of initial holes. See table for values of objective etc. and figure references.

SOME APPROACHES FOR SHAPE FINDING IN OPTIMAL STRUCTURAL DESIGN

H.P. Mlejnek, U. Jehle and R. Schirrmacher

Institute for Computer Applications, University of Stuttgart
Federal Republic of Germany

1. Introduction

Shape optimization by moving boundaries is now a well established technology, which enters commercial programs. Following the philosophy of Schmit [1], Fleury [2] and many other contributions, the movable shape problem is described by means of a blending function, which is governed by a comparatively small number of variables. This property disposes the approach favourable to mathematical progamming (MP), which can handle all types of objectives and constraints very easily. However since we have to preselect the position of the moving boundary as well as the type of the blending function, the process of finding an optimal shape is quite predetermined. We are for instance not able to create automatically voids.

In [3] Bendsoe and Kikuchi introduced a novel approach of distributing mass within a specified design domain utilizing a stiffness-density relation. The design space is discretized and the element densities form the variables of an topological design, which can be treated as easy as sizing (fig. 1). This procedure relates almost total freedom to the design process, involves however a large number of variables. Nevertheless we are able to create an efficient algorithm by working under simplified conditions. In [3] the compliance is minimized for a specified amount of structural mass and constrained densities and optimality criteria methods OC are employed to solve that problem. It became immediately apparent to the authors, that the generated designs seemed to be very useful as starting designs for optimal shape moves. Consequently we established that procedure, developed a modified scheme for the material stiffness-density relation, introduced multiple loading cases and spacious applications [4],[5],[6],[7]. Investigations for quite general sets of objectives and constraints employing MP showed however, that the large number of density variables is for a general application disadvantageous.

Nevertheless the designs provided for the objective compliance, specified displacement or eigenvalue seem to be very useful as starting designs for a more general MP-approach and therefore the optimal mass distribution (OMD) was implemented as a preoptimizer PREOPT to the general shape optimizer OPTIMA-S, which works by boundary moves (OBM).

The mathematical formulation of both approaches is effectively the same. We minimize an objective function

$$W(x) = min!$$ (1)

which denotes in the simplest case of OMD the max-compliance-function over the loading cases and may be of arbitrary nature for OBM. The minimization is subject to constraints

$$g(x) \leq o$$ (2)

of an arbitrary number and character for OBM (stress, displacements, eigenfrequencies, buckling values). In the simplest case of OMD we have only one active constraint namely the amount of mass assigned to the design space. In both approaches direct variable limits are involved.

$$\underline{x} \leq x \leq \overline{x} \tag{3}$$

Objective function W and generally also the constraints g are implicit and highly non-linear functions of the design variables x.

Therefore the working scheme for OBM as well as for OMD involves three essential steps, which are based on L. Schmit's philosophy [1]. First an analysis model (finite element discretized structure) is used to compute values and design derivatives of implicit behaviour functions. Second we use these informations to construct an explicit approximation (behaviour model). Third we feed this explicit approximation into an optimizer. The optimum of the behaviour model is used for a refresh operation, which closes the design loop. In the following we discuss this approach with special considerations of second order sensitivities and approximations of behaviour functions.

2. Aspects of second order sensitivity computation

We start our considerations with an exemplary discussion of displacement dependent functions as

$$g\left(x, r\left(x\right)\right) \tag{4}$$

which depends either explicitly or implicitly via the displacement vector $r(x)$ on the design variables x.

$$Kr = R \tag{5}$$

describes the static equilibrium and K denotes the stiffness matrix, R the load. As relation for the first derivative we find in textbooks

$$\frac{dg}{dx_i} = \frac{\partial g}{\partial x_i} + \frac{\partial g}{\partial r} K^{-1} \left[\frac{dR}{dx_i} - \frac{dK}{dx_i} r \right] \tag{6}$$

where 'd' indicates here a complete differentiation with respect to a design variable and '∂' denotes a differentiation to the explicit variables only.

The adjoint approach makes use of the solution

$$s_a = K^{-1} \frac{\partial g}{\partial r^t} \tag{7}$$

and the socalled direct approach utilizes the result

$$\frac{dr}{dx_i} = K^{-1} \left[\frac{dR}{dx_i} - \frac{dK}{dx_i} r \right] \tag{8}$$

Correspondingly we need for m functions and n variables in the adjoint method m forward-backward-substitutions (FBS) and for the direct approach n FBS. If we proceed to second derivatives of (4) we receive

$$
\begin{aligned}
\frac{d^2 g}{dx_i dx_j} = \frac{\partial^2 g}{\partial x_i \partial x_j} &+ \frac{d\boldsymbol{r}^t}{dx_j} \frac{\partial^2 g}{\partial \boldsymbol{r} \partial \boldsymbol{r}^t} \frac{d\boldsymbol{r}}{dx_i} \\
&+ \frac{\partial^2 g}{\partial \boldsymbol{r} \partial x_j} \frac{d\boldsymbol{r}}{dx_i} + \frac{\partial^2 g}{\partial \boldsymbol{r} \partial x_i} \frac{d\boldsymbol{r}}{dx_j} \\
&+ \boldsymbol{s}_a^t \left[\frac{d^2 \boldsymbol{R}}{dx_i dx_j} - \frac{d^2 \boldsymbol{K}}{dx_i dx_j} \boldsymbol{r} - \frac{d\boldsymbol{K}}{dx_i} \frac{d\boldsymbol{r}}{dx_j} - \frac{d\boldsymbol{K}}{dx_j} \frac{d\boldsymbol{r}}{dx_i} \right]
\end{aligned}
\tag{9}
$$

Eq. (9) may be regarded as second order adjoint approach. We detect immediately, that no additional FBS are necessary to compute (9), provided we know both the adjoint and direct solutions of the first order computations, which involve $n + m$ FBS (Haftka [8]). With reference to (7) we may also rewrite (9) to yield the second order direct approach, which utilizes the second displacement vector derivatives

$$
\frac{d^2 \boldsymbol{r}}{dx_i dx_j} = \boldsymbol{K}^{-1} \left[\frac{d^2 \boldsymbol{R}}{dx_i dx_j} - \frac{d^2 \boldsymbol{K}}{dx_i dx_j} \boldsymbol{r} - \frac{d\boldsymbol{K}}{dx_i} \frac{d\boldsymbol{r}}{dx_j} - \frac{d\boldsymbol{K}}{dx_j} \frac{d\boldsymbol{r}}{dx_i} \right]
\tag{10}
$$

and requires the direct first order results (n FBS) plus $n(n+1)/2$ FBS.

Hitherto the discussion was based on the tacid assumption, that derivatives of load vector $\boldsymbol{R}$ and stiffness matrix $\boldsymbol{K}$ with respect to the design variables are available. In OMD having element densities as design variables an analytic derivative computation is easy to implement. We simply substitute the material stiffness by the known material stiffness derivatives and compute formally element stiffnesses. Moreover all second mixed derivatives are zero. If we regard the special function compliance, the adjoint solution is given by the displacement vector $\boldsymbol{r}$ and the adjoint approach provides trivial and cheap first order results. For second derivatives of compliance we need however the direct first order solutions (displacement derivatives) and the expense is increased considerable.

In OBM we work with shape variables. The implementation of analytic stiffness derivatives is nontrivial and affects also element dependent routines. If we are not willing to take this burden, we may use numerical schemes as we do in this paper. We can either use the second adjoint approach replacing unknown derivatives by the corresponding difference expression. This method can be denoted as second order semi-analytic computation and its accuracy is subject to current investigations. We require for this approach at least approximations for first order displacement sensitivities. The simple first order semianalytic result is in some cases quite inaccurate and may be not admissable for this purpose. In the simplest case we need $n + m$ FBS. The second order direct approach involves the same problems and requires at least n plus $n(n+1)/2$ FBS. A third choice is given by a second order overall difference scheme (see e.g. fig. 2). This scheme provides first and second derivatives with an cut-off error of order Δ^2 where Δ denotes the size of variable change (modification). If we develop an iterative computation of the modified states giving incremental function changes we may use quite small modification values. We use the triangularized stiffness matrix of the un-modified state 0, but need per modification some FBS (4-5 with initial state 0). We may however utilize computed information to reduce this expense as indicated e.g. in fig. 2. The complete setup of second derivatives by the overall difference approach is quite expensive involving $2n^2$ modifications. For other than experimental investigations the implementation of analytic derivatives is therefore even more urgent if second order sensitivities are demanded. As we will see in the next chapter we use for economic reasons preferable separable approximations. In this case the number of necessary mod-ifications is reduced to $2n$. We turn our attention now to a second class of functions, namely eigenvalue functions as buckling values or eigenfrequencies.

$$g(\lambda(x)) \tag{11}$$

where

$$(B - \lambda K_e)\Phi = o \tag{12}$$

describes the corresponding eigenvalue equation. The matrix B denotes either the mass matrix or the geometrical stiffness, K_e is the elastic stiffness matrix and the eigenvalue λ embraces either the inverse of square circular eigenfrequency or the negative inverse of the load scale factor. A side condition fixes the length of the eigenvector Φ, e.g. for stiffness orthogonal eigenvectors

$$\Phi^t K_e \Phi = 1 \tag{13}$$

The relation for first derivatives requires as prerequisite only B respectively K_e matrix derivatives.

$$\frac{d\lambda}{dx_i} = \Phi^t \left(\frac{dB}{dx_i} - \lambda \frac{dK_e}{dx_i} \right) \Phi \tag{14}$$

This aspect was already discussed in the first part of this section. We may however point out, that dynamic problems in the OMD approach involve only constant mass derivatives, since the mass matrix is linear in the density variables. Buckling problems employ the displacement dependent geometrical stiffness matrix and are more difficult to handle. We proceed now to second derivatives of eigenvalues.

$$\begin{aligned}
\frac{d^2\lambda}{dx_i dx_j} = \Phi^t &\left(\frac{d^2 B}{dx_i dx_j} - \frac{d\lambda}{dx_i} \frac{dK_e}{dx_j} - \frac{d\lambda}{dx_j} \frac{dK_e}{dx_i} \right) \Phi \\
+ \Phi^t &\left(\frac{dB}{dx_i} - \frac{d\lambda}{dx_i} K_e - \lambda \frac{dK_e}{dx_i} \right) \frac{d\Phi}{dx_j} \\
+ \Phi^t &\left(\frac{dB}{dx_j} - \frac{d\lambda}{dx_j} K_e - \lambda \frac{dK_e}{dx_j} \right) \frac{d\Phi}{dx_i}
\end{aligned} \tag{15}$$

Unfortunately this relation includes also eigenvector derivatives which can be deter-mined from

$$(B - \lambda K_e)\frac{d\Phi}{dx_i} = - \left(\frac{dB}{dx_i} - \frac{d\lambda}{dx_i} K_e - \lambda \frac{dK_e}{dx_i} \right) \Phi \tag{16}$$

Obviously this system is singular. A necessary side condition is obtained by differentiating (13).

$$2\boldsymbol{\Phi}^t K_e \frac{d\boldsymbol{\Phi}}{dx_i} = -\boldsymbol{\Phi}^t \frac{dK_e}{dx_i}\boldsymbol{\Phi} \tag{17}$$

We may solve this set of equation either by Nelson's approach [9,10] or by the generalized inverse procedure [11]. Both methods are expensive, since they work on a large set of equations.

A first choice for numerical computation of derivatives would embrace again the replacement of matrix derivatives by differences (semianalytical approach). This scheme would however not circumvent the expensive solution of (16),(17). A second choice is to employ the overall difference approach. Since the corresponding modification should be kept small, we may engage a vector iteration procedure for modified systems. This method will be discussed in a forthcoming paper. A third possibility is to gain (approximate) second order information from the difference of first order derivatives, utilizing the first order derivatives of two neighboured designs.

3. Second order explicit approximations

In this section we discuss some possible explicit approximation schemes, which utilize second order information.

Half quadratic scheme
This approach was already used in [12]. A typical function g is described by

$$g = a_0 + \sum_j \left[a_j\,(x_j - x_{j0}) + \frac{1}{2}b_j\,(x_j - x_{j0})^2 \right] \tag{18}$$

and the coefficients are determined by first order

$$a_j = \left(\frac{dg}{dx_j} \right)_0 \tag{19}$$

respectively by second order sensitivities

$$b_j = \left(\frac{d^2g}{dx_j^2} \right)_0 \tag{20}$$

We require only direct second derivatives since the approximation is separable. Also we may establish a convex function by taking into account only positive second derivatives (fig. 3a). In the latter case we drop of course already computed information and receive only linear contributions. This model behaves in application very well. Since the curvature is constant, we observe a somewhat too optimistic behaviour.

Generalized power approach
This scheme was also applied in [12]. The explicit approximation reads now

$$g = a_0 + \sum_j a_j x_j^{p_j} \tag{21}$$

and the powers are determined by second derivatives

$$p_j = 1 + \left(\frac{d^2 g}{dx_j^2}\right)_0 x_{j0} / \left(\frac{dg}{dx_j}\right)_0 \tag{22}$$

and the coefficients by first order sensitivities.

$$a_j = \left(\frac{dg}{dx_j}\right)_0 x_{j0}^{1-p_j} / p_j \tag{23}$$

Again this scheme is separable and we need only second derivatives. We receive in general ($p_j \neq 2$) also variable curvature. For positive second order sensitivities ($d^2 g / dx_j^2)_0 \geq 0$ we obtain convex contributions (fig. 2b). Negative first order sensitivities may result in negative powers, which limits the range of application to positive variables. This limitation provides no specific problems. The model exhibits a well balanced behaviour. It may be converted to a first order scheme by prescribing fixed powers, e.g. the hybrid approach [13,14] selects $p_j = \pm 1$ to establish a convex approximations. The use of other powers (generalized hybrid scheme) is possible.

Generalized method of moving asymptotes
This approach was initiated by Svanberg [15] as convex approximation scheme. It uses either an upper or an lower asymptote dependent from the sign of the first derivative (fig. 3c)

$$g_i = a_{i0} + \sum_j a_{ij} < \begin{array}{c} 1/(U_{ij} - x_j) \\ 1/(x_j - L_{ij}) \end{array} \tag{24}$$

The original version did not consider function dependent asymptotes, but worked only with variable dependent asymptotes. This restriction has the advantage that a minimizing variable set in the dual approach can be determined analytically, whereas the more general scheme above needs one dimensional line searches. The individual asymptotes may be fixed by utilizing second derivatives as follows:
If $(dg_i/dx_j)_0 \geq 0$

$$U_{ij} = x_{j0} + 2\left(\frac{dg_i}{dx_j}\right)_0 / \left(\frac{d^2 g_i}{dx_j^2}\right)_0 \tag{25}$$

$$a_{ij} = (U_{ij} - x_{j0})^2 \left(\frac{dg_i}{dx_j}\right)_0 \tag{26}$$

If $(dg_i/dx_j)_0 < 0$

$$L_{ij} = x_{j0} + 2\left(\frac{dg_i}{dx_j}\right)_0 / \left(\frac{d^2 g_i}{dx_j^2}\right)_0 \tag{27}$$

$$a_{ij} = -(x_{j0} - L_{ij})^2 \left(\frac{dg_i}{dx_j}\right)_0 \tag{28}$$

We receive again a separable scheme, which needs only second derivatives. Since $U_{ij} > x_{j0}$ respectively $L_{ij} < x_{j0}$ are natural requirements, only positive second derivatives can be admitted. This scheme is therefore by definition convex. As we easily see from (25) respectively (27) large second derivatives will lead to tightly to x_{j0} neighboured asymptotes. The approximation assumes therefore quickly very large values, which makes this approach under this conditions quite conservative. This property leads to delayed design steps and to many refreshment iterations.

Full second order Taylor
The full second order Taylor approximation is nonseparable and requires all second derivatives.

$$g = a_0 + a^t (x - x_0) + \frac{1}{2} (x - x_0)^t B (x - x_0) \tag{29}$$

$$a_j = \left(\frac{dg}{dx_j} \right)_0 \tag{30}$$

$$b_{jk} = \left(\frac{d^2 g}{dx_j dx_k} \right)_0 \tag{31}$$

The large amount of data prohibits a general application in that sense, that objective and constraints are all approximated by this scheme. It may be a useful tool for some experiments discussing the separability of a real behaviour. One possible application suggested by Fleury [16] is the approximation of the Lagrangian in a dual approach employing sequential quadratic programming. The above scheme is convex for positive definite B. It is well behaved, employs constant curvature and tends to be too optimistic in structural applications.

4. Optimizer and second order approximations

We turn our attention in shortness to the third working step, namely to the optimization working on the explicit approximations discussed in the previous section. The most efficient procedure tested uptodate is the dual solution scheme for separable convex approximations (SCP). A special CP-version in OMD (objective compliance, constraint specified mass) was even faster as simple OC-methods. Nevertheless we have to accept, that we loose valuable second order information by applying convex schemes. The possible impact of second order improvements will be diminished. In some cases we had to drop 50% of all second order derivatives. Primal procedures (e.g. extended penalty) do not inherit this precondition and work with all types of approximation. The economy is however inferiour to the dual methods. If separibility is admitted and second order affects are of minor importance we may therefor rely on the dual schemes as CP or SQP.

5. Examples for optimal boundary moves

Triangular plate
We start our series of example with a triangular plate (fig. 4) submitted to a line load at each corner. We perform optimization with four behaviour models. The generalized hybrid scheme with a fixed exponent of 2 belongs to the first order family. As second order models we use the generalized power scheme, the generalized method of moving asymptotes and the half quadratic scheme.

As we may see in fig. 5 the first order generalized hybrid model exposes a very optimistic approximation. We gain a high mass reduction after the first loop for the price of 5 violated constraints. Via the refreshment steps the model is capable to remove these violations. The second order models perform almost identical and are describing the structural behaviour with higher quality leading to less significant violations. In the final end all approximation candidates are leading to the same optimum design (fig. 6).

Plate with hole
Our next example is the well known plate with hole (fig. 7). Making use of the double symmetry we optimize only one quarter. All methods perform like in our first example with the exception of the generalized method of moving asymptotes. Since in this example the second derivatives are comparatively large, the computed asymptotes are tightly neighboured to the generation points. Consequently we quickly approach the positions of infinite constraint values. This involves a highly conservative approximation and thus a slow convergence following a series of feasable designs (fig. 8). The optimal design is shown in fig. 9.

Cube with cavity
The next presented example is a three dimensional problem test namely a cube with cavity (fig. 10). We also take advantage of the double symmetry. The Fig. 11 shows the behaviour of a typical constraint in a progress direction and the approximation with different approximation schemes. All models are build up with the same information and correspond to the same curvature. As in all test example the generalized MMA is also here too conservative. The generalized power and half quadratic schemes are here too optimistic which is however not typical for this approximation types. In the most cases these two candidates show a similar behaviour being sometimes too optimistic and sometimes too conservative. The weight history (fig. 12) demonstrates a high mass reduction of the first order model after the first loop, involves however an infeasible design. The optimum design (fig. 13) fullfills the Kuhn-Tucker condition and reduces the weigth about 21 per cent.

Efficency
To compare the first order generalized hybrid with the second order generalized power scheme for the presented examples we introduce the efficieny factor.

$$e = \frac{n_{s1}}{n_{s2}} * \frac{t_{s1}}{t_{s2}} \tag{32}$$

A value greater than 1 indicates the superiority of second order scheme. We receive for our examples
  triangular plate $e = 0.9375$
  plate with hole $e = 1.01$
  cube with cavity $e = 1.08$

Obviously the first example is solved more efficient by first order scheme. In the other two cases we have a marginable advantage of the second order schemes.

6. Examples for optimal material distribution

Triangular plate

In linear static the examples of the genesis of structures correspond to those of OBM in the previous section. For the triangular plate we start with identical dimensions and a mean density of 70%. The first and second order approximation models lead to the same optimal layout showing a convex outer boundary (fig. 14). Additionally a void in the center of the plate is generated. Obviously the solutions provided by second order models are more accurate as the Kuhn-Tucker-check proves. For the computation of the second order derivatives we need one FBS per variable. This extra expense would be only balanced, if the number of refreshment loops is reduced at least by one third. In the convex second order scheme a high portion of non-positiv second order sensitivities had to be dropped. The necessary linearizations lead to a badly conditioned redesign. The efficient reduction of reanalysis loops could not be achieved (fig. 15).

Cube with cavity

We use again the same structure as in OBM (see previous section). However for historical reasons the load is $2p_0$ in x-direction and p_0 in y,z-direction. The mean density assigned to the design space is 80%. The discretisation embraced 432 brick elements (HEXE8). The material distribution was computed with first and quasi second order approximations. First order models were the power scheme with fixed power and the moving asymptote procedure with global asymptote position. For the generation of second order information the first order sensitivities of two subsequent design loops were utilized. As demonstrated in fig. 16 the mass distribution is the same for all approximation schemes showing the formation of an elliptic cavity. The redesign step with 432 variables was very quick. The reduction of objective compliance (fig. 17) is similar for all approximation models. The quasi-second order model has no positive effect on a better convergence. One reason for that behaviour may be, that local exponents or asymptotes do not differ much from the assumed global values in the first order models. Another reason may be the high portion of assigned structural mass of 80%.

Barrier (eigenfrequency)

Our aim is the design of a barrier foundation (fig. 18), which maximizes the first bending eigenfrequency of the system. It is assumed, that the shape of the bar is fixed (density 1). The foundation design space is filled with a mean density of 80%. In fig. 19 we give the designs obtained for first and second order MMA. Both designs are essentially the same. The second order model supplies a design with a better massiv/empty separation. The optimizer needs only one to two iterations to find the maximum for both approximations (fig. 20). The second order model will however be only superiour, if it reduces the number of iterations by one sixth due to the costly computation of the eigenvector derivative.

Column buckling

In linear buckling we investigated a flat truss (fig. 21) subjected to a single pressure load, whose buckling value had to be maximized. The computation of first order sensitivities requires already a lot of CPU-time, because the geometrical stiffness derivative is stress dependent. Thus the second order approach again was based on the cheaper quasi-second order scheme utilizing differences of first order results.As demonstrated in fig.

22, the optimized structure accumulates mass at the outside bottom and narrows up to the top. At the top we observe some mass accumulation for load diffusion, which is marked clearer by the quasi-second order model. In this example the first order power scheme shows an oscillating behaviour of the objective function and it has not converged after 10 iterations, whereas the quasi-second order power scheme and the first order moving asymptotes converge very fast.

7. Conclusions and outlook

Analytic (density variables, OMD) and iterative (shape variables, OBM) schemes for the computation of second order sensitivities have been developed and tested. Numerical second order computations require roughly twice the expense compared with that for first order sensitivities, providing also more accurate first order information. As assumed second order approximation schemes stabilize the refreshment iteration and reduce the number of loops and reanalysis cycles. Since for optimal material distribution also first order schemes work stable, the benefit is meager. The reduction of model loops does not balance the expense for second order sensitivities. We gain only in accuracy and in stability for optimal boundary moves, not in efficiency. Therefore second order approaches remain a domain for a subset of difficult problems, which tend to be unstable or which cannot be approximated separable. For future developments concerning the less sensitive realm of problems, it seems more promising to utilize cheap approximate second order information exploiting e.g. design history.

Acknowledgments

A portion of the work reported in this paper was undertaken with the support of the Deutsche Forschungsgemeinschaft (DFG), research project 'Verhaltensmodelle'. The authors would like to express their gratitude for this generous support in research.

References

[1] L. A. Schmit, Structural optimization - some key ideas and insights, New directions in optimum structural design, ed. by E. Atrek, R. H. Gallagher, K. M. Ragsdell and O. C. Zienkiewicz, John Wiley, 1984

[2] L.A. Schmit and C. Fleury, Structural synthesis by combining approximation concepts and dual methods, AIAA, 18, 1252-1260, 1980

[3] M. P. Bendsoe and N. Kikuchi, Generating optimal topologies in structural design using a homogenization method, Comp. Meth. Appl. Mech. Eng., 71, 197-224, 1988

[4] H. P. Mlejnek and R. Schirrmacher, An engineers approach to optimal material distribution and shape finding, submitted to Comp. Meth. Appl. Mech. Eng. in 1989

[5] H. P. Mlejnek, R. Schirrmacher and U. Jehle, Strategies and potential of modern optimization, International FEM-Congress Baden-Baden, FRG, nov. 20-21, 1989, proceedings ed. by IKOSS GmbH Stuttgart, 1989

[6] H. P. Mlejnek, Optimale Materialverteilungen, COMETT-Seminar Bayreuth, June 18-22, 1990

[7] U. Jehle and R. Schirrmacher, Softwareaspekte der Programme OPTIMA-S und PREOPT, COMETT-Seminar, Bayreuth, June 18-22, 1990

[8] R. T. Haftka, Second-order sensitivity derivatives in structural analysis, AIAA, Vol. 20, No.12, 1765-1766, 1982

[9] R. B. Nelson, Simplified calculation of eigenvector derivatives, AIAA, Vol.14, 1201-1206, Sept. 1976

[10] X. Cao and H.P.Mlejnek, Second order eigensensitivity analysis of discrete structural systems, Second World Conference on Computational Mechanics, Stuttgart, FRG, Aug. 27-31, 1990

[11] R. Penrose, A generalized inverse for matrices, proceedings of the Cambrigde Philosophical Society, Vol. 51, 406-413, 1955

[12] H. P. Mlejnek and P. Schmolz, Some contribution of optimal design using explicit behaviour models, Eng. Opt., Vol. 11, 121-139, 1987, also presented as lecture in ASI computer aided optimal design, Troia, Portugal, June 29 - July 11, 1986

[13] J. H. Starnes,Jr and R. T. Haftka, Preliminary design of composite wings for buckling stress and displacement constraints

[14] V. Braibant and C. Fleury, An approximation concepts approach to shape optimal design, Comp. Meth. Appl. Mech. Eng., 53, 119-148, 1985

[15] K. Svanberg, The method of moving asymptotes - a new method for structural optimization, Int. J. Num. Meth. in Eng., Vol. 24, 359-373, 1987

[16] C. Fleury and H. Smaoui, Convex approximation strategies in structural optimization, Proc. GAMM Seminar Discretization methods and structural optimization - procedures and applications, Oct. 5-7, 1988, Springer

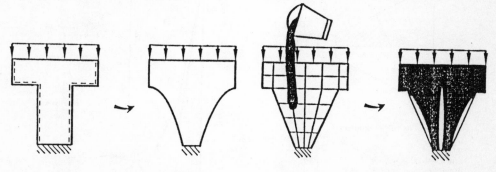

Optimal boundary moves   Optimal material distribution

Fig. 1: Alternative approaches in optimal shape design

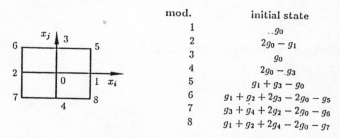

	mod.	initial state
	1	$..g_0$
	2	$2g_0 - g_1$
	3	g_0
	4	$2g_0 - g_3$
	5	$g_1 + g_3 - g_0$
	6	$g_1 + g_2 + 2g_3 - 2g_0 - g_5$
	7	$g_3 + g_4 + 2g_2 - 2g_0 - g_6$
	8	$g_1 + g_2 + 2g_4 - 2g_0 - g_7$

Fig. 2: Modification scheme for second order derivatives

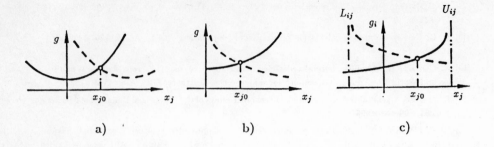

a)   b)   c)

Fig. 3: Typical convex approximations : a) half quadratic b) power scheme c) moving asymptotes

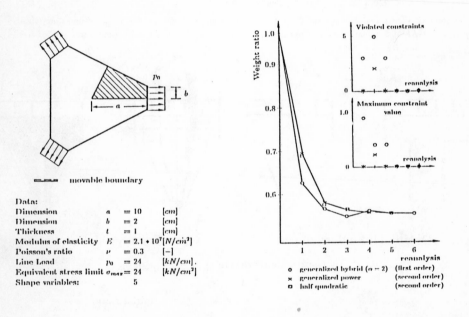

movable boundary

Data:
Dimension a $= 10$ $[cm]$
Dimension b $= 2$ $[cm]$
Thickness t $= 1$ $[cm]$
Modulus of elasticity E $= 2.1 * 10^7 [N/cm^2]$
Poisson's ratio ν $= 0.3$ $[-]$
Line Load p_0 $= 24$ $[kN/cm]$.
Equivalent stress limit $\sigma_{max} = 24$ $[kN/cm^2]$
Shape variables: 5

o   generalized hybrid ($\alpha - 2$)   (first order)
x   generalized power   (second order)
□   half quadratic   (second order)

Fig. 4: Triangular plate: problem description   Fig. 5: Triangular plate: weight convergence

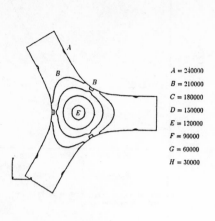

A = 240000
B = 210000
C = 180000
D = 150000
E = 120000
F = 90000
G = 60000
H = 30000

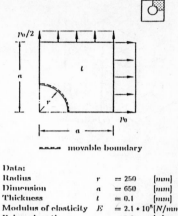

movable boundary

Data:
Radius	r	$= 250$	[mm]
Dimension	a	$= 650$	[mm]
Thickness	t	$= 0.1$	[mm]
Modulus of elasticity	E	$= 2.1 * 10^5$	$[N/mm^2]$
Poisson's ratio	ν	$= 0.3$	[-]
Line Load	p_0	$= 100/650$	$[N/mm]$
Equivalent stress limit	$\sigma_{max}=7$		$[N/mm^2]$
Shape variables:		7	

Fig. 6: Triangular plate: optimal design and equivalent stresses

Fig. 7: Plate with hole: problem description

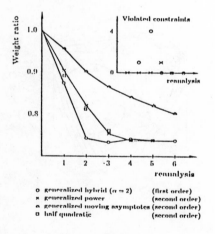

o generalized hybrid ($\alpha = 2$)   (first order)
× generalized power   (second order)
• generalized moving asymptotes (second order)
□ half quadratic   (second order)

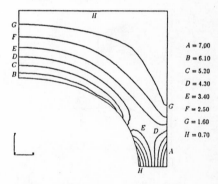

A = 7.00
B = 6.10
C = 5.20
D = 4.30
E = 3.40
F = 2.50
G = 1.60
H = 0.70

Fig. 8: Plate with hole: weight convergence

Fig. 9: Plate with hole: optimal design and equivalent stresses

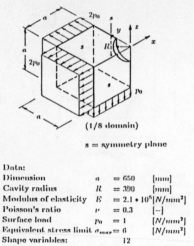

(1/8 domain)

s = symmetry plane

Data:
Dimension a = 650   [mm]
Cavity radius R = 390   [mm]
Modulus of elasticity E = $2.1 * 10^5$ [N/mm^2]
Poisson's ratio ν = 0.3   [−]
Surface load p_0 = 1   [N/mm^2]
Equivalent stress limit σ_{max} = 6 [N/mm^2]
Shape variables: 12

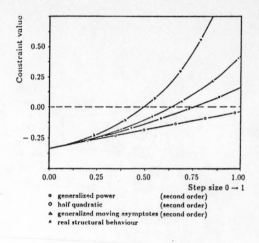

Fig. 10: Cube with cavity: problem description

Fig. 11: Cube with cavity: behaviour of constraints in progress direction

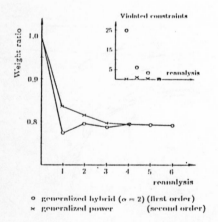

o generalized hybrid ($\alpha = 2$) (first order)
× generalized power   (second order)

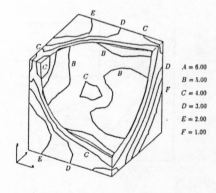

$A = 6.00$
$B = 5.00$
$C = 4.00$
$D = 3.00$
$E = 2.00$
$F = 1.00$

Fig. 12: Cube with cavity: weight convergence

Fig. 13: Cube with cavity: optimal design and equivalent stresses

49

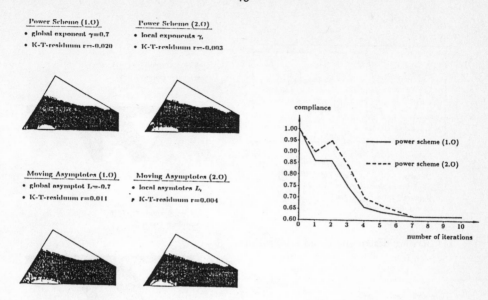

Fig. 14: Triangular plate: optimal mass distribution

Fig. 15: Triangular plate: compliance history

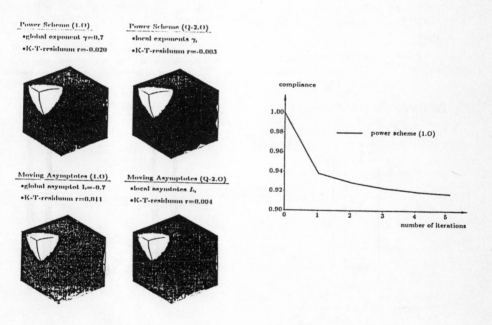

Fig. 16: Cube with cavity: optimal mass distribution

Fig. 17: Cube with cavity: compliance history

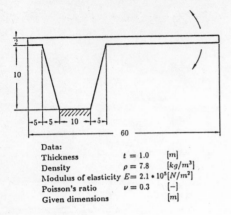

Data:
Thickness $t = 1.0$ $[m]$
Density $\rho = 7.8$ $[kg/m^3]$
Modulus of elasticity $E = 2.1 \cdot 10^5 [N/m^2]$
Poisson's ratio $\nu = 0.3$ $[-]$
Given dimensions $[m]$

Fig. 18: Barrier: design space and problem description

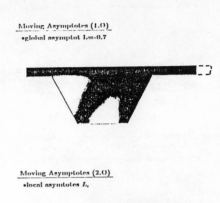

Moving Asymptotes (1.0)
• global asymptot L=-0.7

Moving Asymptotes (2.0)
• local asymtotes L_i

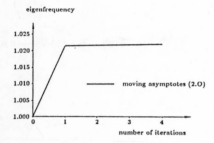

eigenfrequency

Fig. 20: Barrier: optimal mass distribution

Power Scheme (1.0)   Power Scheme (Q-2.0)
• global exponent $\gamma=0.7$ • local exponents γ_i
• K-T-residuum r=-0.600   • K-T-residuum r=-0.081

Fig. 19: Barrier: frequency history

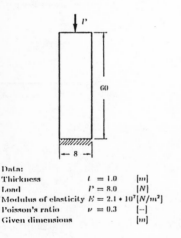

Data:
Thickness $t = 1.0$ $[m]$
Load $P = 8.0$ $[N]$
Modulus of elasticity $E = 2.1 \cdot 10^7 [N/m^2]$
Poisson's ratio $\nu = 0.3$ $[-]$
Given dimensions $[m]$

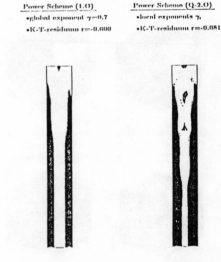

Fig. 22: Plane truss: optimal mass distribution

Fig. 21: Plane truss: design space and problem description

SHAPE OPTIMIZATION AND IDENTIFICATION OF 2-D ELASTIC
STRUCTURES BY THE B E M

Grzegorz Krzesinski
Institute of Aircraft Technology and Applied Mechanics
Technical University of Warsaw
ul.Nowowiejska 22/24 , 00-665 Warsaw

Abstract

The shape optimization of two-dimensional elastic structures and some inverse problems of elasticity are treated in the paper. Boundary element method, spline representation of unknown boundary and nonlinear programming techniques are used. Advantages and disadvantages of BEM compared with FEM for shape design are discussed. The technique of cubic splines boundary approximation provides good shape flexibility using relatively small number of design variables. Some numerical results illustrating different problems are presented.

Introduction

Most of the works on numerical shape optimization are based on the FEM formulation ([1,2,3,4]). The BEM has been used in this field only for the last few years (e.g.[5,6]). The shape design requires a FEM model which changes in the course of optimization and this may cause some particular difficulties in the solution process ([1,7]). To avoid the sometimes drastic distortion of the FEM grid sophisticated automated mesh generation techniques may be required ([1,8]). These difficulties with FE formulation can be partially overcome by using the BEM. This method usually requires modelling of the boundary only and it seems to be well suited for the boundary shape optimization .

In shape design problems description of a moving boundary plays a central role. In many works the coordinates of the boundary nodes are used as design variables. In this case the accuracy of the FE model can deteriorate and may lead to unrealistic, jagged shape ([1,7]). This technique results additionally in many design variables in the optimization process. Another approach is the representation of the boundary as a linear combination of shape functions with coefficients being the design variables. Splines, composed of low-order polynomial pieces are especially suitable for shape design. A spline representation of the boundary was applied in some works using FEM (e. q. [3]).

In this paper the BEM together with spline approximation of the boundary is employed. The presented general statement of the shape optimization problem allows us to solve shape optimization and shape identification problems as well as some inverse questions of elasticity.

Formulations of the shape optimization problem

In many engineering problems a boundary shape function f_s (Fig. 1) optimal due to a chosen criterion is required. The problem is described by a functional:

$$J_s: \quad f_s \longrightarrow c \in R \tag{1}$$

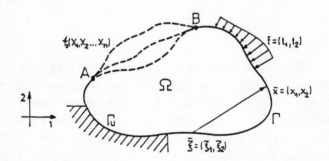

Fig. 1. The body under consideration.

The value of c can represent, for example, maximum Von Mises stress or compliance of the structure. To solve the problem of minimization of J_s f_s is represented as a function of unknown parameters (design variables) $\bar{X} = (X_1, X_2, \ldots X_n)$.

Hence the shape design problem can now be written as a nonlinear programming problem for a function F: $\bar{X} \longrightarrow c$.

$$\min F(\bar{X}) \tag{2}$$

subject to given geometrical constraints:

$$g_i \geq 0 \quad , \quad i = 1, 2 \ldots, m ,$$
$$h_j = 0 \quad , \quad j = 1, 2 \ldots, k .$$

The formulation presented above can be generalized. Let a problem be defined by a function transformation:

$$J_T: \quad f_s \longrightarrow v, \tag{3}$$

where v is a function describing e.g. a distribution of reactions or boundary displacements.

Then an inverse problem can be considered:

Find f_s^*, such that $J_T(f_s^*)=v^*$, where v^* is a given, prescribed function. The function v^* can be obtained from experimental measurement (identification problem) or required by the designer.

Because such the function v^* may not exist we solve the modified problem:

$$\min_{f_s} d(v, v^*) = \min_{f_s} d(J_T(f_s), v^*) = \min_{f_s} d(f_s) = \min_{\bar{X}} d(\bar{X}) \tag{4}$$

where d denotes the distance between v and v^* defined by a metric.

Using the formulation (4) some identification problems as well as optimum design problems in junction and contact can be solved. For example, elements introducing a concentrated load into a thin walled elastic structures (Fig. 2) can be designed so as to avoid strong concentration of stresses (force applied in the form of assumed, known advantageously distributed load).

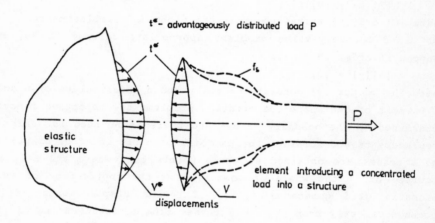

Fig. 2. Designing of an element introducing a concentrated load into the elastic structure.

An integrated shape design program including BEM analysis, spline boundary approximation, nonlinear programming techniques and interactive graphic presentation was developed ([9]). Some special techniques were examined e.g. computation with an increasing number of design variables.

Boundary element method

The analytical basis of the BEM in 2D elasticity (without body forces) is the integral equation ([10]):

$$c_{kj}(\bar{\xi})u_j(\bar{\xi}) = \int_\Gamma [U_i^{(k)}(\bar{x},\bar{\xi})t_i(\bar{x}) - T_i^{(k)}(\bar{x},\bar{\xi})u_i(\bar{x})]d\Gamma(\bar{x}) \qquad (5)$$

where $\bar{x}, \bar{\xi} \in \Gamma$, $\bar{u}$ and $\bar{t}$ denote displacement and traction vectors respectively (Fig. 1) and c_{kj} is the tensor dependent on the geometry of the boundary at point $\bar{\xi}$. If the boundary is smooth at $\bar{\xi}$ then $c_{kj}(\bar{\xi})=\delta_{kj}/2$ and $c_{kj}=\delta_{kj}$ if $\bar{\xi}$ is an interior point of Ω.
The kernel functions $U_i^{(k)}$ and $T_i^{(k)}$ are the fundamental Kelvin solutions for displacements and tractions due to the unit concentrated force in an elastic infinite space.

In order to solve (5) we transform the integral equation to a set of linear algebraic relations. The boundary is approximated by a set of boundary elements over which u_i and t_i vary in some assumed manner ([10]). The result is given as the system of 2N equations where N denotes the number of nodes on the boundary:

$$[T]\{u\} = [U]\{p\} \qquad (6)$$

where $\{u\}$ and $\{p\}$ represent nodal tractions and displacements.
For mixed boundary-value problems appropriate columns in (6) may be swapped in order to obtain :

$$[A]\{y\} = \{b\} \qquad (7)$$

where the vector $\{y\}$ consists of the unknown nodal values of u_i and t_i.

In most of the shape optimization problems the objective function is determined by the boundary values and there is no need for calculating displacements and stresses in the domain. In BEM interior displacements and stresses are obtained at those points only where the solution is required or in the whole of domain after the application of adequate automatic grid generator ([11]). Interior displacements ($\bar{\xi}\in\Omega$) are obtained directly from (5) and stresses from differentiating of (5) and application of Hooke's law.

The comparison BEM to FEM shows that the BEM usually delivers more accurate solutions than the FEM for the same level of computational costs, specially if we restrict the results to the boundary ([9],[12]). The BEM is well suited to problems with infinite domains and the procedures of the method require relatively small operating memory in spite of not banded, unsymmetric matrix [A].
This method is however less versatile for structural analysis and its applicability and efficiency are limited to some particular problems.

Representation of the boundary

A feature of good representation of the moving boundary is to provide sufficient shape flexibility using small number of design parameters. Shape description with cubic splines, successfully used in computer graphics ([13]), seems to be the adequate technique. In the simplest way the approximation of the unknown boundary curve is treated as the spline interpolation problem (Fig. 3) with the knots (z_i) and the angles at points A i B (α, β) being design variables X_i^o:

$$f_s(s) = X_i^o \, C_i(s) \tag{8}$$

However , this method corresponds with searching for the shape function f_s in the basis of functions $C_i(s)$, which have a disadvantageous oscillatory character.

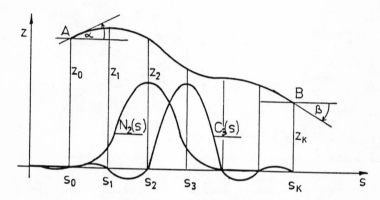

Fig. 3. Approximation of the moving boundary.

Application of B-splines is more efficient. It enables us to obtain the representation:

$$f_s(s) = X_i \, N_i(s) \tag{9}$$

where functions $N_i(s)$ are local and non-oscillatory.
A local property of the moving boundary (local modifications do not propagate) is advantageous and often required in the shape optimization techniques . Representation of the boundary (9) can be realized in cartesian or polar coordinates. The boundary curve is independently divided into a set of boundary elements.

Examples of application

Optimization of a shape of a hole in a plate (Fig. 4)
The aim of the design is to find the best possible shape of a hole in

an infinite plate under a biaxial stress field (σ_1, σ_2) minimizing the maximum Von Mises equivalent stress σ_{red} in the structure. This classical test problem was solved to examine the effectiveness and accuracy of the numerical algorithms.

Only the quarter of the plate ($x_1 \geq 0$, $x_2 \geq 0$) is modelled. The hole is described by 4 design variables in polar coordinates. The moving part of the boundary is divided into 24 elements. The distribution of σ_{red} along the boundary for the initial shape is showed in Fig. 4(a) and for the final shape in Fig. 4(b). In the presented case $(\sigma_2 = 0.5\sigma_1)$ the result of computation is $\sigma_{red}^{max} = 1.52 \, \sigma_1$ (known analytical result $1.50\sigma_1$).

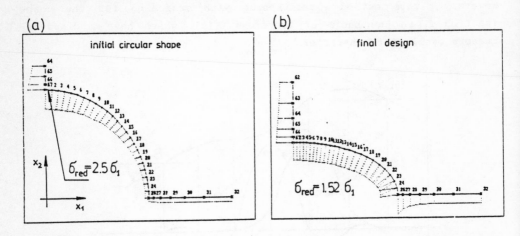

Fig. 4. Optimization of a hole in a biaxial stress field.

Designing of a reinforcement of the circular hole in a plate (Fig. 5)

This example is based on the generalized formulation (4). The aim of the design is to find a shape of external boundary of a plate reinforcement. The reinforcement is joined on its internal circular boundary with the plate in order to avoid the stress concentration in the plate around the hole. The ideal solution would be such a boundary shape (f_s^*) for which in the plate there was no concentration of stresses (constant state of stress σ_1, σ_2). In this case we can define corresponding tractions t_i^* and displacements u_i^* on the circular line AB which define the best interaction between the plate and the reinforcement.

The optimization consists in finding the shape function f_s, which produces displacements $\bar{u}$ on the curve AB and satisfies the condition $\min(d(\bar{u}, \bar{u}^*))$. In the case presented in Fig. 5(c) (3 design variables) the result is $d_{min} = 0.012d_o$, where d_o denotes the value of d for the initial circular shape. The thickness of the reinforcement was assumed as $h_{re} = 4h_{pl}$. It was verified that the corresponding stress concentration factor was reduced from $1.37\sigma_1$ to $1.07\sigma_1$.

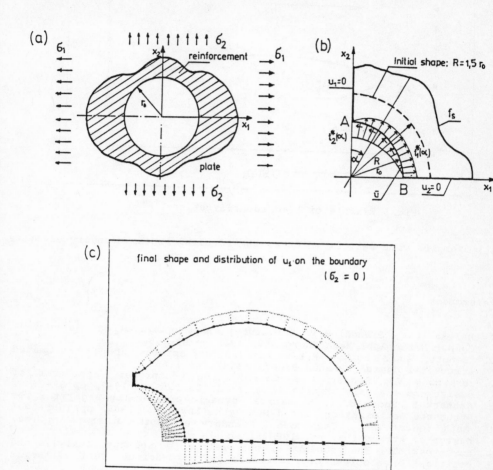

Fig. 5. Optimum design of the reinforcement of the hole.

Identification of unknown load applied to a structure (Fig. 6)

In this case, based on the formulation (4), design variables control unknown distribution of load which produces known distribution of displacements. In the presented example (Fig. 6) the identified (parabolic) load acts on the upper part of the boundary. The displacement $\overset{\bullet}{u}_2$ on the lower part of the boundary is known. Fig. 6 presents the identified load and its distribution which results from optimization (5 design variables).

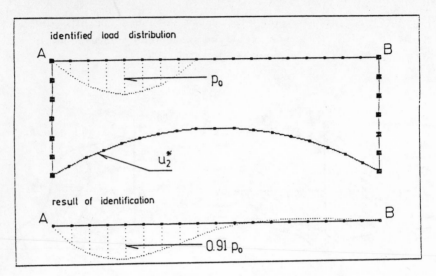

Fig. 6. Example of load identification.

All examples were calculated on an IBM-PC using linear boundary elements.

References

1. Haftka R. T. , Grandhi R. V. : Structural shape optimization - a survey, Comput. Meths. Appl. Mech. Engrg. 57, 1986, pp. 91-106.
2. Benett J. A. , Botkin M. E. (Eds.): The optimum shape. Automated structural design, Plenum Press, New York, 1986.
3. Braibant V. , Sander G. , : Optimization techniques: synthesis of design and analysis, Finite Elem. Anal. Design 3, 1987, pp. 57-78.
4. Schnack E. , Sporl U. : A mechanical dynamic programming algorithm for structure optimization, Int. J. Num. Meths. Engrg. 23, 1986, pp. 1985-2004.
5. Mota Soares C. A. , Choi K. K. : Boundary elements in shape optimal design of structures, pp. 199-231 in [2]
6. Burczynski T. : The boundary element method for selected analysis and optimization problems of deformable systems, Scient. Bull. Silesian Tech. Univ. , 1989 (in Polish)
7. Taylor J. E. : Anomalies arising in analysis and computational procedures for the prediction of optimal shape, pp. 353-363 in [2].
8. Kikuchi N. , Chung K. Y. , Torigaki T. , Taylor J. E. : Adaptive finite element methods for shape optimization of linearly elastic structures, Comput. Meths. Appl. Mech. Engrg. 57, 1986, pp. 67-89.
9. Krzesinski G. : Application of BEM to optimum shape design of elastic structures. Ph. D. Thesis, Warsaw Techn. Univ. , 1989 (in Polish).
10. Brebbia C. A. , Telles J. C. F. , Wrobel L. C. : Boundary element techniques, Springer Verlag, 1984.
11. Krzesinski G. , Zmijewski K. H: Automatic generation of triangular meshes in arbitrary complex 2D domains, Mechanics & Computer, Polish Ac. Sci. , V 10 in the press (in Polish).
12. Radaj D. , Mohrmann W. , Schilberth G. : Economy and convergence of notch stress analysis using boundary and finite element methods. Internat. J. Num. Meths. Engrg. 20, 1984, pp. 565-572.
13. Pavlidis T. : Graphics and image processing. Springer Verlag, 1982.

Applications of the COC Algorithm in Layout Optimization

G.I.N. Rozvany and M. Zhou
Fachbereich 10 (Bauwesen), Universität Essen, Postfach 10 37 64,
D-4300 Essen 1, Germany

ABSTRACT. Whilst the advantages of iterative continuum-based optimality criteria (COC) methods in cross-section optimization ("sizing") were discussed elsewhere (e.g. [4, 5]), this paper is devoted to applications of the above technique in layout optimization. Earlier studies of the latter field by others used a two-stage process consisting of separate topological and geometrical optimization and considered only a very small number of members. It will be shown here that the COC algorithm achieves a simultaneous optimization of the topology and geometry in layout problems with many thousand potential members. Moreover, some observations on shape optimization by the "homogenization" method are offered and an alternative approach to global shape optimization is suggested.

1. Introduction

One of the basic difficulties in structural optimization at present is the considerable *discrepancy between the analysis capability* ($10^4 - 10^5$ DF's) *and optimization capability* (a few hundred DF's) of currently available soft- and hardware, if primal-type mathematical programming (MP) methods are employed. It was shown recently that an *optimizer* based on iterative continuum-type optimality criteria (COC) methods could handle *several million variables* if a suitable *analyser* were available for systems of that magnitude. This means that the COC technique not only eliminates but even reverses the above discrepancy. In the present paper, applications of the COC methods to layout optimization are considered and potential extensions to global shape optimization are outlined.

2. The Theory of Optimal Layouts

The above theory was developed in the late seventies by W. Prager and the first author, and extended considerably by the latter in the eighties. It is based on two underlying concepts, namely:
- *continuum-based optimality criteria* (COC), which are necessary (in convex problems also sufficient) conditions for cost minimality, and
- the *structural universe* (also called in the literature "basic structure" or "ground structure"), which is the union of all potential members.

   Some of the above optimality criteria are usually reinterpreted in terms of a fictitious or *"adjoint structure"* as equilibrium, compatibility and strain-stress relations. Since they also give an adjoint strain requirement (usually an inequality) for vanishing or non-optimal members (of zero cross-section), satisfaction of all optimality criteria for the entire structural universe ensures layout optimality in convex problems.

The original, or *classical layout theory* considered structures of low material-volume/domain-volume ratio (termed "volume fraction"), such as trusses, grillages, arch grids and cable nets. In these structures the effect of the member intersections on strength, stiffness and structural weight is neglected and the specific cost per unit area or volume is therefore the sum of the specific costs of the members in various directions.

In the so-called *advanced layout theory*, the volume fraction is relatively high and hence the above simplifying assumption is unjustified. Considering a two- or three-dimensional solid, the optimal solution usually consists of three types of regions, namely: (a) *solid (black) regions* filled with material, (b) *empty (white) regions*, and (c) *perforated (grey) regions* having a very fine system of cavities. The *microstructure* of the latter is usually first optimized locally and then the *layout* of the corresponding ribs or fibres is optimized on a macroscopic scale [1, 2].

3. Iterative COC Methods Based on the Classical Layout Theory

In order to illustrate this procedure with a simple example, the equations of optimal elastic design for a single deflection constraint are given in Fig. 1, in which $Q = (Q_1, \ldots, Q_n)$ are generalized stresses (stresses or stress resultants), q are generalized strains, p are loads, u are displacements, $x \in D$ are spatial coordinates where D is the structural domain containing all potential member centerlines, S_1 and S_2 are subsets of D, z are cross-sectional dimensions, $\psi(z)$ is the "specific cost" where the "total cost" to be minimized is $\Phi = \int_D \psi(z) \, dx$. Quantities having an overbar are associated with the adjoint system. The symbol $\mathcal{G}$ denotes a generalized gradient which, however, reduces to the usual gradient operator for differentiable functions, e.g. $\mathcal{G}_{,z}[\psi(z)] = \mathrm{grad}\psi = (\partial\psi/\partial z_1, \ldots, \partial\psi/\partial z_r)$.

The above optimality conditions have been derived using the calculus of variations and represent an exact optimal solution. However, for large, real systems an analytical solution of the equations in Fig. 1, in general, cannot be obtained.

The iterative COC method solves the above equations using the following two steps in each iteration:

(a) *analysis* of the real and adjoint structures in a discretized form using an FE program;

(b) *resizing* (i.e. evaluation of the cross-sectional dimensions z_{ji}) for given values of the generalized stresses Q_i and $\overline{Q}_i$.

A comprehensive set of optimality criteria for various combinations of design constraints and geometrical restrictions is given in a recent book [3] by the first author.

In the iterative COC procedure, it is necessary to prescribe a minimum value for each cross-sectional parameter z_{ji} of each element i. In layout optimization, a very small value (usually 10^{-12}) is adopted for these lower limits.

For a detailed description of the COC procedure, the reader is referred to previous publications of the authors [4, 5].

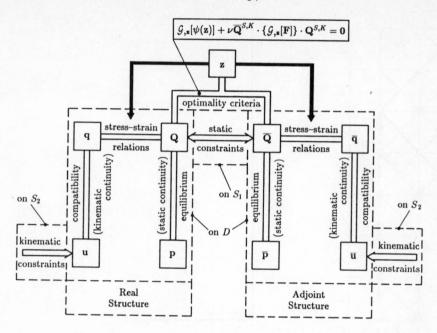

$$\mathcal{G}_{,\mathbf{z}}[\psi(\mathbf{z})] + \nu \overline{\mathbf{Q}}^{S,K} \cdot \{\mathcal{G}_{,\mathbf{z}}[\mathbf{F}]\} \cdot \mathbf{Q}^{S,K} = 0$$

Fig. 1 Fundamental relations of optimal elastic design with a deflection constraint.

4. Applications of the Classical Layout Theory

4.1 Analytical Results

The layout of least-weight *trusses* was studied already around the turn of the century by Michell [6], but relatively few exact analytical solutions were established until recently. On the other hand, the layout of *least-weight beam systems (grillages)* has been determined for most boundary and loading conditions in a closed analytical form (e.g. [7, 8]) and even a computer program has been developed for deriving by purely analytical operations and plotting the optimal grillage layout [9]. In addition, closed form analytical solutions were determined for arch-grids and cable nets (e.g. [10]), even for the combination of external load and selfweight. Quite recently, a systematic exploration of least-weight truss layouts for various boundary conditions was carried out (e.g. [11]).

4.2 Numerical Results by the COC Method: Test Examples

4.2.1 Vertical and Horizontal Supporting Lines with a Point Load

The first truss layout investigated by the iterative COC method is shown in Fig. 2a in which thick lines indicate optimal members. All other members took on a cross-sectional area of 10^{-12}, except for members "a" and "b" which were $< 3 \times 10^{-12}$. The COC solution agreed with the analytical solution to a *twelve digit accuracy*. The adjoint field associated with the exact solution, which was found later [11], is shown in Fig. 2b.

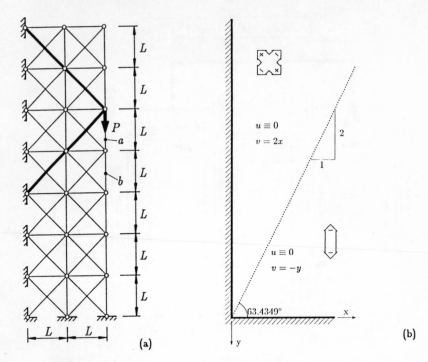

Fig. 2 The first layout problem investigated by the iterative COC method.

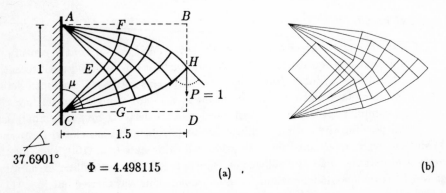

Fig. 3 Exact and iterative COC solutions for a layout consisting of a Hencky net.

4.2.2 Vertical Point Load and Vertical Supporting Line of Limited Length

The analytical solution for this problem was discussed by Hemp [12] and is shown in Fig. 3a. The optimal solution obtained using the iterative COC method with 5055 potential members is shown in Fig. 3b which represents 0.76% greater structural weight than the exact solution. With 12992 members, this error reduced to 0.46%.

4.2.3 Simply Supported Truss with a Point Load

Figures 4a–d show one half of a COC solution for a truss with a central point load and

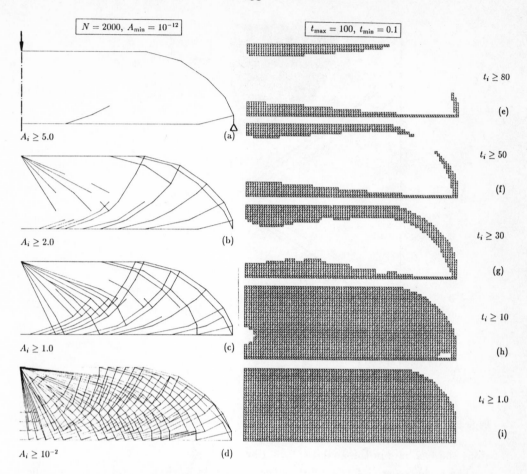

$N = 2000,\ A_{\min} = 10^{-12}$

$A_i \geq 5.0$ (a)

$A_i \geq 2.0$ (b)

$A_i \geq 1.0$ (c)

$A_i \geq 10^{-2}$ (d)

$t_{\max} = 100,\ t_{\min} = 0.1$

$t_i \geq 80$ (e)

$t_i \geq 50$ (f)

$t_i \geq 30$ (g)

$t_i \geq 10$ (h)

$t_i \geq 1.0$ (i)

Fig. 4   A comparison of a truss layout and a plate of variable thickness obtained by the COC method.

two simple supports and various ranges of cross-sectional areas. For a comparison, Figs. 4e-i show the COC solution for a plate of variable thickness (t) with the same support conditions and loading.

4.2.4 Optimal Truss for Two Alternate Load Conditions

Figure 5 shows the COC solution for two alternate loading conditions and compliance constraints using 7170 potential members in the structural universe, part of which is shown in the top right corner. The analytical solution consists of two bars (broken lines in Fig. 5) which the numerical solution tries to attain with a limited number of admissible member directions. While the above solution represents a 3.495% weight error in comparison with the analytical solution, an improved COC run with 12202 members gave an error of only 0.0198% and consisted mostly of heavy cross-sections

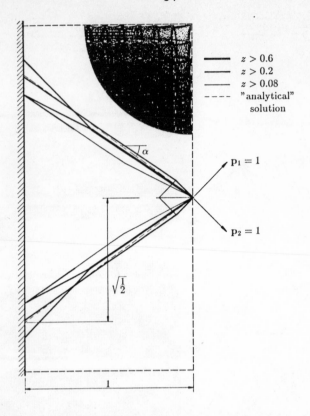

Fig. 5   Iterative COC solution for two alternate loads and part of the structural universe.

along the two bars in the analytical solution.

The authors were informed during the meeting in Karlsruhe that in an unpublished symposium paper Bendsøe and Ben-Tal [18], who used a different method, also presented optimal truss layouts involving several thousand potential members.

5. Shape Optimization by Homogenization

It was mentioned in Section 2 under *advanced layout theory* that in shape optimization one type of region, termed *perforated (grey) region*, contains a fine system of cavities or, theoretically, an infinite number of internal boundaries. A similar result was obtained in plate optimization where optimal solid and perforated plates were found ([17, 19], see also [1, 2, 13]) to contain systems of ribs of infinitesimal spacing and hence the thickness function has an infinite number of discontinuities over a finite width. Although the material in these problems is isotropic, the "homogenization" method consists of replacing ribbed or fibrous elements with homogeneous but anisotropic elements whose stiffness or strength is direction- but not location-dependent within the element (e.g. [19]).

From a historical point of view, the basic idea of homogenization was introduced already by Prager and the first author (e.g. [7, 8]), although they used the terms

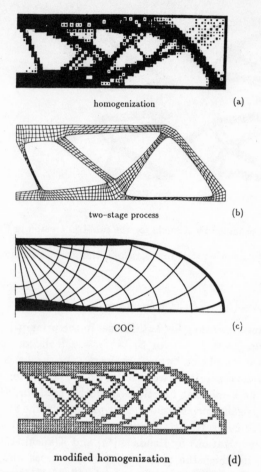

homogenization   (a)

two-stage process   (b)

COC   (c)

modified homogenization   (d)

Fig. 6 (a, b) Simplified topology derived by Olhoff and Rasmussen for the problem treated in Fig. 4, (c) the exact topology suggested by COC layout solutions, and (d) solution obtained by a modified homogenization method (see Section 5.1).

"grillage-like continua" and "truss-like continua". In these structures, a theoretically infinite number of bars or beams occur over a unit width but the above authors replaced this system with a continuum, whose specific cost, stiffness and strength depended on the "lumped" width of the bars or beams over a unit width. This concept was clearly equivalent to homogenization, but without considering its mathematical implications. Applications of homogenization in global shape optimization were discussed in milestone contributions by Bendsøe (e.g. [14]) which represent one of the most important recent developments in structural optimization.

Returning now to the shape optimization of a perforated plate in bending or in plane stress, various mathematical studies suggested that the optimal microstructure consists of ribs of first- and second-order infinitesimal width in the two principal directions, if the structure is optimized for a given compliance. The stiffness and cost properties of this microstructure were discussed in papers by the first author, Olhoff, Bendsøe *et al.*

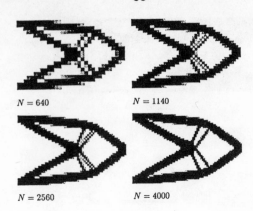

$N = 640$ $N = 1140$

$N = 2560$ $N = 4000$

Fig. 7 Simplified topology obtained by Kikuchi for the problem treated in Fig. 3.

[1, 2]. For a zero value of Poisson's ratio, for example, the non-dimensionalized specific cost becomes

$$\psi = (S_1 - 2S_1 S_2 + S_2)/(1 - S_1 S_2) , \tag{1}$$

where S_1 and S_2 are the non-dimensionalized stiffnesses in the principal directions. The above formula, for example, gives $\psi = 1$ for $S_1 = S_2 = 1$ (solid or "black" regions) and $\psi = 0$ for $S_1 = S_2 = 0$ (empty or "white" regions). On the basis of the specific cost function in (1), *analytical* solutions were obtained for axisymmetric perforated plates [1, 2]. Naturally, the same cost functions could be used for *numerical* shape optimization, and should give the correct solution for plates in plane stress or bending. This development is pursued currently by the authors.

The main aim of the investigations by Bendsøe [14] and Kikuchi [15] is to come up with a practical topology, in which the perforated ("grey") areas disappear and the optimal structure consists of solid ("black") and empty ("white") regions only. This procedure can be seen from Figs. 6a and b, in which Olhoff, Bendsøe and Rasmussen [16] first determined the approximate optimal topology (Fig. 6a) for the support condition and loading considered in Fig. 4, and then carried out a separate shape optimization (Fig. 6b) for the more detailed design conditions. This "homogenization" method, indeed, gives negligibly small perforated (grey) areas, as can be observed in Fig. 6a, and also in Fig. 7 which was obtained by Kikuchi [15] for the load and support conditions in Fig. 3. The latter seems to give the same topology, irrespective of the number of elements (N) employed.

The following circumstantial evidence seems to indicate that the *exact* optimal topologies differ from those obtained by the homogenization method (e.g. Figs. 6b, 7):

- In the analytical solutions obtained for perforated plates [1, 2], a high proportion of the plate area is covered by perforated (grey) areas.
- It was shown previously [1, 2] that the solution for very low volume fractions tends to that for grid-type structures (Michell frames or least-weight grillages). This was also observed by Prager who commented on some optimal solid plate designs by

Cheng and Olhoff [17]. It was also noted [1] that, as the volume fraction increased progressively in analytical solutions, solid (black) regions developed in areas where the ribs in the perforated regions had the greatest density. Making use of this observation, the grid-type solutions in Figs. 4a-d imply the topology in Fig. 6c (graphics by Dr. Gollub), in which the width of the solid regions is based on the cross-sectional areas of the "concentrated" bars along the top and bottom chords of the truss. The spacing of the members in the perforated region is theoretically infinitesimal, but a finite number of members would have to be used in any practical solution. In the neighbourhood of the top right corner, we have an empty (white) region. It can also be observed that the "homogenized" solution in Fig. 6a tries to achieve the solution in Fig. 6c, except that in areas of low rib density (e.g. right bottom region inside the chord) it comes up with empty regions. As Bendsøe pointed out in Karlsruhe, this can be attributed to the fact that here non-optimal microstructures were used for the perforated (grey) regions. Whilst this does not change the cost of solid (black) and empty (white) regions, it does increase artificially the cost of perforated (grey) regions and hence it tends to suppress the latter.

- The solutions for plates of varying thickness represent an "isotropized" version of the exact solution. This can be observed by comparing Figs. 4e-i with Fig. 6c. The plate thickness in the former is roughly proportional to the average material density over the latter, with ribs occurring in Figs. 4e-i along the solid regions of Fig. 6c. This is a further confirmation of the improved topology in Fig. 6c. Moreover, a modified isotropized homogenization method (Section 5.1) fully confirmed the solution in Fig. 6c, as can be seen from Fig. 6d.

Remark. The contention that existing homogenization methods give a *simplified* topology compared to the exact solution *by no means represents a criticism of these extremely important techniques.* Such "condensed" topologies are in fact very practical because, naturally, it is impossible to use an infinitesimal bar spacing in real structures. However, the exact optimal topology can also be of practical significance, because the client could be told by the designer that further weight savings can always be achieved by increasing the number of "holes" in the design and then former could decide as to how far he can go within realistic manufacturing capabilities.

5.1 An Alternative Method for Suppressing Perforated (Grey) Regions

Since the use of *non-optimal* microstructures homogenized into an *anisotropic* continuum introduces some *unknown* penalty for "grey" regions into shape optimization, the perforated (grey) regions could also be suppressed by using an *isotropic* microstructure but with a suitable penalty function for such regions. This can also be justified on practical grounds, as can be seen from the argument that follows.

As a first approximation, we could assume that the specific material cost (i.e. weight) is roughly proportional to the specific stiffness of perforated regions (Fig. 8a, which is also valid for plates of variable thickness). On the other hand, the extra manufacturing cost of cavities would increase with the size of the cavities if we consider a casting process requiring some sort of formwork for the cavities (Fig. 8b). Note that for empty (white) regions with $s = 0$ the manufacturing cost also becomes zero. The specific

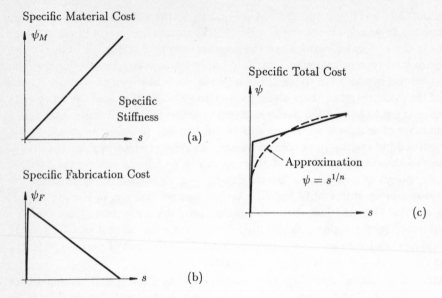

Fig. 8   Suppression of "grey" regions in isotropized solutions.

total cost and its suitable approximation is shown in Fig. 8c. The use of the above type of cost function promotes the suppression of perforated (grey) areas in isotropized designs which require *only one design parameter* (s) per element. The introduction of orthogonal cavities in the usual homogenization process [14, 15] requires 3 design variables per element for two-dimensional systems and 6 variables for three-dimensional ones. The solution in Fig. 6d was obtained with an n-value of 1.86 in Fig. 8c, and represents a topology closer to the "exact" optimal design than the design in Fig. 6b. As expected, simpler topologies can be obtained by adopting a higher n-value (i.e. by increasing the penalty for grey regions).

Conclusions

- The iterative COC method enables us, probably for the first time in the history of structural optimization
- (a) to optimize *simultaneously* the topology and geometry of grid-type structures (trusses, grillages, shell-grids, etc.),
- (b) for any combination of the usual design conditions (stresses, displacement, natural frequency, stability, etc. constraints),
- (c) using a fully automatic method capable of handling many thousand potential members.
- Shape optimization by "homogenization" is essentially a numerical approximation of the solutions furnished by the "advanced layout theory". The discretization errors in topologies obtained by this method can be assessed by a comparison with COC solutions for grid-like systems (e.g. Figs. 4a-d) or for "isotropized" systems (e.g. Figs. 4e-i). With a suitable penalty formulation, the latter could also be used for suppressing perforated (grey) regions in shape optimization.

References

1. Rozvany, G.I.N.; Olhoff, N.; Bendsøe, M.P.; Ong, T.G.; Sandler, R.; Szeto, W.T.: Least-Weight Design of Perforated Elastic Plates, I and II. *Int. J. of Solids Struct.* **23**, 4, 521-536, 537-550, (1987).

2. Ong, T.G.; Rozvany, G.I.N.; Szeto, W.T.: Least-Weight Design of Perforated Elastic Plates for Given Compliance: Nonzero Poisson's Ratio. *Comp. Meth. Appl. Mech. Eng.* **66**, 3, 301-322, (1988).

3. Rozvany, G.I.N.: *Structural Design via Optimality Criteria.* Kluwer Acad. Publ., Dordrecht, (1989).

4. Rozvany, G.I.N.; Zhou, M.; Rotthaus, M.; Gollub, W.; Spengemann, F.: Continuum-Type Optimality Criteria Methods for Large Finite Element Systems with a Displacement Constraint, Part I. *Struct. Optim.* **1**, 1, 47-72, (1989).

5. Rozvany, G.I.N.; Zhou, M.; Gollub, W.: Continuum-Type Optimality Criteria Methods for Large Finite Element Systems with a Displacement Constraint, Part II. *Struct. Optim.* **2**, 2, 77-104, (1990).

6. Michell, A.G.M.: The Limits of Economy of Material in Frame-Structures. *Phil. Mag.* **8**, 47, 589-597, (1904).

7. Prager, W.; Rozvany, G.I.N.: Optimal Layout of Grillages. *J. Struct. Mech.* **5**, 1, 1-18, (1977).

8. Rozvany, G.I.N.: Optimality Criteria for Grids, Shells and Arches. In: Haug E.J.; Cea, J. (Eds.), *Optimization of Distributed Parameter Structures.* (Proc. NATO ASI held in Iowa City), pp. 112-151, Sijthoff and Noordhoff, Alphen aan der Rijn, The Netherlands, (1981).

9. Hill, R.H.; Rozvany, G.I.N.: Prager's Layout Theory: A Nonnumeric Computer Method for Generating Optimal Structural Configurations and Weight-Influence Surfaces. *Comp. Meth. Appl. Mech. Engrg.* **49**, 1, 131-148, (1985).

10. Rozvany, G.I.N.; Wang, C.M.: On Plane Prager Structures, I and II. *Int. J. Mech. Sci.* **25**, 7, 519-527, 529-541, (1983).

11. Rozvany, G.I.N.; Gollub, W.: Michell Layouts for Various Combinations of Line Supports, Part I. *Int. J. Mech. Sci.*, proofs returned.

12. Hemp, W.S.: *Optimum Structures.* Clarendon, Oxford, (1973).

13. Rozvany, G.I.N.; Olhoff, N.; Cheng, J.-T.; Taylor, J.E.: On the Solid Plate Paradox in Structural Optimization. *J. Struct. Mech.* **10**, 1, 1-32, (1982).

14. Bendsøe, M.P.: Optimal Shape Design as a Material Distribution Problem. *Struct. Optim.* **1**, 4, 193-202, (1989).

15. Kikuchi, N.; Suzuki, K.: Shape and Topology Optimization by the Homogenization Method. In: Rozvany, G.I.N. (Ed.), *Proc. CISM Advanced School, Shape and Layout Optimization of Structural Systems*, (held in Udine, July 1990), Springer-Verlag, Vienna, (1991).

16. Olhoff, N.; Bendsøe, M.P.; Rasmussen, J.: On CAD-Integrated Structural Topology and Design Optimization. *Proc. 2nd World Congress on Computational Mechanics*, pp. 95-99, Int. Assoc. Comp. Mech., Stuttgart, 1990.

17. Cheng, K.-T.; Olhoff, N.: An Investigation Concerning Optimal Design of Solid Elastic Plates. *Int. J. Solids Struct.* **17**, 3, 305-323, (1981).

18. Bendsøe, M.P.; Ben-Tal, A.: Truss Topology Optimization, *Symp. Polish Acad. Sci. Optimal Design and Control of Structures*, Jablonna, Poland, June 1990

19. Olhoff, N.; Lurie, K.A.; Cherkaev, A.V.; Fedorov, A.V.: Sliding Regimes and Anisotropy in Optimal Design of Vibrating Axisymmetric Plates. *Int. J. Solids Struct.* **17**, 931-948 (1981)

Shape Optimization in Machine Tool Design

M. Weck, W. Sprangers
WZL, Chair for Machine Tools
RWTH Aachen
Steinbachstr. 53/54
5100 Aachen
Germany

__Abstract__: Combining finite element analysis with optimization algorithms in optimization systems allows the design of machine tools to be distinguished by a maximum stiffness at the point of processing, optimum stress distribution on fillets, or a minimum weight for highly accelerated machine components. This lecture is intended to give an insight into the specific problems of optimizing the shape of machine tool components and the current work being done in this area at the chair of machine tools at the University of Aachen.

Introduction

In the field of machine tool design, depending on the application of the machine, different design objectives must be fulfilled. The finishing accurancy is the most important quality feature of a cutting machine and, therefore the stiffness at the point of processing becomes the main objective of design. In the design of machines finite element analysis is generally used today. Structural optimization is also becoming more and more important as the requirements for finishing precision and productivity increases. In addition, the pressure on the market and the lack of development time give structural optimization systems growing importance to preserve the competition of a company.

Structural optimization in machine tool design

In machine tool design the most important objectives are/1/:

- maximum stiffness at the point of processing
- minimal component weight
- smooth stress distribution on fillets or notches
- maximum natural frequency

The optimization aim of maximum stiffness at the point of processing of a machine tool is the most important objective function. A machine tool's main components usually have thin wall thickness and, therefore, can be built up with shell elements. In an optimization process the thickness of these elements are changed continuously or discretly depending on the production technology of the component (casting or welding). The design aim for highly accelerated machine parts is minimal weight in order to reduce the loads on the drive units. Here, fiber reinforced materials with low weight and high stiffness are becoming more and more important. By optimizing the angles of the fibers and the thickness of the layers a design with maximal stiffness and minimal weight can be generated.
Beside these sizing problems, shape optimization is also used in machine tool design: for example, the optimization of the stress distribution in fillets and notches and the minimum weight design of solid parts (eg. a machine tool base made of reaction resin concrete). This lecture will concentrate on the shape optimal design of machine tool components along with discribing current systems and optimization results.

Shape optimal design

In shape optimal design systems the description of the geometry that must be changed plays a very important role. The following requirements are imposed on these descriptions:
- a small number of design variables
- allowance for large changes in shape
In the case of the minimization of stress peaks in notches and

following these requirements cubic splines are used to describe the curve which has to optimized.
In the three dimensional case, the system developed at the WZL uses geometry elements for the optimization.

Minimization of notch stress in plain and rotationally symmetric structures.

The aim of this optimization is the minimisation of stress peaks on fillets and notches. <u>Fig. 1</u> gives some optimization examples and shows the method in more detail.

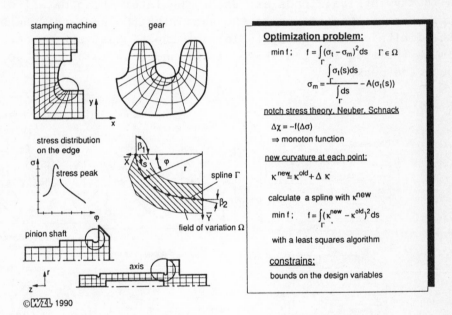

<u>Fig. 1:</u> Notch stress minimization, plain and rotationally symmetric case

The interesting curve is modelled by a spline with the possibility of defining a slope at the end knots to fit the curve in the structure. The formulation of a polar coordinate system where the angles to the points are fixed, leads to a parametric form for the optimization algorithm.

Based on the notch stress theory and the work of Neuber and Schnack, the mechanical problem is changed to a geometrical problem /2,3,4/. The solution to the optimization problem is done with the least squares algorithm /5/.

Minimization of the stress peaks in a pinion shaft.

In this example, a pinion shaft (a rotationally symmetric part) is optimized. <u>Fig. 2</u> shows the CAD representation of the shaft and the finite element mesh. Only half of the structure has to be modelled because rotationally symmetric elements with asymmetrical loads are used. The interesting detail of this part is the offset of the axes with the undercut where the finite element analysis locates the highest stress peaks. In an optimization the shape of the undercut is changed to reduce the peaks.

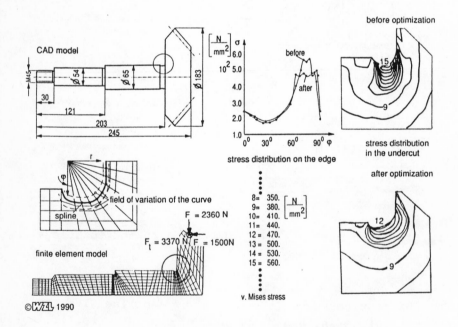

<u>Fig. 2:</u> Optimization of the undercut of a pinion shaft

Fig. 2 compares the stress distribution on the edge before and after the optimization. The change in shape is detailed in the right part of the figure. The stress peaks were reduced by 16%.

Optimization of solid structures

In the second part of the lecture a system for the shape optimum design of solid structures is presented. The aim of this optimization system is the weight minimisation or stiffness maximization under displacement, stress or geometrical constraints. In this system geometric elements are used for description of the shape.

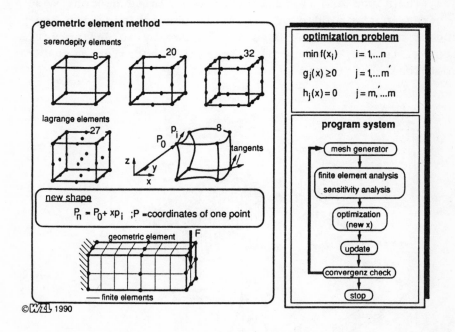

Fig. 3: Shape optimal design of solids

Fig. 3 shows the geometric elements available in the system. Four of these elements are defined by the coordinates of their

points and the shape function, the fifth element also needs
the slope directions at the points /6,7/. The coordinates and
the slope directions may be design parameters in the optimization which is done with the SQP method from Powell /8/.

Weight minimization of a machine tool base made of reaction
resin concrete

Machine tool parts made of reaction resin concrete, e.g. a machine tool base (fig. 4), must be described with hexaedron elements in order for carry out the finite element analysis. In
this optimization the structural volume is to be minimized.
Fig. 4 shows the model of the base build up with geometry elements. The loads and restraints of the finite element model
are given in the middle of fig 4. In the optimization the
shape of the hole of the base is changed by 10 design variables, x1 to x10, whereby displacement constraints and
geometrical constraints must be fulfilled.

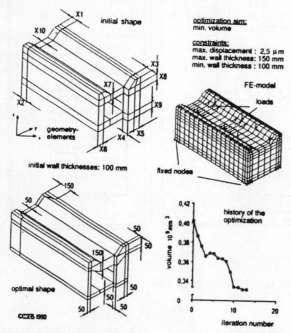

Fig. 4: Optimization of a machine tool base made of
        reaction resin concrete

The result of this optimization is shown in the lower part of
fig. 4. The thickness of the side walls, the front and back
plates and the ground plate were reduced. The top plate has
increased in thickness. The structural volume was reduced
by 18%.

Conclusion

The motivations for using optimization systems in the design
of machine tools are the increasing precision requirements and
productivity and the decreasing of development time. Shape
optimization is one of the manifold possibilities for the solution of a design problem. Problem oriented systems are a
good choice for effective optimization calculations as the
examples presented in this lecture have shown.

References

1. Weck, M.   Werkzeugmaschinen Band 2.
                              Konstruktion und Berechnung.
                              VDI-Verlag Düsseldorf, 1985.

2. Neuber, H.   Kerbspannungslehre Grundlage
                              für genaue Spannungs-
                              berechnung. Springer-Verlag,
                              Berlin 1937.

3. Schnack, E.   Ein Iterationsverfahren zur
                              Optimierung von Kerbober-
                              flächen. VDI-Forschungsheft
                              589, VDI-Verlag Düsseldorf,
                              1978.

4. Förtsch, F.   Entwicklung und Anwendung von
                                  Methoden zur Optimierung des
                                  mechanischen Verhaltens von
                                  Bauteilen. VDI-Z Reihe 1,166,
                                  VDI-Verlag, Düsseldorf, 1988.

5. Lindström, P.   A new linesearch algorithm for
   Wendin, P.   nonlinear least squares prob-
                                  lems. Rep. UMINF-82.81, issn
                                  0348-0542 Univ. Umea, 1983.

6. Mortenson, E.M.   Geometric modelling. John Wi-
                                  ley & Sons 1985.

7. Wassermann, K.   Three dimensional shape op-
                                  timization of arch dams with
                                  prescribed shape functions.
                                  Jour. struc. Mech. pp. 465-
                                  489, 1983-1984.

8. Powell, M.J.D.   A fast algorithm for non-
                                  linearly constrained opti-
                                  mization calculations. Lec-
                                  tures notes in Mathematics
                                  No.630. Ed. G.A. Watson
                                  Springer-Verlag Berlin 1987.

SHAPE OPTIMIZATION AND DESIGN ELEMENT SELECTION FOR PROBLEMS WITH CONSTRAINED STATIC AND DYNAMIC RESPONSE

B. Specht, H. Baier
Dornier GmbH
Friedrichshafen, W. Germany

SUMMARY AND INTRODUCTION

In many cases, design problems are governed by both static and dynamic constraints, where for the latter case engineering experience and feeling is often less evolved. In addition, the more challenging design variables are those for structural shape or the selection of proper types of design elements instead of only cross-sectional areas or plate thicknesses. In this paper some strategies and applications of optimal shape design will be presented which specifically include dynamic response constraints.

Concept of Software

The underlying software approach aims at the solution of general problems in structural optimization. Objective and constraint functions may depend on any results from statics, eigendynamics, buckling and combinations (e.g. statics + dynamics). Potential design variables are not only dimensioning parameters, but also shape parameters and material property parameters including orientation angles and layer thicknesses of laminates. The related software package OPOS can be identified as a data management system to provide the communication between software of optimization algorithms, stand along structural analysis and sensitivity, objective and constraints evaluation and potential further support routines (see Fig. 1).

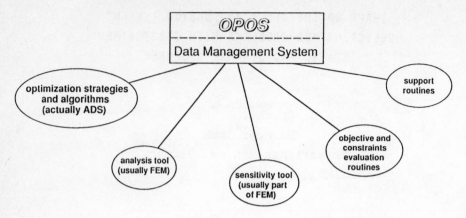

<u>Figure 1:</u>   The general software approach of OPOS

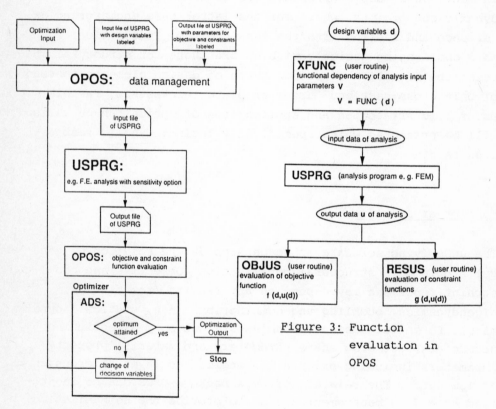

<u>Figure 2:</u> Data flow of OPOS

<u>Figure 3:</u> Function
            evaluation in
            OPOS

The interfacing problem between OPOS and the analysis program
(usually a FEM code) is solved in a peripheral way. That means
OPOS does not require interface data in special formats but
interprets the formatted I/O files of the analysis program
(USPRG in Fig. 2). Actually in use are the FEM programs ASKA,
PERMAS and SAP and an analysis code for axisymmetric shells
called BOSOR.

OPOS reads in an initial step a typical input and output file
of USPRG which fit to the actual structural problem. The user
has to lable in these files all places where variable values
have to be inserted or where response values have to be read
from, respectively. Using those information OPOS generates the
actual input file for USPRG with the variable values inserted
at all function evaluation steps within the optimization loop.
The required response values for objective and constraints evaluation are then read from the actual output file (see Fig. 2).
This data communication concept yields a very high degree of
generality in the definition of an optimization problem because
any analysis (e.g. FEM) input data is a potential variable.
Their dependency on the design parameters is defined in the
user routine XFUNC (see Fig. 3) in terms of an arbitrary algebraic relations. On the output side any analysis response data
can be used to evaluate the objective and constraint functions
via the user routines OBJUS and RESUS (see also Fig. 3).

With respect to a general approach the gradients are calculated
via finite differences. For that OPOS requires perturbed function values, which can be provided via either a complete reanalysis or a perturbation approximation. The latter approach is
based on a semianalytic approximation in statics or a modal
subspace approximation in dynamics. These perturbation analysis
are used with the FEM codes ASKA or PERMAS to provide the sensitivity data.

The optimizer implemented in OPOS is the program ADS [1] which
offers a wide spectrum of optimization strategies (e.g. direct
solution, sequential linearization, sequential quadratic programming) and algorithms (e.g. Flatcher-Reeves, Davidon-
Fletcher-Powell, BFGS for unconstrained problems and the method
of feasible directions for constrained problems).

Method of Shape Update

It is obvious for a number of reasons that a general approach
for shape optimization with FEM should not take directly the
nodal coordinates of the mesh as independent design variables
but vary them dependent on a relative small number of shape
parameters s_i. Each of these shape parameters is related to a
nodal displacement set which will be called a shape mode ψ_i.
The shape parameter values together define a linear combination
of the shape modes which is the resulting shape variation $\underline{u}$.

$$\underline{u} = \sum_i \underline{\psi}_i s_i \qquad\qquad (1)$$

The new nodal coordinates $\underline{x}$ of the modified mesh result from an
addition of the shape variation $\underline{u}$ to the old coordinates $\underline{x}^\circ$

$$\underline{x} = \underline{x}^\circ + \underline{u} \qquad\qquad (2)$$

The crucial point of the outlined approach is the evaluation of
the shape modes $\underline{\psi}_i$. This is done in OPOS by standard FEM algorithms based on an idea of Rajan and Belegundu [2]. For this
purpose the actual FEM mesh of the structure is copied into a
so called dummy mesh with nodal point geometry and topology unchanged. A shape mode is now generated from this dummy mesh
(see Fig. 4) by a static deformation analysis with boundary
conditions and loads defined by the user. Each of these dummy
load cases generates one shape mode. It is further obvious that
the dummy static analysis has to be performed only once and for
all in an initial step. Inside of the optimization loop one has
simply to combine the modes linearly according to formula (1).

The shape mode method as described above is a general approach
for updating of FEM meshes. Internal nodes are moved automatically in a natural way and hence mesh degenerations are at
least postponed. The limitations are no changes in mesh topo-

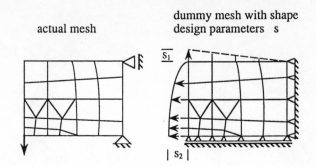

actual mesh

dummy mesh with shape design parameters s

Fig. 4: Dummy load case for generation of shape modes

logy and moderate shape variation until extreme element distorsions will spoil the results. In these cases the optimization has to be interrupted and restarted after remeshing.

Applications

The applications with OPOS cover a wide range of problems due to the already outlined generality of the software. Spacecraft automobile and mobile bridges are the main fields of application and the related optimization problems to be solved required a large flexibility in the problem definition.

Two examples with shape design variables are described in the following.

Example 1

The automobile parking latch which is outlined in Fig. 5 represents a 2D shape optimization problem. Variable parts of the contour are plotted in dashed lines. The aim is to minimize the stress peak in the notch due to a load at the contacts surface of the tooth. The constraints are geometrical limits of contour variations.

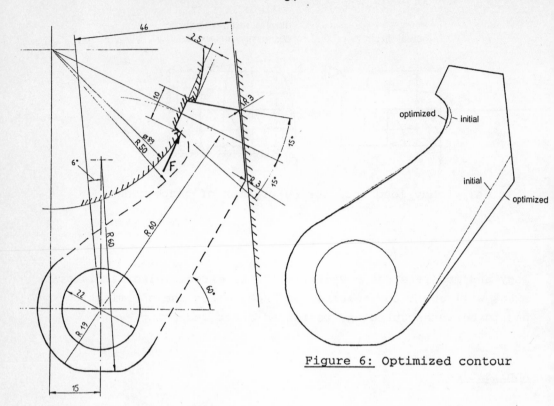

Figure 5: Geometry and static load of the parking latch

Figure 6: Optimized contour

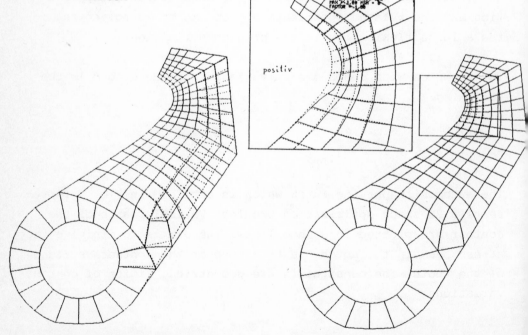

Figure 7: Shape Mode No. 1

Figure 8: Shape Mode No. 3

Fig. 7 and Fig. 8 show two of totally 5 shape modes for mesh
updating purpose. The mode in Fig. 7 represents a linear variation of a variable boundary part and has been generated via
prescribed boundary displacements as dummy load case. The mode
in Fig. 7 is one of 4 variations of the curved boundary part.
These modes have been generated by distributed normal boundary
forces as dummy load cases. The initial and the optimized contours are outlined in Fig. 6. The applied optimization strategy
was a direct solution with the method of feasible directions
[1]. The stress peak could be reduced by 25 % in 7 optimization
steps.

Example 2

A 3D shape optimization is requested in the design of rubber
parts of an automobile bearing (see Fig. 9). The objective is
to maximize the fundamental eigenfrequency of such a rubber
part with stiffness properties to be held fixed. But the constraints turned out to be too restrictive to allow a significant change in the objective. Hence only upper bounds for the
stiffness were defined for demonstration purpose.

Fig. 10 shows the 6 shape modes which were all generated by
dummy load cases in terms of prescribed boundary displacements.
The resulting optimized shape of the rubber part is outlined in
Fig. 11. The related change of the frequency objective can be
read from the the diagram in Fig. 12. There are two jumps in
the objective curve which indicate an approximation update for
the frequency evaluation. This evaluation is based on a modal
subspace approximation within each of the optimization cycles.
The approximation subspace is defined by the 10 lowest eigenmodes evaluated at the beginning of each optimization cycle.

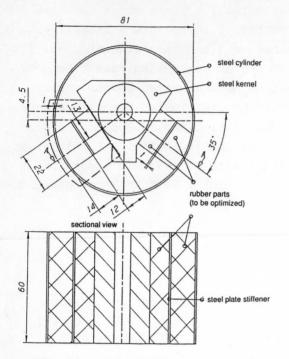

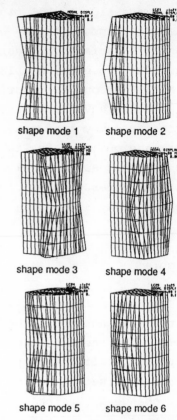

shape mode 1   shape mode 2

shape mode 3   shape mode 4

shape mode 5   shape mode 6

Figure 9: Rubber parts in an automobile bearing

Figure 10: Shape modes of FEM model of a rubber part

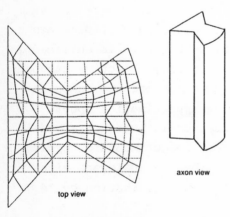

top view

axon view

Figure 11: Final shape of a rubber part

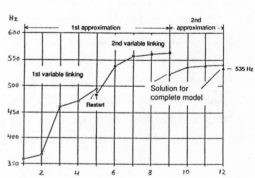

Figure 12: Fundamental frequency [Hz] over iteration steps

REFERENCES

[1] G.N. Vanderplaats; ADS - A Fortran Program for Automated
     Design Synthesis, Version 3.00, March 1988, Engineering
     Design Optimization, INC.

[2] S.D. Rajan, A.D. Belegundu, Shape Optimal Design Using
     Fictitious Loads. AIAA Journal, Vol. 27, No. 1, 1989,
     pp. 102-107

SRTUCTURAL OPTIMIZATION
WITH ADVANCED MATERIALS

OPTIMAL DESIGN WITH ADVANCED MATERIALS

Pauli Pedersen

Department of Solid Mechanics

The Technical University of Denmark, Lyngby, Denmark

Abstract — Recent results from sensitivity analysis for strain energy with anisotropic elasticity are applied to thickness and orientational design of laminated membranes. Primarily the first order gradients of the total elastic energy are used in an optimality criteria based method. This traditional method is shown to give slow convergence with respect to design parameters, although the convergence of strain energy is very good. To get a deeper insight into this rather general characteristic, second order derivatives are included and it is shown how they can be obtained by first order sensitivity analysis. Examples of only thickness design, only orientational design and combined thickness—orientational design will be presented.

1. INTRODUCTION

Design with advanced materials, such as anisotropic laminates, is a challenging area for optimization. We shall here restrict ourselves to plane problems, as in the early work of BANICHUK [1] (which includes further early references). Recent work by the author [2],[3] was conducted independently and the formulations are rather parallel. Similar research is carried out by SACCHI LANDRIANI & ROVATI [4]. In the present paper we combine these orientational optimizations with thickness optimization. The further goal is to get a deeper insight into the redesign procedures based on optimality criteria.

The sensitivity analysis that proves local gradient determination relative to a fixed strain field is presented. The physical understanding of these results have many aspects outside the scope of the present paper. The early paper by MASUR [5] includes valuable information about this sensitivity analysis.

For orthotropic materials, a single optimization parameter controls the orientational design. This parameter includes information about material as well as about the state of strain. It is used as an optimization criterion and in principle, the optimization procedure is a non—gradient technique. In this way local extrema are avoided.

When the principal axes of an orthotropic material are equal to, say, the principal strain axes, it follows directly that principal stress axes also equal those of material and strain. However, optimal orientations exist for which the principal axes of material differ from those of the principal strains. Even for this case it is proved in [3] that the principal axes of stress equal those of the principal strains.

The sensitivity analysis for thickness change is extended to include the mutual sensitivities, i.e. change in energy density with respect thickness changes not at the same point. A symmetry relation is proven.

A number of actual examples will be shown and discussed, but are not included in this short Conference paper.

2. SENSITIVITY ANALYSIS FOR ENERGY IN NON–LINEAR ELASTICITY

Let us start with the **work equation**

$$W + W^C = U + U^C \tag{2.1}$$

where W, W^C are physical and complimentary work of the external forces, and U, U^C are physical and complementary elastic energy, also named strain and stress energy, respectively.

The work equation (2.1) holds for any **design** h and therefore for the total differential quotient wrt. h

$$\frac{dW}{dh} + \frac{dW^C}{dh} = \frac{dU}{dh} + \frac{dU^C}{dh} \tag{2.2}$$

Now in the same way as h represents the design field generally, ϵ represents the strain field and σ represents the stress field. Remembering that as a function of h, ϵ we have W, U , while the complementary quantities W^C, U^C are functions of h, σ . Then we get (2.2) more detailed by

$$\frac{\partial W}{\partial h} + \frac{\partial W}{\partial \epsilon}\frac{\partial \epsilon}{\partial h} + \frac{\partial W^C}{\partial h} + \frac{\partial W^C}{\partial \sigma}\frac{\partial \sigma}{\partial h} = \frac{\partial U}{\partial h} + \frac{\partial U}{\partial \epsilon}\frac{\partial \epsilon}{\partial h} + \frac{\partial U^C}{\partial h} + \frac{\partial U^C}{\partial \sigma}\frac{\partial \sigma}{\partial h} \tag{2.3}$$

The **principles of virtual work** which hold for solids/structures in equilibrium are

$$\frac{\partial W}{\partial \epsilon} = \frac{\partial U}{\partial \epsilon} \tag{2.4}$$

for the physical quantities with strain variation and for the complimentary quantities with stress variation we have

$$\frac{\partial W^C}{\partial \sigma} = \frac{\partial U^C}{\partial \sigma} \tag{2.5}$$

Inserting (2.4) and (2.5) in (2.3) we get

$$\frac{\partial U^c}{\partial h} - \frac{\partial W^c}{\partial h} = -\left[\frac{\partial U}{\partial h} - \frac{\partial W}{\partial h}\right] \tag{2.6}$$

and for design independent loads

$$\left[\frac{\partial U^c}{\partial h}\right]_{\substack{\text{fixed} \\ \text{stresses}}} = -\left[\frac{\partial U}{\partial h}\right]_{\substack{\text{fixed} \\ \text{strains}}} \tag{2.7}$$

as stated by MASUR [5]. Note that the only assumption behind this is the design independent loads $\partial W/\partial h = 0$, $\partial W^c/\partial h = 0$.

To get further into a **physical interpretation** of $(\partial U/\partial h)_{\text{fixed strains}}$ (and by (2.7) of $(\partial U^c/\partial h)_{\text{fixed stresses}}$) we need the relation between external work W and strain energy U. Let us assume that this relation is given by the constant c

$$W = cU \tag{2.8}$$

For linear elasticity and dead loads we have $c = 2$ and in general we will have $c > 1$.

Parallel to the analysis from (2.1) to (2.3) we based on (2.8) get

$$\frac{\partial W}{\partial h} + \frac{\partial W}{\partial \epsilon}\frac{\partial \epsilon}{\partial h} = c\frac{\partial U}{\partial h} + c\frac{\partial U}{\partial \epsilon}\frac{\partial \epsilon}{\partial h} \tag{2.9}$$

that for design independent loads $\partial W/\partial h = 0$ with virtual work (2.4) gives

$$\frac{\partial W}{\partial \epsilon}\frac{\partial \epsilon}{\partial h} = \frac{\partial U}{\partial \epsilon}\frac{\partial \epsilon}{\partial h} = \frac{c}{1-c}\frac{\partial U}{\partial h} \tag{2.10}$$

and thereby

$$\frac{dU}{dh} = \frac{\partial U}{\partial h} + \frac{\partial U}{\partial \epsilon}\frac{\partial \epsilon}{\partial h} = \frac{1}{1-c}\left[\frac{\partial U}{\partial h}\right]_{\substack{\text{fixed} \\ \text{strains}}} \tag{2.11}$$

Note, in this important result that with $c > 1$ we have different signs for dU/dh and $(\partial U/\partial h)_{\text{fixed strains}}$.

For the case of **linear elasticity and dead loads** we have with $c = 2$ and adding (2.7)

$$\frac{dU}{dh} = -\left[\frac{\partial U}{\partial h}\right]_{\substack{\text{fixed} \\ \text{strains}}} = \left[\frac{\partial U}{\partial h}\right]_{\substack{\text{fixed} \\ \text{stresses}}} \tag{2.12}$$

For the case of **non–linear elasticity** by

$$\sigma = E\epsilon^n \tag{2.13}$$

and still dead loads $(W^C = 0)$ we get $c = 1+n$ and thereby

$$\frac{dU}{dh} = -\frac{1}{n}\left[\frac{\partial U}{\partial h}\right]_{\substack{fixed \\ strains}} = \frac{1}{n}\left[\frac{\partial U}{\partial h}\right]_{\substack{fixed \\ stresses}} \tag{2.14}$$

3. OPTIMALITY CRITERIA

We want to minimize the elastic strain energy U

$$\text{Minimize}\left[U = \sum_{e=1}^{N} U_e\right] \tag{3.1}$$

which is obtained as the sum of the element energies U_e for $e = 1,2,...,N$. Two groups of design parameters are considered. The material orientations θ_e for $e = 1,2,...,N$ assumed constant in each element, and the element thicknesses t_e for $e = 1,2,...,N$, also constant in each element. The constraint of our optimization problem is a given volume $\bar{V}$, i.e., by summation over element volumes V_e for $e = 1,2,...,N$

$$V - \bar{V} = \sum_{e=1}^{N} V_e - \bar{V} = 0 \tag{3.2}$$

The **gradients of volume** are easily obtained for thicknesses

$$\frac{\partial V}{\partial t_e} = \frac{\partial V_e}{\partial t_e} = \frac{V_e}{t_e} \tag{3.3}$$

and volume do not depend on material orientation

$$\frac{\partial V}{\partial \theta_e} = 0 \tag{3.4}$$

The **gradients of elastic strain energy** is simplified by the results of section two and thereby localized

$$\frac{\partial U}{\partial h_e} = -\left[\frac{\partial U}{\partial h_e}\right]_{fixed\ strains} = -\left[\frac{\partial(u_e\ V_e)}{\partial h_e}\right]_{fixed\ strains} \tag{3.5}$$

valid for $h_e = \theta_e$ as well as for $h_e = t_e$. The strain energy density u_e is introduced by $U_e = u_e V_e = u_e a_e t_e$ with a_e for element area.

With fixed strains, the thickness has no influence on the strain energy density u_e and thus with (3.3) and (3.5), we directly get

$$\frac{\partial U}{\partial t_e} = -\frac{U_e}{t_e} = -\frac{u_e V_e}{t_e} \tag{3.6}$$

With respect to material orientation the gradient is more complicated, because even with fixed strains will the energy density u_e depend on θ_e. A rather simple formula is derived in [2], in terms of principal strains ϵ_I, ϵ_{II} $\left(|\epsilon_I| > |\epsilon_{II}|\right)$ — angle ψ from direction of ϵ_I to principal material direction — and material parameters C_2 and C_3

$$\frac{\partial U}{\partial \theta_e} = \left[V(\epsilon_I - \epsilon_{II})^2 \sin2\psi \left[C_2 \frac{\epsilon_I + \epsilon_{II}}{\epsilon_I - \epsilon_{II}} + 4C_3 \cos2\psi\right]\right]_e \tag{3.7}$$

With the gradients determined by (3.3), (3.4), (3.6) and (3.7) we can now formulate optimality criteria. For the **thickness optimization** the well–known criterion of proportional gradients gives $-u_e V_e/t_e \sim V_e/t_e$ which means constant energy density, equal to the mean strain energy density $\bar{u}$

$$u_e = \bar{u} \text{ for all } e \tag{3.8}$$

See also the early paper by MASUR [5] for this optimality criterion.

For the **material orientation optimization** we have an unconstrained problem, and thus from (3.7) the optimality criterion

$$\sin2\psi \left[C_2 \frac{\epsilon_I + \epsilon_{II}}{\epsilon_I - \epsilon_{II}} + 4C_3 \cos2\psi\right]_e = 0 \text{ for all } e \tag{3.9}$$

How is a thickness distribution that fulfill (3.8) obtained? We shall discuss a practical procedure, cf. ROZVANY [6], which is based on a number of approximations. Firstly, we neglect the mutual influences from element to element, i.e. each element is redesigned independently (but simultaneously)

$$(t_e)_{next} = t_e + (\Delta t_e) \tag{3.10}$$

Secondly, the optimal mean energy density $\bar{u}$ is taken as the present mean energy density $\tilde{u}$. Thirdly, the element energy U_e is assumed constant through the change Δt_e and then from (3.8) we get

$$\frac{U_e}{V_e(1 + \Delta t_e/t_e)} = \tilde{u} \quad , \text{i.e.}$$

(3.11)

$$\Delta t_e = t_e(u_e - \tilde{u})/\tilde{u} \quad \text{or} \quad (t_e)_{next} = t_e\, u_e/\tilde{u}$$

It is natural to ask, why the gradient of element energy is not taken into account

$$(U_e)_{next} = U_e + \frac{\partial U_e}{\partial t_e} \Delta t_e$$

(3.12)

but this is explained by the fact that although $\partial U/\partial t_e$ is known by (3.6) the gradient of the local energy (the element strain energy)

$$\frac{\partial U_e}{\partial t_e} = \left[\frac{\partial U_e}{\partial t_e}\right]_{\text{fixed strain}} + \left[\frac{\partial U_e}{\partial \epsilon}\right] \frac{\partial \epsilon}{\partial t_e}$$

(3.13)

is more difficult to determine. The two terms in (3.13) have different signs, and also the other neglected terms $\partial U_e/\partial t_i$ for $e \neq i$ may be of the same order. Although the procedure (3.11) mostly work rather satisfactory, we shall extend our analysis to the coupled problem.

4. MUTUAL SENSITIVITIES

The redesign procedure by (3.11) neglect the mutual sensitivities, i.e. the change in element energy due to change in the thickness of the other elements. These sensitivities can be calculated by classical sensitivity analysis. Assume the analysis is related to a finite element model

$$[S]\{D\} = \{A\}$$

(4.1)

where $\{A\}$ are the given nodal actions, $\{D\}$ the resulting nodal displacements and $[S] = \sum_e [S_e]$ the system stiffness matrix accumulated over the element stiffness matrices $[S_e]$ for $e = 1,2,...,N$.

Let h_e be an element design parameter without influence on $\{A\}$, then we get

$$[S] \frac{\partial \{D\}}{\partial h_e} = -\frac{\partial [S]}{\partial h_e} \{D\} = \{P_e\}$$

(4.2)

where the right-hand side $\{P_e\}$ is a pseudo load, equivalent to design change. Knowing $\partial\{D\}/\partial h_e$ it is straight forward to calculate $\partial U_i/\partial h_e$. Generally the computational efforts correspond to one additional load for each design parameter.

Then with all the gradients $\partial U_e/\partial t_i$ available we can formulate a procedure for simultaneously redesign of all element thicknesses

When maximum value criteria are used, feasibility of the constraint functionals $\tilde{g}_i$ and $\hat{h}_j$ at every time within the given time-intervall is desired. I.e. the most critical value has to be examined. Together with the first two steps the optimization problem results in a min-max parameter optimization problem:

$$\text{Min} \left\{ \bar{f}[\underline{a}(\underline{x})] = \max_t f[\underline{u}(t), \tilde{\underline{\xi}}\{\underline{a}(\underline{x}), t\}, t] \right|$$

$$\bar{g}_i[\underline{a}(\underline{x})] = \max_t g_i[\underline{u}(t), \tilde{\underline{\xi}}\{\underline{a}(\underline{x}), t\}, t] \leq 0, \ i = 1,\ldots,n_G; \tag{6}$$

$$\bar{h}_j[\underline{a}(\underline{x})] = \max_t |\hat{h}_j[\underline{u}(t), \tilde{\underline{\xi}}\{\underline{a}(\underline{x}), t\}, t]| = 0, \ j = 1,\ldots,n_H-n_u; \ t \in [t_0, t_1] \right\}.$$

The calculation of the maximum values is a subordinated optimization problem and can be treated separately. For this purpose a special adaptive search technique is used which is an extension of the algorithm described in [6].

The three steps of the direct optimization strategy are an integrated part of the optimization procedure SAPOP [7-9]. Fig. 1 contains the extensions of the optimization loop. After determination of the form parameters by a design model the form functions are calculated at time t_0. A subsequent structural analysis produces the actual values of the state functions. The results of the evaluation model are the values of objective and constraint functions. Applying the adaptive search technique the inner optimization loop is repeated several times until the maximum values of the functions are found. The latter are subsequently delivered to the optimization algorithm, which calculates a new set of design variables and starts the global optimization loop once more until a convergency criteria responds.

3. DESIGN AND EVALUATION MODEL

Following the optimization strategy suitable form functions for composites have to be defined. For this the fibre angles and layer thicknesses of the laminate are interpreted as form parameters of the form functions "laminate stiffnesses" and "hygrothermal force resultants" [10]. The design variables are attached to layer angles α_i and layer thicknesses t_i by using a simple linear transformation model:

$$\underline{a}(\underline{x}) = \begin{bmatrix} \alpha(\underline{x}) \\ t(\underline{x}) \end{bmatrix} = \underline{\underline{A}} \ \underline{x} + \underline{x}_o . \tag{7}$$

This allows variable linking and variable fixing, so that symmetric and antimetric layers or a special laminate configuration can be modelled. Subsequently at given time $t = t_i$ the diffusion equation is solved [1, 12]

$$\frac{c(\zeta,t)}{\partial t} = \frac{\partial}{\partial \zeta}\left(D_{33}\frac{c(\zeta,t)}{\partial \zeta}\right) \quad \text{with diffusion coefficient } D_{33}. \tag{8}$$

The calculation of laminate stiffnesses and hygrothermal force resultants is based on the quasi elastic approximation of the hygrothermal viscoelastic material law [2,9]:

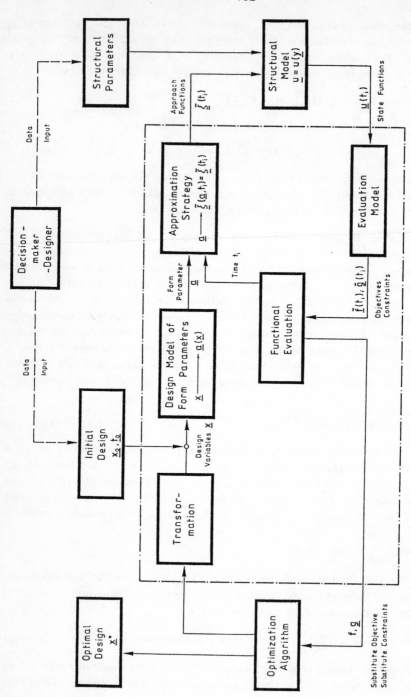

Fig. 1: Optimization loop for time-variant optimization problems

$$A^{\alpha\beta\gamma\delta}(\tau) = \sum_{k=1}^{n_s} {}_k E^{\alpha\beta\gamma\delta}(\tau) \, [h_{k+1} - h_k], \qquad \text{(strain stiffness)} \qquad (9a)$$

$$B^{\alpha\beta\gamma\delta}(\tau) = \sum_{k=1}^{n_s} {}_k E^{\alpha\beta\gamma\delta}(\tau) \, \frac{1}{2}[h_{k+1}^2 - h_k^2], \qquad \text{(coupling stiffness)} \qquad (9b)$$

$$K^{\alpha\beta\gamma\delta}(\tau) = \sum_{k=1}^{n_s} {}_k E^{\alpha\beta\gamma\delta}(\tau) \, \frac{1}{3}[h_{k+1}^3 - h_k^3], \qquad \text{(bending stiffness)} \qquad (9c)$$

$$S^{\alpha 3\gamma 3}(\tau) = \frac{5}{4}\sum_{k=1}^{n_s} {}_k C^{\alpha 3\gamma 3}(\tau) \, \frac{1}{3}\left[h_{k+1} - h_k - \frac{4}{3t^2}(h_{k+1}^3 - h_k^3)\right]. \quad \text{(shear stiffness)} \quad (9d)$$

$$\left\{N_T^{\alpha\beta}(t), M_T^{\alpha\beta}(t)\right\} = \sum_{k=1}^{n_s} {}_k E^{\alpha\beta\gamma\delta}(\tau) \, {}_k\alpha_{T\gamma\delta}(\tau) \int_{h_k}^{h_{k+1}} \Theta(t)\left\{1,\zeta\right\} d\zeta \qquad (9e)$$
$$\text{(hygrothermal force resultants)}$$

$$\left\{N_c^{\alpha\beta}(t), M_c^{\alpha\beta}(t)\right\} = \sum_{k=1}^{n_s} {}_k E^{\alpha\beta\gamma\delta}(\tau) \, {}_k\alpha_{c\gamma\delta}(\tau) \int_{h_k}^{h_{k+1}} c(t)\left\{1,\zeta\right\} d\zeta \qquad (9f)$$

with the plain relaxation tensor ${}_k E^{\alpha\beta\gamma\delta}(\tau)$ of layer k at reduced time τ due to hygro-thermal time-shift for thermorheologic simple material behaviour [11] and distance h_k of layer bottom surface k to laminate mid surface. The actual stiffnesses and hygrothermal stress resultants are the input of the structural analysis module, which provides the evaluation model with laminate stresses, strains and displacements as state functions to calculate failure criteria, objectives and constraints.

4. STRESS AND WEIGHT OPTIMIZATION OF A COMPOSITE CANTILEVER TRUSS

A cantilever truss made of CFC-tubes (Fig. 2) shall be improved so that during a certain time period the objectives "laminate failure criteria" and "weight" are minimized.

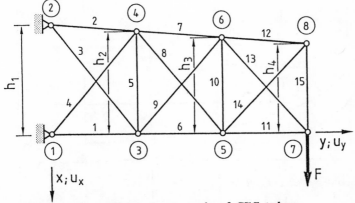

Fig. 2: Composite cantilever truss made of CFC-tubes
Geometry: height $h_i = 1$ m, length $l = 5$ m, mean tube diameter $d = 80$ mm
Loads: dead weight, single load $F = 50$ kN, moisture diffusion at surface concentration $c_o = 1\%$, time $t \leq 10^5$ h, temperature $\vartheta = 23\,°C$
Material: C-fibre 300, epoxy resin 934C, 4 layers $(0°, 90°, 90°, 0°)$

The total laminate thickness of each tube is a design variable and the mean tube diameter is held constant during optimization. The multiple objective problem reads:

$$\tilde{\underline{f}}(\underline{x},t_i) = \begin{bmatrix} \tilde{f}_1(\underline{x},t_i) \equiv W \\ \tilde{f}_2(\underline{x},t_i) \equiv \max_a \max_k [\,_kB_{FBa}(\underline{x},t_i),\,_kB_{ZFBa}(\underline{x},t_i)] \end{bmatrix}. \tag{10}$$

Tube stiffnesses, hygrothermal loads and tube cross sections are provided by the design model at time t_i and are given to the structural analysis modul. From the analysis results fibre break failure $_kB_{FBa}$ and bonding break failure $_kB_{ZFBa}$ are calculated at each single layer k of each tube a [13]. The objectives are reduced to a scalar preference function using the maximum value criterion (6) and the method of constraint oriented transformation [7]. The failure criteria of each truss member built an additional constraint:

$$g_a(\underline{x}) = \max_{t_i} \max_k \{[\,_kB_{FBa}(\underline{x},t_i),\,_kB_{ZFBa}(\underline{x},t_i)] - 1\} \le 0, \; t_i \in [t_o, t_j] \tag{11}$$

In a first optimization calculation a stationary moisture distribution achieved at the final state of the diffusion process is considered (Fig. 3). Second the structure is optimized with time-variant moisture distribution. At a same utilization of the failure criteria the optimal weight differs up to 28 %. Fig. 4 shows the design variables of the weight-optimal design at utilization rate 1. Neglecting the time-dependency results in a weak underdimensioned structure. Truss members with small laminate thickness achieve a constant moisture profile earlier as larger dimensioned members. Accordingly internal forces are varying with time and constraints can get infeasible. During the process of diffusion considerable constraint violations occur although all criteria are feasible in the final stationary state (Fig. 5).

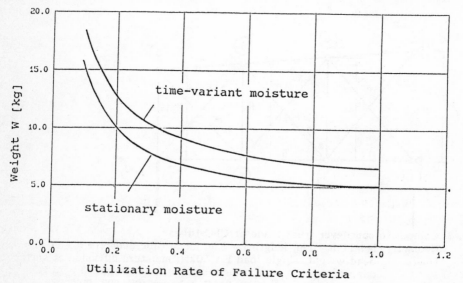

Fig. 3: Functional efficient solutions of a composite cantilever truss

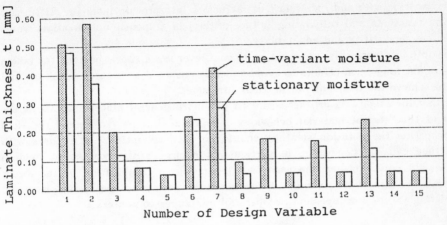

Fig 4: Design variables of the weight optimal design

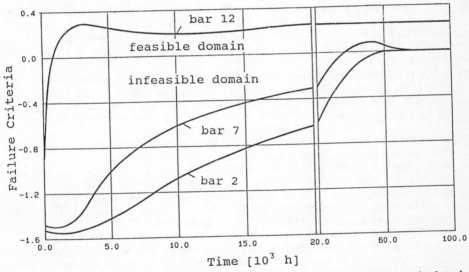

Fig. 5: Time-variation of failure criteria of the weight optimal design, calculated with stationary moisture distribution

6. CONCLUSION

A new direct optimization strategy for structures with time-variant material behaviour is introduced. The optimization can be carried out using the well-known and efficient algorithms of mathematical programming. The general formulation of this strategy allows an application on further types of time-dependent problems. For laminated composite structures a special optimization model has been developed. The long-term material behaviour is described by the quasi-elastic-approximation of linear viscoelasticity and Fick's law of diffusion. Interpreting laminate stiffnesses as form functions and calculating failure criteria in a special optimization model guarantees modularity and adaptability.

The optimization results of a composite cantilever truss show the need for taking into account long-term effects to avoid underdimensioned structures which are not able to achieve the demands during their life-time.

Further investigations are necessary to get more general statements about the effect of time-variant material behaviour on optimal designs. To extend the field of application the solution strategy should be used to optimize structures with short-time dynamic behaviour. In particular for large scale systems of modern light-weight constructions and aerospace technology a combination of the direct solution strategy with decomposition methods seems to be very promising considering future developements of parallel and super computers.

References

[1] SPRINGER, G.S.: Environmental Effects on Composite Materials. Westport: Technomic Publishing Co. Vol. 1: 1981. Vol. 2: 1984. Vol 3: 1988

[2] SCHAPERY, R.A.: Viscoelastic Behavior and Analysis of Composite Materials. In: SENDECKIJ, G.P. (Ed.): Composite Materials Vol. 2. New York: Academic Press (1974) 85-168

[3] SIMS, D.F.: Viscoelastic Creep and Relaxation Behavior of Laminated Composite Plates. Ph. D. Diss., Southern Methodist University, Dallas, Texas 1972

[4] WILSON, D.W.; VINSON, J.R.: Viscoelastic Buckling Analysis of Laminated Composite Columns. In: VINSON, J.R.; TAYA, M. (Edt.): Recent Advances in Composites. ASTM STP 864, Philadelphia (1985) 368-383

[5] VINSON, J.R.; SIERAKOWSKY, R.L.: The Behaviour of Structures Composed of Composite Materials. Netherland: Martinus Nijhoff Publ. 1986

[6] GRANDHI, R.V.; HAFTKA, R.T.; WATSON, L.T.: Design-Oriented Identification of Critical Times in Transient Responce. AIAA J. 24 (1986) 649-656

[7] ESCHENAUER, H.: Rechnerische und experimentelle Untersuchungen zur Strukturoptimierung von Bauweisen. DFG-Forschungsbericht, Universität- GH Siegen, Technische Informationsbibliothek TIB Hannover 1985

[8] ESCHENAUER, H.; POST, P.U.: Optimization Procedure SAPOP Applied to Optimal Layouts of Complex Structures. In KARIHALOO, B,L,; ROZVANY, G.I.N. (Edts.): Structural Optimization. Dordrecht: Kluwer Academic Publishers 1988

[9] ESCHENAUER, H.; POST, P.U.; BREMICKER, M: Einsatz der Optimierungsprozedur SAPOP zur Auslegung von Bauteilkomponenten. Bauingenieur 63(1988)515-526

[10] POST, P.U.: Optimierung von Verbundbauweisen unter Berücksichtigung des zeitabhängigen Materialverhaltens. Dissertation Universität-GH Siegen. VDI-Fortschrittbericht, Reihe 1, Nr. 172, Düsseldorf: VDI-Verlag 1989

[11] CROSSMAN, F.W., MAURI, R.W., WARREN. W.J.: Moisture Altered Viscoelastic Response of Graphite/Epoxy Composites. In: VINSON, J.R. (Edt.): Composite Materials - Environmental Effects. ASTM-STP 658. 1987, pp. 205-220

[12] GITSCHNER, H.-W.: Diffusionsbedingte Verformungs- und Spannungszustände in glasfaserverstärkten Verbundwerkstoffen. Diss. RWTH- Aachen 1980

[13] PUCK, A.: Festigkeitsberechnungen an Glasfaser/Kunststoff-Laminaten bei zusammengesetzter Beanspruchung. Kunststoffe 59 (1969) 780-787

[14] ESCHENAUER, H.; POST, P.U.: Zur Optimierung von Bauteilen aus linear-viskoelastischem Verbundwerkstoff. ZAMM 68 (1988) T176-T178

DESIGN OF LAMINATED COMPOSITES UNDER TIME DEPENDENT LOADS AND MATERIAL BEHAVIOUR

Joachim Bühlmeier
Institute for Computer Applications,
Pfaffenwaldring 27, 7000 Stuttgart 80 , F R G

1. Problem description

In the design of laminated composites it is necessary to define a large number of design variables and manifold objective functions or multicriteria objective functions. The question is: are we able to solve all the arising problems economically on the FEM-level, or should we split some tasks so that we can handle them on a local sublevel (fig.1)? The advantage is that we can reduce the effort significantly when operating on a sublevel; the weak point is, however, the definition of FEM-level-corresponding constraints and the design of corresponding objective functions. This means we try to introduce decoupling conditions between the FEM- and local level of optimisation. The analysis procedure on the local level may be defined according to the used elements, the aspect ratio or the location of the load measurement points.

Here, we consider in a first step only the local level optimisation based on the classical thin laminate theory with extensions to the in-plane and transversal behaviour that we use to have the ability to compute strains caused by temperature and moisture loads, or to improve the results in the case of low aspect ratios (fig. 2). One may regard the stand-alone application of the local level optimisation also as a predesign procedure.

The standard loads as normal and shear loads, moments and twist are extended by time-dependent temperature and moisture loads in the short and medium time range caused by ply temperature and moisture strains $\alpha_k \Delta T_k(t), \beta_k \Delta m_k(t)$ as well as by initial loads caused by creep and relaxation strains $\eta_k = \eta_k(t, \sigma_k, \gamma_k)$ in the long time range. Optimisation normally results in an extremely high ratio of actual stresses to limit stresses, so these time effects have to be taken into account when there is a given exposure time.

The additional loads in the formulation of the classical thin laminate theory are

$$\begin{bmatrix} N \\ M \end{bmatrix}_T = \sum_{k=1}^{n} \int_{h_{k-1}}^{h_k} \lceil \hat{\kappa}_k \quad \hat{\kappa}_k \rfloor \begin{bmatrix} \alpha_k \\ \alpha_k z \end{bmatrix} \Delta T_k(t) dz$$

$$\begin{bmatrix} N \\ M \end{bmatrix}_m = \sum_{k=1}^{n} \int_{h_{k-1}}^{h_k} \lceil \hat{\kappa}_k \quad \hat{\kappa}_k \rfloor \begin{bmatrix} \beta_k \\ \beta_k z \end{bmatrix} \Delta m_k(t) dz$$

$$\begin{bmatrix} N \\ M \end{bmatrix}_t = \sum_{k=1}^{n} \int_{h_{k-1}}^{h_k} \lceil \hat{\kappa}_k \quad \hat{\kappa}_k \rfloor \begin{bmatrix} \eta_k(t, \sigma_k, \gamma_k) \\ \eta_k(t, \sigma_k, \gamma_k) z \end{bmatrix} dz$$

2. Optimisation

The standard optimisation problem

$$W(\boldsymbol{x}) = min!$$

$$g_i(\boldsymbol{x}) \geq 0 \qquad i = 1, m$$

$$x_{j_{min}} \leq x_j \leq x_{j_{max}} \qquad j = 1, n$$

is solved iteratively by introducing approximations for the weight function and the implicit constraints such that

$$\tilde{w} = w_B + \sum_{j=1}^{n} x_{jB} \left(\frac{\partial w}{\partial x_j}\right)_B \left[\left(\frac{x_j}{x_{jB}}\right)^{\beta_{0j}} - 1\right] \frac{1}{\beta_{0j}}$$

$$\beta_{0j} = \alpha_0 sign \left(\frac{\partial W}{\partial x_j}\right)_B$$

$$\tilde{g}_i = g_{iB} + \sum_{j=1}^{n} x_{jB} \left(\frac{\partial g_i}{\partial x_j}\right)_B \left[\left(\frac{x_i}{x_{jB}}\right)^{\beta_{ij}} - 1\right] \frac{1}{\beta_{ij}}$$

$$\beta_{ij} = \alpha_i \, sign \left(\frac{\partial g_i}{\partial x_j}\right)_B$$

The approximated problem is changed to an unconstrained problem with the concept of the interior penalty function (Fiacco, McCormick):

$$\phi(\boldsymbol{x}, r_k) = \widetilde{W}(\boldsymbol{x})$$

$$+ r_k \left(\sum_{i=1}^{m} \frac{1}{\tilde{g}_i} + \sum_{j=1}^{n} \frac{x_{j_{min}}}{x_j - x_{j_{min}}} + \sum_{j=1}^{n} \frac{x_{j_{max}}}{x_{j_{max}} - x_j}\right)$$

Due to the different demands in the design of laminated composites, the objective function may be figured or designed in numerous ways. We reduced the possibilities for the user to 5 basic forms: the single value

$$W_{tot}(\boldsymbol{x}) = W(\boldsymbol{x})$$

the objective weighting

$$W_{tot}(\boldsymbol{x}) = \sum_{j=1}^{l} c_j W_j(\boldsymbol{x}) = \boldsymbol{c}^t \boldsymbol{W}$$

the objective function ratios

$$W_{tot}(\boldsymbol{x}) = \sum_{j=1}^{l} c_j \frac{W_j(\boldsymbol{x})}{W_{j+1}(\boldsymbol{x})}$$

the objective weighting with demand levels d_j

$$W_{tot}(\boldsymbol{x}) = \sum_{j=1}^{l} c_j (W_j(\boldsymbol{x}) - d_j)$$

and objective function differences

$$W_{tot}(\boldsymbol{x}) = \sum_{j=1}^{l} c_j(W_j(\boldsymbol{x}) - W_{j+1}(\boldsymbol{x}))$$

One can use all forms in the minimisation and maximisation mode. Possible elements of the objective functions are the laminae weight, the elements of stiffness and flexibility matrices, thermal coefficients of expansion and curvature, moisture coefficients of expansion and curvature, thermal conduction, thermal and moisture strains or total strains and total curvatures.

The constraints consist of implicit constraints g_i as given from a selected failure criterion (Tsai-Wu, ZTL or Puck) and explicit constraints h_i for the design variables. First-ply-failure only is regarded. The reserve factors $R_{LF,min}$ to longitudinal fracture and $R_{TF,min}$ to transverse fracture must be given. Both types of constraints then read

$$g_i = 1 - \frac{R_{LF\,min}}{R_{i,LF}} \geq 0$$

$$g_i = 1 - \frac{R_{TF\,min}}{R_{i,TF}} \geq 0$$

$$h_i = 1 - \frac{t_i}{t_{max,i}} \geq 0$$

$$h_i = 1 - \frac{t_{min,i}}{t_i} \geq 0$$

where angles α_i, sums of ply thicknesses with determined orientations due to minimum content requirements from manufacturing experiences or elements in the stiffness-respectively flexibility matrix may also occur in h_i. Several restriction steps to the ply-sequences can be selected: no restriction, arbitrary coupled plies, symmetrical laminate, symmetrical laminate plus arbitrary coupled plies.

3. Time-dependent loads

The application of composites based on resin-fiber materials is often restricted because of their bad long-time properties caused by moisture absorption and creep/relaxation effects. We try to include this behaviour in the optimisation procedure and consider first the moisture diffusion.

The Fickian moisture absorption law gives

$$\frac{\partial m}{\partial t} = div(\boldsymbol{D}\,\mathrm{grad}\,\mathrm{m})$$

which can be reduced in our case to the transversal direction

$$\frac{\partial m}{\partial t} = \frac{\partial}{\partial z}\left(D_{zz}\frac{\partial m}{\partial z}\right)$$

Analytical solutions of this differential equation exist for an unlayered cross section. In the case of layered material, the diffusion coefficient D_{zz} may change from layer to layer and we are forced to use a finite difference method. In the following Δ_t means the time interval and Δ_z means the thickness interval, so the differential equation can be written as

$$\frac{\Delta_t m}{\Delta t} = D_{zz}(z)\frac{\Delta_z^2 m}{\Delta z^2}$$

$$\Delta_t m = m_{n,k+1} - m_{n,k}$$

$$\Delta_z m = m_{n+1,k} - m_{n,k}$$

$$\Delta_z^2 m = (m_{n+1,k} - m_{n,k}) - (m_{n,k} - m_{n-1,k})$$
$$= m_{n+1,k} + m_{n-1,k} - 2m_{n,k}$$

$$m_{n,k+1} = m_{n,k} + D_{zz}\frac{\Delta t}{\Delta z^2}(m_{n+1,k} + m_{n-1,k} - 2m_{n,k})$$

When we compare the analytical (fig. 3a) with the numerical solution (fig. 3b), we see that a subdivision of one ply into 5 strips already gives the analytical solution .
In the case of temperature loading, we get a similar differential equation

$$\frac{\partial T}{\partial t} = \frac{\partial}{\partial z}\left(\lambda_{zz}\frac{\partial T}{\partial z}\right)$$

where λ_{zz} is the thermal conductivity. With the help of the finite difference procedure, we are able to compute the moisture and temperature per ply in dependence of a given starting state and exposure time. The corresponding loading cases are built up at relevant time steps (fig. 4).

4. Time-dependent material behaviour

Secondly, we regard the creep/relaxation effects with the help of a simplified constitutive model. In reality and in a comprehensive inelastic analysis, there are numerous behaviour properties that we will not regard in a first analysis step. Among these are the viscoelastic, the viscoplastic and elastoplastic strains. In general, the corresponding flows affect one another, they are load history and load velocity dependent. We reduce our consideration to the aging flow strains in

$$\eta = \eta_{\substack{aging \\ flow}} + \eta_{\substack{visco \\ elastic}} + \eta_{\substack{visco- \\ plastic}} + \eta_{\substack{elasto- \\ plastic}}$$

Fig. 5 gives the general one dimensional behaviour in time. From material testing under several load levels and prestressing time intervals one gets relevant constitutive models for the numerical prediction of structural behaviour in the inelastic response regime. It is usual to simplify the general one dimensional creep law in the rate form

$$\dot{\eta}_{af} = f(\sigma, t, T, \eta_{af})$$

by the time-hardening creep model to

$$\dot{\eta}_{af} = f(\sigma, t) = \alpha \sigma^P t^m$$

and determine α, p, m with the method of least squares.

For a unidirectional ply no exact two-dimensional law was available, so we assumed in the transversal direction nearly the resin values, in the longitudinal direction an approximation of the fiber values and for the off-diagonal terms a weak coupling of both. In general, the single ply in a laminate is neither kinematically nor statically determined, that means the plies are exposed to neither pure creep nor pure relaxation and we have to find the correct state by the linear expansion

$$\Delta\boldsymbol{\eta} = \Delta t(\dot{\boldsymbol{\eta}} + \boldsymbol{A}_\eta \Delta\boldsymbol{\sigma} + \boldsymbol{B}_\eta \Delta\boldsymbol{\eta})$$

where

$$
\boldsymbol{A}_\eta =
\begin{bmatrix}
\dfrac{\partial\dot{\eta}_{xx}}{\partial\sigma_{xx}} & \dfrac{\partial\dot{\eta}_{xx}}{\partial\sigma_{yy}} & \dfrac{\partial\dot{\eta}_{xx}}{\partial\sigma_{xy}} \\[2ex]
\dfrac{\partial\dot{\eta}_{yy}}{\partial\sigma_{xx}} & \dfrac{\partial\dot{\eta}_{yy}}{\partial\sigma_{yy}} & \dfrac{\partial\dot{\eta}_{yy}}{\partial\sigma_{xy}} \\[2ex]
\dfrac{\partial\dot{\eta}_{xy}}{\partial\sigma_{xx}} & \dfrac{\partial\dot{\eta}_{xy}}{\partial\sigma_{yy}} & \dfrac{\partial\dot{\eta}_{xy}}{\partial\sigma_{xy}}
\end{bmatrix}
$$

and

$$
\boldsymbol{B}_\eta =
\begin{bmatrix}
\dfrac{\dot{\eta}_{xx}(t) - \dot{\eta}_{xx}(t+\Delta t)}{\Delta\eta_{xx}} & \dfrac{\dot{\eta}_{xx}(t) - \dot{\eta}_{yy}(t+\Delta t)}{\Delta\eta_{yy}} & \dfrac{\dot{\eta}_{xx}(t) - \dot{\eta}_{xy}(t+\Delta t)}{\Delta\eta_{xy}} \\[2ex]
\dfrac{\dot{\eta}_{yy}(t) - \dot{\eta}_{xx}(t+\Delta t)}{\Delta\eta_{xx}} & \dfrac{\dot{\eta}_{yy}(t) - \dot{\eta}_{yy}(t+\Delta t)}{\Delta\eta_{yy}} & \dfrac{\dot{\eta}_{yy}(t) - \dot{\eta}_{xy}(t+\Delta t)}{\Delta\eta_{xy}} \\[2ex]
\dfrac{\dot{\eta}_{xy}(t) - \dot{\eta}_{xx}(t+\Delta t)}{\Delta\eta_{xx}} & \dfrac{\dot{\eta}_{xy}(t) - \dot{\eta}_{yy}(t+\Delta t)}{\Delta\eta_{yy}} & \dfrac{\dot{\eta}_{xy}(t) - \dot{\eta}_{xy}(t+\Delta t)}{\Delta\eta_{xy}}
\end{bmatrix}
$$

Between $\Delta\boldsymbol{\sigma}$ and $\Delta\boldsymbol{\eta}$ the elastic strain- stress relation exists in incremental form

$$\Delta\boldsymbol{\sigma} = \hat{\boldsymbol{\kappa}}(\Delta\boldsymbol{\gamma} - \Delta\boldsymbol{\eta})$$

that leads to

$$[\boldsymbol{I} - \boldsymbol{B}_\eta \Delta t]\Delta\boldsymbol{\eta} = \Delta t\dot{\boldsymbol{\eta}} + \Delta t \boldsymbol{A}_\eta \Delta\boldsymbol{\sigma}$$

$$\begin{bmatrix} I & \hat{\kappa} \\ -\Delta t A_\eta & I - B_\eta \Delta t \end{bmatrix} \begin{bmatrix} \Delta\sigma \\ \Delta\eta \end{bmatrix} = \begin{bmatrix} \hat{\kappa}\Delta\gamma \\ \Delta t\dot{\eta} \end{bmatrix}$$

This equation separates $\Delta\gamma$ in an elastic part $\Delta\epsilon$ (corresponding to $\Delta\sigma$) and in an inelastic part $\Delta\eta$.

For the iteration, $\hat{\kappa}\Delta\gamma$ is the driving force, convergence is achieved when the force steps are small enough, that means the norm of the stress ratios has to be less than 1

$$\left\| \frac{\partial\sigma_{n+1}^{i+1}}{\partial\sigma_{n+1}^{i}} \right\| < 1$$

From this, one derives the time-step Δt. Results are given in fig. 6.

5. Extensions

To the constant strains ϵ^0 and linear strains $z_k\kappa^0$ in the ply k we add a quadratic and a cubic strain distribution $\epsilon^0 + z_k\kappa^0 + z_k^2\epsilon^1 + z_k^3\kappa^1$. These strains have no external resultants and we call the corresponding stresses "self-equilibrating stress groups". Integrating about all plies, we get

$$\begin{bmatrix} N^0 \\ M^0 \\ N^1 \\ M^1 \end{bmatrix} = \sum_{k=1}^{n} \int_{h_{k-1}}^{h_k} \begin{bmatrix} \sigma \\ z\sigma \\ z^2\sigma \\ z^3\sigma \end{bmatrix} \mathrm{d}z = \sum_{k=1}^{n} \int_{h_{k-1}}^{h_k} \begin{bmatrix} \hat{\kappa}_k & \hat{\kappa}_k & \hat{\kappa}_k & \hat{\kappa}_k \end{bmatrix} \begin{bmatrix} \epsilon^0 \\ z\kappa^0 \\ z^2\epsilon^1 \\ z^3\kappa^1 \end{bmatrix} \mathrm{d}z$$

$$\begin{bmatrix} N^0 \\ M^0 \\ N^1 \\ M^1 \end{bmatrix} = \begin{bmatrix} A & B & D - \frac{t^2}{12}A & G - \frac{3}{20}t^2 B \\ & D & G - \frac{t^2}{12}B & C - \frac{3}{20}t^2 D \\ & & C - \frac{t^2}{12}D & E - \frac{3}{20}t^2 G \\ sym & & & F - \frac{3}{20}t^2 C \end{bmatrix} \begin{bmatrix} \epsilon^0 \\ \kappa^0 \\ \epsilon^1 \\ \kappa^1 \end{bmatrix}$$

where ϵ^1 and κ^1 give the amount of the quadratic or cubic contributions and further

$$A = \sum_{k=1}^{n} \hat{\kappa}_k(h_k - h_{k-1}) \qquad B = \frac{1}{2}\sum_{k=1}^{n} \hat{\kappa}_k(h_k^2 - h_{k-1}^2) \qquad D = \frac{1}{3}\sum_{k=1}^{n} \hat{\kappa}_k(h_k^3 - h_{k-1}^3)$$

$$G = \frac{1}{4}\sum_{k=1}^{n} \hat{\kappa}_k(h_k^4 - h_{k-1}^4) \qquad C = \frac{1}{5}\sum_{k=1}^{n} \hat{\kappa}_k(h_k^5 - h_{k-1}^5) \qquad E = \frac{1}{6}\sum_{k=1}^{n} \hat{\kappa}_k(h_k^6 - h_{k-1}^6)$$

$$F = \frac{1}{7}\sum_{k=1}^{n} \hat{\kappa}_k(h_k^7 - h_{k-1}^7)$$

For the determination of the load vector $\{N^0 M^0 N^1 M^1\}$, we first compute the layer temperatures and moisture values ΔT_k, Δm_k from the finite difference analysis. The integration about the thickness gives

$$
\begin{bmatrix} N_0 \\ M_0 \\ N_1 \\ M_1 \end{bmatrix}_t = \sum_{k=1}^{n} \int_{h_{k-1}}^{h_k} \lceil \hat{\boldsymbol{\kappa}}_k \quad \hat{\boldsymbol{\kappa}}_k \rfloor \left(\begin{bmatrix} \alpha_k \\ z\alpha_k \\ z^2\alpha_k \\ z^3\alpha_k \end{bmatrix} \Delta T_k(t) + \begin{bmatrix} \beta_k \\ z\beta_k \\ z^2\beta_k \\ z^3\beta_k \end{bmatrix} \Delta m_k(t) \right) \mathrm{d}z
$$

A further extension is the inclusion of transversal shear and normal loads. We consider first monoclinic anisotropic material

$$
\begin{bmatrix} \sigma'_{xx} \\ \sigma'_{yy} \\ \sigma'_{xy} \\ \sigma'_{zz} \\ \sigma'_{yz} \\ \sigma'_{zx} \end{bmatrix} = \begin{bmatrix} \hat{x}'_{11} & \hat{x}'_{12} & \hat{x}'_{16} & \hat{x}'_{13} & 0 & 0 \\ \hat{x}'_{21} & \hat{x}'_{22} & \hat{x}'_{26} & \hat{x}'_{23} & 0 & 0 \\ \hat{x}'_{61} & \hat{x}'_{62} & \hat{x}'_{66} & \hat{x}'_{63} & 0 & 0 \\ \hat{x}'_{31} & \hat{x}'_{32} & \hat{x}'_{36} & \hat{x}'_{33} & 0 & 0 \\ 0 & 0 & 0 & 0 & \hat{x}'_{44} & \hat{x}'_{45} \\ 0 & 0 & 0 & 0 & \hat{x}'_{54} & \hat{x}'_{55} \end{bmatrix} \begin{bmatrix} \epsilon'_{xx} \\ \epsilon'_{yy} \\ \epsilon'_{xy} \\ \epsilon'_{zz} \\ \epsilon'_{yz} \\ \epsilon'_{zx} \end{bmatrix}
$$

and assume a constant transversal stress distribution, so that $N_{\perp k} = N_\perp$ and introduce a mean transverse stiffness $\hat{\boldsymbol{\kappa}}_{22}$. With the abbreviation

$$
\begin{bmatrix} \boldsymbol{\sigma}_1 \\ \boldsymbol{\sigma}_2 \end{bmatrix} = \begin{bmatrix} \hat{\boldsymbol{\kappa}}_{11} & \hat{\boldsymbol{\kappa}}_{12} \\ \hat{\boldsymbol{\kappa}}_{12}^t & \hat{\boldsymbol{\kappa}}_{22} \end{bmatrix} \begin{bmatrix} \boldsymbol{\epsilon}_1 \\ \boldsymbol{\epsilon}_2 \end{bmatrix}
$$

for the above stress-strain relation we arrive at

$$
\begin{bmatrix} N_0 \\ N_\perp \\ M_0 \end{bmatrix} = \begin{bmatrix} A & C & B \\ C & G & E \\ B & E & D \end{bmatrix} \begin{bmatrix} \boldsymbol{\epsilon}^0 \\ \boldsymbol{\epsilon}^\perp \\ \boldsymbol{\kappa}^0 \end{bmatrix}
$$

where

$$
A = \sum_{k=1}^{n} (\hat{\boldsymbol{\kappa}}_{11} - \hat{\boldsymbol{\kappa}}_{12} \hat{\boldsymbol{\kappa}}_{22}^{-1} \hat{\boldsymbol{\kappa}}_{12}^t)_k (h_k - h_{k-1}) \qquad B = \frac{1}{2} \sum_{k=1}^{n} (\hat{\boldsymbol{\kappa}}_{11} - \hat{\boldsymbol{\kappa}}_{12} \hat{\boldsymbol{\kappa}}_{22}^{-1} \hat{\boldsymbol{\kappa}}_{12}^t)_k (h_k^2 - h_{k-1}^2)
$$

$$D = \frac{1}{3}\sum_{k=1}^{n}(\hat{\kappa}_{11} - \hat{\kappa}_{12}\hat{\kappa}_{22}^{-1}\hat{\kappa}_{12}^{t})_k(h_k^3 - h_{k-1}^3) \qquad G = \sum_{k=1}^{n}\hat{\kappa}_{22k}(h_k - h_{k-1})$$

$$C = \sum_{k=1}^{n}(\hat{\kappa}_{12}\hat{\kappa}_{22}^{-1}\hat{\kappa}_{22})_k(h_k - h_{k-1}) \qquad E = \frac{1}{2}\sum_{k=1}^{n}(\hat{\kappa}_{22}\hat{\kappa}_{22}^{-1}\hat{\kappa}_{21})_k(h_k^2 - h_{k-1}^2)$$

This procedure is equivalent to the weakening of the in-plane stiffness and derives physically from the opening of the transversal degrees of freedom.

6. Results

In general for good-natured problems, one needs 5 to 6 iteration loops. For constraint-cornered tasks, one has to suppose 15 to 20 iteration loops and bad constraint- or weight-function-behaviour may lead to zig-zagging.

A typical task which may be solved without difficulties is the preliminary simple weight optimisation with ply thicknesses as variables and reserve factors as constraints as well as a following parameter study about the number of layers (fig. 7, application for a cross-ply). One may find for different ratios t_{max}/t_{min} a different optimal number of layers. Fig. 8 shows the increasing weight for decreasing ratios t_{max}/t_{min}.

Fig. 9 gives a first representative example of a multicriterion optimisation. The curves show the relation between weight W and curvature κ_{xy}, when we vary the weighting in

$$W_{tot} = c_1 W + c_2 \kappa_{xy}$$

or

$$W_{tot} = c_1 F_{12} + c_2 \kappa_{xy} + c_3 W$$

The given limits W_{min} and $\kappa_{xy,min}$ deliver the solution of best possible compromise.

In the same way, we get in a normal load case the optimum design vector for a weight function, which contains the weight, the thermal conduction and the normal strains in x-direction

$$W_{tot} = c_1 W + c_2 \dot{Q}_{xx} + c_3 \epsilon_{xx}$$

For low strain values we get an undesired high thermal conduction, which is in this case linearly coupled with weight and vice versa, so the compromise is also given here by the limit values (fig. 10).

Fig. 11 shows the possibility to tailor a laminae to given physical properties. There is no load and we design a material that has a zero thermal coefficient of expansion in y-direction. With the growing constraint ($\alpha_{yy} \Rightarrow 0$) weight increases.

Considering the ply-orientations as variables we touch several problems:

(i) the influence of angle-variations on objective functions as well as on constraints is much less than thickness variations. This means numerical differences of two to three digits in the gradient vectors. The disadvantage is that ply-orientation is locked in simultaneous thickness- and orientation- optimisation problems.

(ii) the approximated objectve function does not meet the real objective function and leads to zig-zagging.

(iii) numerous local optima exist.

The dilemma is that given small step-sizes stop the iteration in a local optimum and large step-sizes soon result in zig-zagging. On the other hand, one should see the limited influence of the ply-orientation and the manufacture possibilities, which are not so perfected that one easily produces every given stack sequence. A typical result for pure ply-orientation is given in fig. 12. The actual load is M_{xy} and the objective function is the corresponding curvature κ_{xy}, weight is constant in this case.

Another method seems to promise better results: one starts with all ply- orientations accepted in the result and extremely small lower limits for the thicknesses. Fig. 13 shows the result for a 16-ply laminate with the loading case N_{xx}, M_{xx}, M_{xy} and the weight function

$$W_{tot} = c_1 W + c_2 \kappa_{xx}$$

Explicitly given are the solution of best possible compromise $c_1 = c_2 = 1.0$ and the solution with the least deviation from a mean thickness.

Most of these examples can be handled with additional time dependent loads; the varying loads are represented by a finite number of loading cases over the given time interval. Fig. 14 shows the end state of an optimisation in time. All plies have the same orientation, so one can see the expected low transverse fracture reserve factors for the outer and the uncritical loading for the inner plies. The optimal design therefore presents increased thicknesses in the cover plies and decreased ones in the core plies. When the temperature profile has reached constant values, the reserve factors are equal and higher than the critical ones. Fig. 15 contains the conditions when a standard laminate $\{0^0 \quad 90^0 \quad 45^0 \quad -45^0\}$ is used.

Inelastic aging flow results are given in the last two figures in a normal load and a momentum load case. Fig. 16 shows different objective functions about exposure time. For $W_{tot} = W$ or $W_{tot} = W + A'_{11}$ without any strain constraint, we have decreasing weight with time, but the critical time instant is $t = 0$ h.

For $W_{tot} = W + 2A'_{11}$, the critical time instant is $t = 60\ 000$ h, the corresponding design is given in the figure. For a prescribed strain γ_{xx} , weight increases at much higher rates. We can observe the same tendencies in the momentum load case (fig. 17).

7. References

[1] Schapery R.A., Stress Analysis of Viscoelastic Composite Materials, J. Composite Materials, Vol. 1 (1967), p. 228

[2] Whitney J.M., Pagano N.J., Shear Deformation in Heterogeneous Anisotropic Plates, Journal of Applied Mechanics, December 1970, p. 1031

[3] Pagano N.J., Hatfield S.J., Elastic Behavior of Multilayered Bidirectional Composites, AIAA Journal, Vol. 10, No. 7, July 1972, p. 931

[4] Khot N.S., Venkayya V.B., Johnson C.D., Tischler V.A., Optimization of Fiber Reinforced Composite Structures, Int. J. Solids Structures, 1973, Vol. 9, p. 1225

[5] Argyris J.H., Doltsinis J.St., Knudson W.C., Vaz L.E., Willam K.J., Numerical Solution of Transient Nonlinear Problems, Proc. FENOMECH 78, Part II, North-Holland Publishing Company, Amsterdam, 1978, p. 341

[6] Schmit L.A., Mehrinfar M., Multilevel Optimum Design of Structures with Fiber-Composite Stiffened-Panel Components, AIAA Journal, Vol. 20, No.1, p. 138

[7] Watkins R.I., Morris A.J., A Multicriteria Objective Function Optimization Scheme for Laminated Composites for Use in Multilevel Structural Optimization Schemes, Comp. Meth. Appl. Mech. Eng. 60, 1987, p. 233

[8] Hsu T.S., Saxena S.K., New Guidelines for Optimization of Finite Element Solutions, Computers and Structures, Vol. 31, No. 2, 1989, p. 203

[9] Hajela P., Shih C.J., Optimal Design of Laminated Composites Using a Modified Mixed Integer and Discrete Programming Algorithm, Computers and Structures, Vol. 32, No. 1, 1989, p. 213

[10] Kam T.Y., Lai M.D., Multilevel Optimal Design of Laminated Composite Plate Structures, Computers and Structures, Vol. 31, No. 2, 1989, p. 197

[11] Yuan F.G., Miller R.E., A New Finite Element For Laminated Composite Beams, Computers and Structures, Vol. 31, No.5, 1989, p. 737

[12] Chen W.H., Huang T., Three Dimensional Interlaminar Stress Analysis at Free Edges of Composite Laminate, Comp. and Struct., Vol. 32, No. 6, 1989, p. 1275

[13] Kam T.Y., Chang R.R., Optimal Design of Laminated Composite Plates with Dynamic and Static Considerations, Computers and Structures, Vol. 32, No. 2, 1989, p. 387

[14] Eschenauer H.A., Multicriteria Optimization Techniques for Highly Accurate Focusing Systems, in Multicriteria Optimization in Engineering and in the Sciences, Plenum Press, New York, edited by Wolfram Stadler, 1988, p. 309

[15] Ehrenstein G.W., Polymerwerkstoffe, Struktur und mechanisches Verhalten, Hanser Verlag, München, 1978

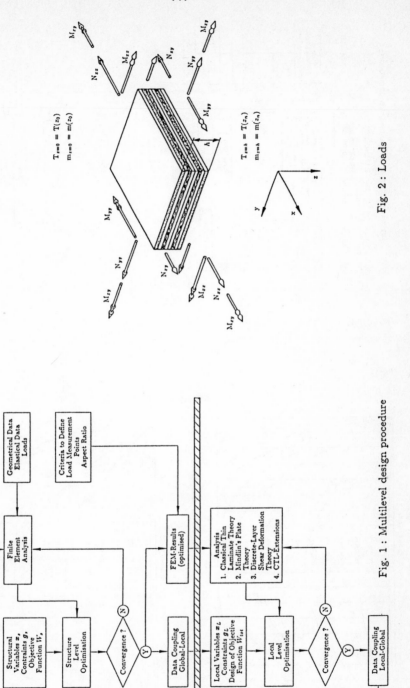

Fig. 2 : Loads

Fig. 1 : Multilevel design procedure

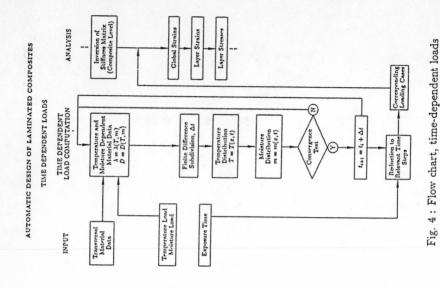

Fig. 4 : Flow chart, time-dependent loads

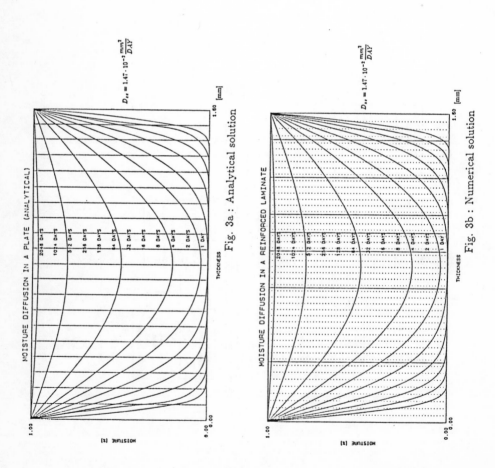

Fig. 3a : Analytical solution

Fig. 3b : Numerical solution

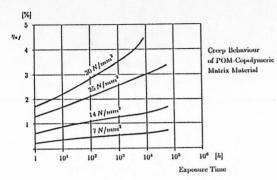

Fig. 5 : Test results [15]

Fig. 6 : Inelastic strains versus time

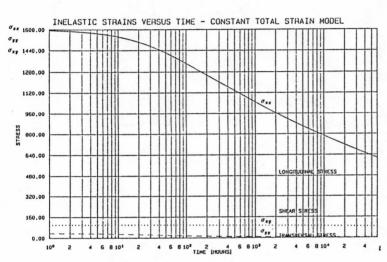

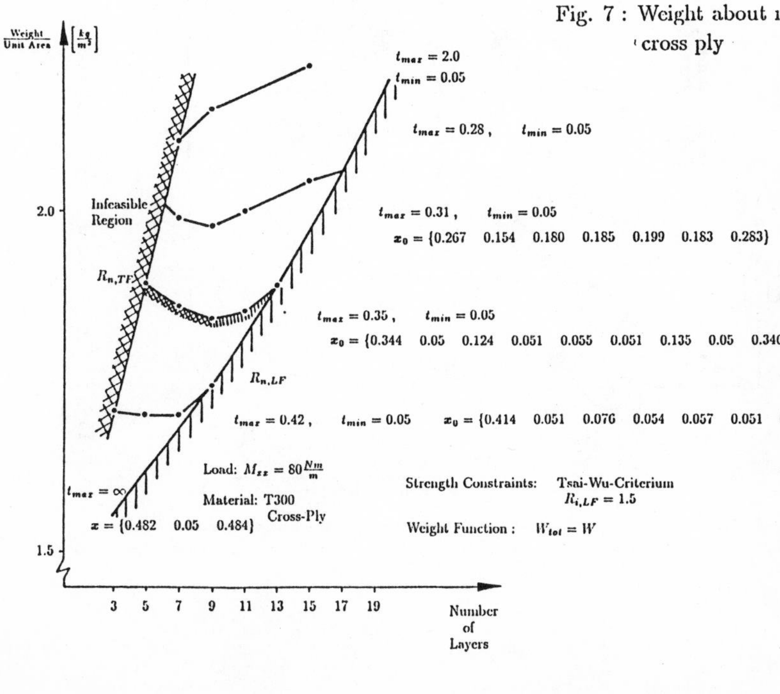

Fig. 7 : Weight about number of layers,
cross ply

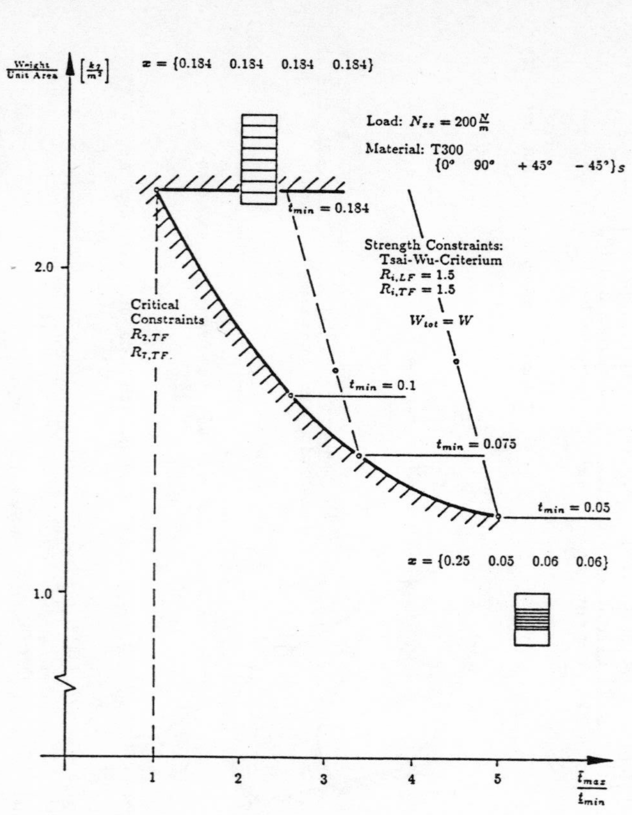

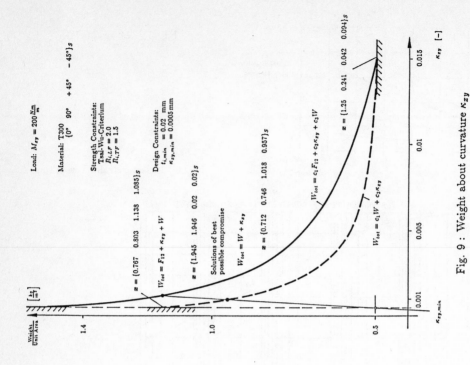

Fig. 9 : Weight about curvature κ_{zy}

Fig. 8 : Weight about design variable ratio

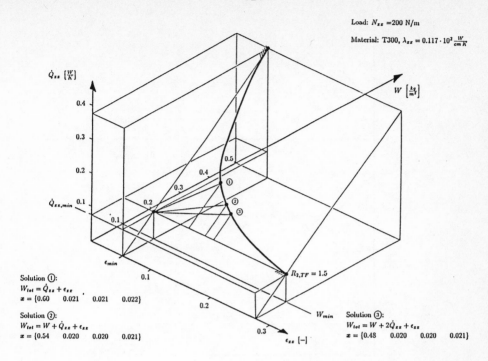

Load: $N_{xx} = 200$ N/m

Material: T300, $\lambda_{xx} = 0.117 \cdot 10^2 \frac{W}{cm\,K}$

Solution ①:
$W_{tot} = \dot{Q}_{xx} + \epsilon_{xx}$
$x = \{0.60 \quad 0.021 \quad 0.021 \quad 0.022\}$

Solution ②:
$W_{tot} = W + \dot{Q}_{xx} + \epsilon_{xx}$
$x = \{0.54 \quad 0.020 \quad 0.020 \quad 0.021\}$

Solution ③:
$W_{tot} = W + 2\dot{Q}_{xx} + \epsilon_{xx}$
$x = \{0.48 \quad 0.020 \quad 0.020 \quad 0.021\}$

Fig. 10 : Thermal conduction about weight and strain ϵ_{xx}

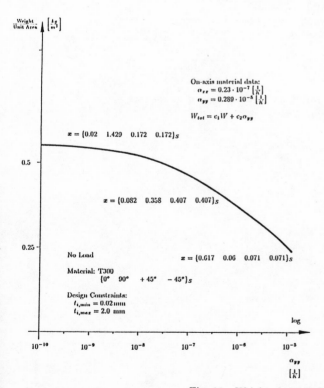

On-axis material data:
$\alpha_{xx} = 0.23 \cdot 10^{-7} \left[\frac{1}{K}\right]$
$\alpha_{yy} = 0.289 \cdot 10^{-5} \left[\frac{1}{K}\right]$

$W_{tot} = c_1 W + c_2 \alpha_{yy}$

$x = \{0.02 \quad 1.429 \quad 0.172 \quad 0.172\}_S$

$x = \{0.082 \quad 0.358 \quad 0.407 \quad 0.407\}_S$

No Load

$x = \{0.617 \quad 0.06 \quad 0.071 \quad 0.071\}_S$

Material: T300
$\{0° \quad 90° \quad +45° \quad -45°\}_S$

Design Constraints:
$t_{i,min} = 0.02$ mm
$t_{i,max} = 2.0$ mm

Fig. 11 : Weight about thermal coefficient of expansion α_{yy}

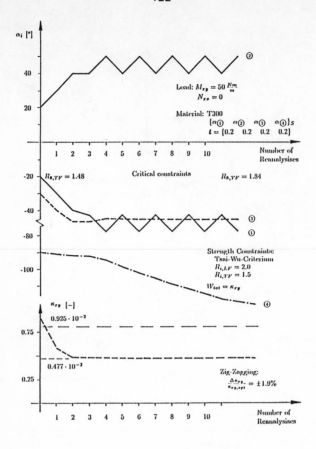

Fig. 12 : Direct optimisation of ply angles

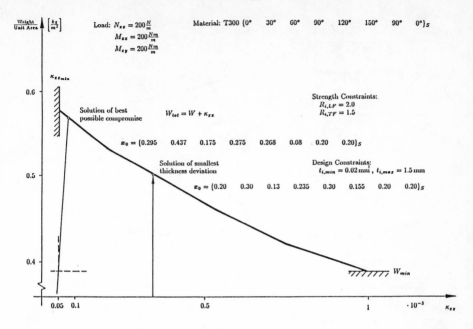

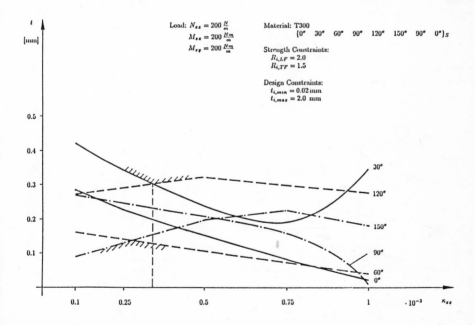

Fig. 13 : Indirect optimisation of ply angles

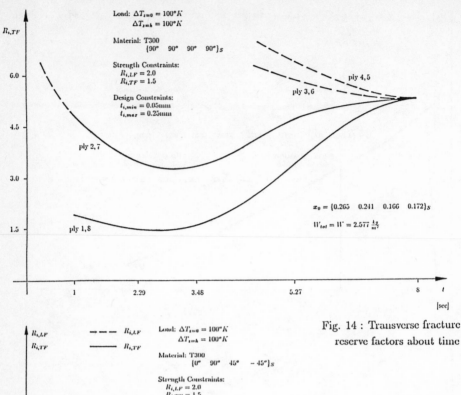

Load: $\Delta T_{z=0} = 100°K$
$\Delta T_{z=h} = 100°K$

Material: T300
$\{90°\quad 90°\quad 90°\quad 90°\}_S$

Strength Constraints:
$R_{i,LF} = 2.0$
$R_{i,TF} = 1.5$

Design Constraints:
$t_{i,min} = 0.05\text{mm}$
$t_{i,max} = 0.25\text{mm}$

ply 4,5

ply 3,6

ply 2,7

ply 1,8

$x_0 = \{0.265\quad 0.241\quad 0.166\quad 0.172\}_S$

$W_{tot} = W = 2.577\,\frac{kg}{m^2}$

Fig. 14 : Transverse fracture reserve factors about time

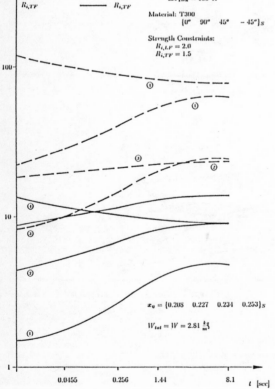

$R_{i,LF}$ $\quad ---\quad R_{i,LF}$
$R_{i,TF}$ $\quad ———\quad R_{i,TF}$

Load: $\Delta T_{z=0} = 100°K$
$\Delta T_{z=h} = 100°K$

Material: T300
$\{0°\quad 90°\quad 45°\quad -45°\}_S$

Strength Constraints:
$R_{i,LF} = 2.0$
$R_{i,TF} = 1.5$

$x_0 = \{0.208\quad 0.227\quad 0.234\quad 0.253\}_S$

$W_{tot} = W = 2.81\,\frac{kg}{m^3}$

Fig. 15 : Transverse fracture reserve factors about time

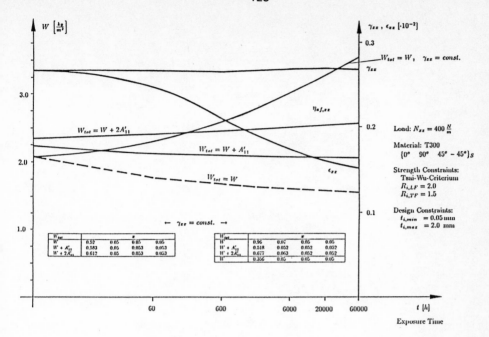

Fig. 16 : Weight and strains about exposure time

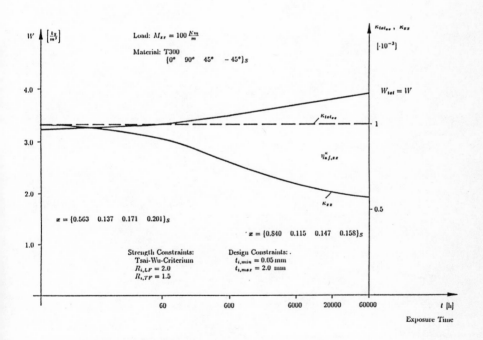

Fig. 17 : Weight and strains about exposure time

OPTIMAL ORIENTATION OF THE SYMMETRY AXES
OF ORTHOTROPIC 3–D MATERIALS

Marco Rovati

Dept. of Structural Mechanics and Design Automation, Univ. of Trento
Via Mesiano 77, 38050 Trento, Italy

Alberto Taliercio

Dept. of Structural Engineering, Politecnico di Milano
Piazza Leonardo da Vinci 32, 20133 Milano, Italy

ABSTRACT

Assuming the elastic energy as a meaningful measure of the global stiffness (or flexibility) of an elastic body, in this paper the interest is paid to the determination of those local orientations of the material symmetry axes in an orthotropic solid which correspond to extreme values of the energy density.

In the general formulation of the problem, and assuming the strain field as given, a linear elastic orthotropic three–dimensional solid is considered and the stationarity conditions are obtained. Such a set of algebraic equations is then explicitly solved referring to the cubic case and the optimal orientations are found as well. It is also pointed out how such orientations depend both on a material parameter and on the strain field.

1. INTRODUCTION

One of the earlier results on optimal orientation of material symmetry axes can be found in a work of Banichuk [1] where necessary conditions for optimal distribution of material properties in orthotropic bodies subjected to plane state of stress are given. Such results have found new impulses in very recent times, mainly due to the growing interest on fiber composite materials. In fact, the optimal orientation of reinforcing fibers seems to be an interesting and promising problem in mechanics of solids.

The aim of the present paper is to give some general results on optimal orientation of the mechanical properties of orthotropic bodies. Up to now, some papers have been published on the subject with reference to plane elastic problems for 2–D structures. After the pioneering work of Banichuk, the same problem was studied by Sacchi & Rovati [4] and Pedersen [2] which independently obtained similar results and gave mechanical interpretations of the optimality conditions. Later on, Pedersen [3] performed a systematic study of the optimal solutions and obtained, for the plane stress problem again, conditions for absolute maxima and minima. Numerical solutions of the problem can be found in Olhoff & Thomsen [5].

In this paper, the optimal orientation of the material symmetry axes in a three–dimensional orthotropic body are sought. The attention is focused on the determination of absolute maxima and minima, i.e., those orientations of the orthotropy axes for which the body exhibits the stiffest (or the most flexible) response. In Section 2, the general formulation of the problem is presented and the optimality conditions are shown. In particular their mechanical interpretation is pointed out, namely the collinearity of principal directions of stress and strain at the optimum. Then two classes of solutions are given as well. In Section 3 the problem is then restricted to the cubic case, for which a complete analytical solution has been performed. It is also highlighted how the optimal solutions depend on the actual strain field and on a parameter depending on the shear stiffness of the material. Moreover it is shown that absolute maxima and minima are attained only in the case of simultaneous collinearity of stress, strain and orthotropy directions or if the longitudinal strains referred to the material symmetry axes are equal. Solutions where only one of the symmetry axes is collinear with one of the principal strains give rise to local maxima or local minima for the energy density.

2. BOUNDS OF ELASTIC STRAIN ENERGY: GENERAL FORMULATION

Consider a linear elastic orthotropic body, defined on the open domain $\mathcal{B} \in \mathbf{R}^3$, with boundary $\partial\mathcal{B} \equiv \partial\mathcal{B}_t \cup \partial\mathcal{B}_u$. Tractions t_i and displacements u_i ($i = 1, 2, 3$) are prescribed on $\partial\mathcal{B}_t$ and on $\partial\mathcal{B}_u$, respectively, while the body forces b_i are given in $\mathcal{B}$.

Denote by x^i, $i = 1, 2, 3$, an orthogonal reference frame with axes aligned, at each point of the body, with the principal directions of orthotropy. Then, indicate with x^α, $\alpha = I, II, III$, the axes of a second reference frame coinciding, at each point, with the principal directions of strain. Let $\underline{g}^{(\alpha)}$ be the unit vectors associated to the frame x^α, with components $g_i^{(\alpha)}$ referred to x^i. Now, the problem to be dealt with consists in determining the local mutual orientations of the frames x^i and x^α, in order to find extreme values (absolute maxima and minima) of the work performed by the external loads. Such optimal orientations can be specified through optimal values of the nine components $g_i^{(\alpha)}$ (which undergo the constraints $g_i^{(\alpha)} g_{(\beta)}^i = \delta_{(\beta)}^{(\alpha)}$, $\alpha, \beta = I, II, III$). For the sake of simplicity, here it seems to be more convenient to characterize the mutual orientations of the two frames through a new set of design variables, namely the three Euler's angles θ_q ($q = 1, 2, 3$), chosen as depicted in Fig. 2.1. The relationship between the direction cosines of the frame x^α and the angles θ_q is

$$\begin{pmatrix} g_1^{(I)} & g_2^{(I)} & g_3^{(I)} \\ g_1^{(II)} & g_2^{(II)} & g_3^{(II)} \\ g_1^{(III)} & g_2^{(III)} & g_3^{(III)} \end{pmatrix} = \begin{pmatrix} c_{\theta_1} c_{\theta_2} & s_{\theta_1} c_{\theta_2} & s_{\theta_2} \\ -c_{\theta_1} s_{\theta_2} c_{\theta_3} - s_{\theta_1} s_{\theta_3} & -s_{\theta_1} s_{\theta_2} c_{\theta_3} + c_{\theta_1} s_{\theta_3} & c_{\theta_2} c_{\theta_3} \\ -c_{\theta_1} s_{\theta_2} s_{\theta_3} + s_{\theta_1} c_{\theta_3} & -s_{\theta_1} s_{\theta_2} s_{\theta_3} - c_{\theta_1} c_{\theta_3} & c_{\theta_2} s_{\theta_3} \end{pmatrix}$$

(2.1)

where, for brevity, the notations $s_{\theta_q} = \sin\theta_q$ and $c_{\theta_q} = \cos\theta_q$ were used. With these assumptions, the problem can be then stated as follows: find

$$\mathcal{F}^0 \equiv \min_{\theta_q} \mathcal{F}(\theta_q) = \min_{\theta_q} \frac{1}{2} \left(\int_{\partial\mathcal{B}_t} t^i \bar{u}_i \, dS + \int_{\mathcal{B}} b^i \bar{u}_i \, d\mathcal{B} \right) \qquad i, q = 1, 2, 3 \qquad (2.2)$$

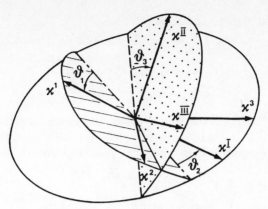

Figure 2.1. *Euler's angles.*

where $\bar{u}_i$ is the actual displacement field for prescribed θ_q, according to equilibrium, compatibility and stress-strain relationship. Now, the functional $\mathcal{F}$ to be minimized (or maximized) is equal, but opposite in sign, to $\mathcal{U}^0 = \min_{u_i} \mathcal{U}$, where $\mathcal{U}(u_i, \theta_q)$ is the total potential energy and the functions u_i belong to the class of the compatible displacement fields. In such a way, the constrained *min* problem (2.2) can be rewritten as the following unconstrained *min max* problem (see [4] for details): find

$$\mathcal{F}^0 = \min_{\theta_q} \max_{u_i} \left(\int_{\partial B_t} t^i u_i \, dS + \int_B b^i u_i \, dB - \int_B \mathcal{E} \, dB \right) \tag{2.3}$$

being $\mathcal{E}$ the specific strain energy

$$2\mathcal{E} = E^{ikrs} \varepsilon_{ik} \varepsilon_{rs}. \tag{2.4}$$

In the reference frame x^i, i.e., in principal directions of orthotropy, the specific strain energy (2.4) can be rewritten in explicit form as

$$\begin{aligned} 2\mathcal{E} = {} & E^{1111} \varepsilon_{11} \varepsilon_{11} + 2E^{1122} \varepsilon_{11} \varepsilon_{22} + 2E^{1133} \varepsilon_{11} \varepsilon_{33} + E^{2222} \varepsilon_{22} \varepsilon_{22} + 2E^{2233} \varepsilon_{22} \varepsilon_{33} \\ & + E^{3333} \varepsilon_{33} \varepsilon_{33} + 4E^{1212} \varepsilon_{12} \varepsilon_{12} + 4E^{2323} \varepsilon_{23} \varepsilon_{23} + 4E^{3131} \varepsilon_{31} \varepsilon_{31}. \end{aligned} \tag{2.5}$$

Now, the stationarity conditions with respect to the Euler's angles θ_q $(q = 1, 2, 3)$ involve only the term $\mathcal{E}$ of the functional $\mathcal{F}$; for this reason, the specific strain energy can be assumed as meaningful parameter of the global stiffness/flexibility of the body. Then, the solution of the problem requires solving the following set of three nonlinear algebraic simultaneous equations

$$\partial \mathcal{E} / \partial \theta_q = 0 \qquad q = 1, 2, 3 \tag{2.6}$$

for the three unknowns θ_q. Denoting $e_{(\beta)}$, $\beta = I, II, III$, the principal strains and recalling that $\varepsilon_{ik} = \sum_{\beta=I}^{III} e_{(\beta)} g_i^{(\beta)} g_k^{(\beta)}$, Eq.s (2.6) can be written as

$$\frac{\partial \mathcal{E}}{\partial \theta_q} = \sigma^{ik} \frac{\partial \varepsilon_{ik}}{\partial \theta_q} = 2 \sum_{\beta=I}^{III} e_{(\beta)} \frac{\partial g_i^{(\beta)}}{\partial \theta_q} g_k^{(\beta)} = 0. \tag{2.7}$$

In explicit form, stationarity conditions (2.7) read

$$\epsilon^k_{3j}\sigma^{ij}\varepsilon_{ik} = 0 \tag{2.8a}$$

$$\left(\epsilon^k_{1j}s_{\theta_1} - \epsilon^k_{2j}c_{\theta_1}\right)\sigma^{ij}\varepsilon_{ik} = 0 \tag{2.8b}$$

$$2\left(e_{(III)} - e_{(II)}\right)\sigma^{ik}g_i^{(II)}g_k^{(III)} = 0 \tag{2.8c}$$

where the symbol ϵ^k_{ij} denotes the Ricci's tensor. By means of suitable linear combinations of Eq.s (2.8), it is possible to rewrite such optimality conditions in the following way, which is more expressive from a physical point of view:

$$\left(e_{(I)} - e_{(II)}\right)\sigma^{ik}g_i^{(I)}g_k^{(II)} = 0 \tag{2.9a}$$

$$\left(e_{(II)} - e_{(III)}\right)\sigma^{ik}g_i^{(II)}g_k^{(III)} = 0 \tag{2.9b}$$

$$\left(e_{(III)} - e_{(I)}\right)\sigma^{ik}g_i^{(III)}g_k^{(I)} = 0. \tag{2.9c}$$

In the most general case, i.e., when $e_{(I)} \neq e_{(II)} \neq e_{(III)}$, the above conditions imply that

$$\sigma^{I\,II} = \sigma^{II\,III} = \sigma^{III\,I} = 0. \tag{2.10}$$

This means that, at the optimum, the principal directions of stress are collinear with the principal directions of strain. Such a mechanical interpretation of the optimality conditions has been already pointed out in [4] for the particular case of plane state of stress and now finds a more general confirm in these results. Collinearity of principal directions of stress and strain can be achieved, in particular, if the principal directions of orthotropy are aligned with those of stress and strain (*trivial* solutions), as an extension of the results obtained for 2–D solids in [2] and [4]. Such a condition is fulfilled when

$$i)\quad \theta_1 = \theta_2 = 0,\ \theta_3 = \pi/2 \quad\Longrightarrow\quad x^1 \equiv x^I,\ x^2 \equiv x^{II},\ x^3 \equiv x^{III} \tag{2.11a}$$

$$ii)\quad \theta_1 = \theta_2 = \theta_3 = 0 \quad\Longrightarrow\quad x^1 \equiv x^I,\ x^2 \equiv x^{III},\ x^3 \equiv -x^{II} \tag{2.11b}$$

$$iii)\quad \theta_1 = \pi/2,\ \theta_2 = 0,\ \theta_3 = \pi/2 \Longrightarrow\ x^1 \equiv -x^{II},\ x^2 \equiv x^I,\ x^3 \equiv x^{III} \tag{2.11c}$$

$$iv)\quad \theta_1 = \pi/2,\ \theta_2 = \theta_3 = 0 \quad\Longrightarrow\quad x^1 \equiv x^{III},\ x^2 \equiv x^I,\ x^3 \equiv x^{II}. \tag{2.11d}$$

The remainder two trivial solutions (i.e., with $x^3 \equiv x^I$) have to be seen as limit cases for $x^I \to x^3$, otherwise the Euler's angles degenerate and can no longer be defined. This yields

$$v)\quad \theta_1 \text{ arbitrary},\ \theta_2 = \pi/2,\ \theta_3 = -\theta_1$$
$$\Longrightarrow\quad x^1 \equiv x^{II},\ x^2 \equiv x^{III},\ x^3 \equiv x^I \tag{2.11e}$$

$$vi)\quad \theta_1 \text{ arbitrary},\ \theta_2 = \pi/2,\ \theta_3 = \pm\pi/2 - \theta_1$$
$$\Longrightarrow\quad x^1 \equiv -x^{III},\ x^2 \equiv x^{II},\ x^3 \equiv x^I. \tag{2.11f}$$

Although the complete analytical solution of the general case seems to be extremely involved, solutions other than the trivial ones can be easily obtained by imposing collinearity of only one of the orthotropy axes and one of the principal direction of strain. For instance, by subsequently aligning x^I with x^1, x^2 and x^3 respectively, one gets the three *almost trivial* solutions

$$x^I \equiv x^1 : \quad \theta_1 = \theta_2 = 0$$

$$\cos 2\theta_3 = \frac{2\left(E^{1122} - E^{3311}\right)e_{(I)} + \left(E^{2222} - E^{3333}\right)\left(e_{(II)} + e_{(III)}\right)}{\left[E^{2222} + E^{3333} - 2\left(E^{2233} + 2E^{2323}\right)\right]\left(e_{(II)} - e_{(III)}\right)} \tag{2.12a}$$

$$x^I \equiv x^2 : \quad \theta_1 = \pi/2 \;\; \theta_2 = 0$$

$$\cos 2\theta_3 = \frac{2\left(E^{1122} - E^{2233}\right)e_{(I)} + \left(E^{1111} - E^{3333}\right)\left(e_{(II)} + e_{(III)}\right)}{\left[E^{1111} + E^{3333} - 2\left(E^{3311} + 2E^{3131}\right)\right]\left(e_{(II)} - e_{(III)}\right)} \tag{2.12b}$$

$$x^I \equiv x^3 : \quad \theta_1 = \theta_2 = \pi/2 \;\; or \;\; \theta_1 = 0, \;\; \theta_2 = \pi/2$$

$$\cos 2\theta_3 = \pm\frac{2\left(E^{3311} - E^{2233}\right)e_{(I)} + \left(E^{1111} - E^{2222}\right)\left(e_{(II)} + e_{(III)}\right)}{\left[E^{1111} + E^{2222} - 2\left(E^{1122} + 2E^{1212}\right)\right]\left(e_{(II)} - e_{(III)}\right)}. \tag{2.12c}$$

Analogous solutions can be obtained by aligning x^{II} or x^{III} with one of the principal directions of orthotropy. Moreover it is not difficult to check that the solutions of the 2-D case (see [2], [3], [4]) can be easily recovered from the general solution just shown.

3. RESTRICTION TO THE CUBIC CASE

Consider now the case of an orthotropic material characterized by three material constants only (cubic case); in this case, the material stiffness coefficients referred to the principal directions of orthotropy are such that

$$E^{1111} = E^{2222} = E^{3333} = k; \;\; E^{1122} = E^{2233} = E^{3311} = \lambda; \;\; E^{1212} = E^{2323} = E^{3131} = \mu. \tag{3.1}$$

Note that, if $k = \lambda + 2\mu$, the isotropic case is recovered. In the cubic case, the specific strain energy takes the form

$$\begin{aligned} 2\mathcal{E} &= \left(k\varepsilon_{11} + \lambda\varepsilon_{22} + \lambda\varepsilon_{33}\right)\varepsilon_{11} + \left(k\varepsilon_{22} + \lambda\varepsilon_{33} + \lambda\varepsilon_{11}\right)\varepsilon_{22} \\ &\quad + \left(k\varepsilon_{33} + \lambda\varepsilon_{11} + \lambda\varepsilon_{22}\right)\varepsilon_{33} + 4\mu\left(\varepsilon_{12}^2 + \varepsilon_{23}^2 + \varepsilon_{31}^2\right) \\ &= \lambda I_\varepsilon^2 + 2\mu II_\varepsilon + (k - \lambda - 2\mu)\left(\varepsilon_{11}^2 + \varepsilon_{22}^2 + \varepsilon_{33}^2\right) \end{aligned} \tag{3.2}$$

where

$$I_\varepsilon = \varepsilon_i^k \delta_k^i = e_{(I)} + e_{(II)} + e_{(III)} \tag{3.3a}$$

$$II_\varepsilon = \varepsilon_i^k \varepsilon_k^i = e_{(I)}^2 + e_{(II)}^2 + e_{(III)}^2 \tag{3.3b}$$

are the first and second invariant of the strain tensor. In this case, it is not difficult to check that the optimality conditions are independent on the material constants and read

$$(\varepsilon_{11} - \varepsilon_{22})\varepsilon_{12} = 0; \;\; (\varepsilon_{22} - \varepsilon_{33})\varepsilon_{23} = 0; \;\; (\varepsilon_{33} - \varepsilon_{11})\varepsilon_{31} = 0. \tag{3.4}$$

The optimality conditions (3.4) are obviously fulfilled if

$$a) \;\; \varepsilon_{12} = \varepsilon_{23} = \varepsilon_{31} = 0 \tag{3.5}$$

which corresponds to the *trivial* solutions $\theta_1 = \theta_2 = \theta_3 = 0$, etc.; in this case, principal directions of orthotropy are collinear with the principal directions of strain and stress, as pointed out in the previous Section.

The second kind of solution

$$b) \quad \bar{\varepsilon}_{11} = \varepsilon_{22} = \varepsilon_{33} \tag{3.6}$$

corresponds to a stationarity point for the specific strain energy at which the body is in an isotropic state in terms of longitudinal strains. The (infinite) values of the Euler's angles corresponding to this solution are such that

$$\cos 2\theta_2 = \frac{2}{3} \frac{2e_{(I)} - \left(e_{(II)} + e_{(III)}\right)}{e_{(I)} - \left(e_{(II)}c_{\theta_3}^2 + e_{(III)}s_{\theta_3}^2\right)} - 1 \tag{3.7a}$$

$$\tan 2\theta_1 = \frac{e_{(I)}c_{\theta_2}^2 + \left(e_{(II)}c_{\theta_3}^2 + e_{(III)}s_{\theta_3}^2\right)s_{\theta_2}^2 - \left(e_{(II)}s_{\theta_3}^2 + e_{(III)}c_{\theta_3}^2\right)}{\left(e_{(III)} - e_{(II)}\right)s_{\theta_2}\sin 2\theta_3}. \tag{3.7b}$$

Finally, the following *almost trivial* solutions have to be considered

$$c) \quad \varepsilon_{11} = \varepsilon_{22}, \varepsilon_{23} = \varepsilon_{31} = 0; \quad \varepsilon_{22} = \varepsilon_{33}, \varepsilon_{31} = \varepsilon_{12} = 0; \quad \varepsilon_{33} = \varepsilon_{11}, \varepsilon_{12} = \varepsilon_{23} = 0, \tag{3.8}$$

where one of the three principal directions of orthotropy (e.g. x^1) is also a principal direction of strain. The corresponding values of the Euler's angles are $\theta_1 = \theta_2 = 0$, $\theta_3 = \pm\pi/4$ (if $x^1 = x^I$ - sol. c_1); $\theta_1 = \pi/2$, $\theta_2 = \pm\pi/4$, $\theta_3 = \pi/2$ (if $x^1 = x^{II}$ - sol. c_2); $\theta_1 = \pi/2$, $\theta_2 = \pm\pi/4$, $\theta_3 = 0$ (if $x^1 = x^{III}$ - sol. c_3).

In order to understand which of these solutions correspond to absolute maxima or minima, it is necessary to compare the corresponding values of the specific strain energy

$$2\mathcal{E}_a = \lambda I_\varepsilon^2 + (k - \lambda)II_\varepsilon \tag{3.9a}$$

$$2\mathcal{E}_b = \lambda I_\varepsilon^2 + 2\mu II_\varepsilon + \frac{1}{3}(k - \lambda - 2\mu)I_\varepsilon^2 \tag{3.9b}$$

$$2\mathcal{E}_{c_1} = \lambda I_\varepsilon^2 + 2\mu II_\varepsilon + (k - \lambda - 2\mu)\left((e_{(II)} + e_{(III)})^2/2 + e_{(I)}^2\right) \tag{3.9c}$$

$$2\mathcal{E}_{c_2} = \lambda I_\varepsilon^2 + 2\mu II_\varepsilon + (k - \lambda - 2\mu)\left((e_{(III)} + e_{(I)})^2/2 + e_{(II)}^2\right) \tag{3.9d}$$

$$2\mathcal{E}_{c_3} = \lambda I_\varepsilon^2 + 2\mu II_\varepsilon + (k - \lambda - 2\mu)\left((e_{(I)} + e_{(II)})^2/2 + e_{(III)}^2\right). \tag{3.9e}$$

It must be noted that the optimal solutions depend both on the strain field and on the material parameter $\frac{k}{\lambda+2\mu}$ which can be greater (material with low shear stiffness) or lower (material with high shear stiffness) than one, as pointed out in [3] for the plane stress case. The particular condition $\frac{k}{\lambda+2\mu} = 1$ corresponds, as already seen, to the isotropic case; in this case the value of the specific strain energy is $2\mathcal{E}_{iso} = \lambda I_\varepsilon^2 + 2\mu II_\varepsilon$. Without loss of generality, assuming $e_{(I)} \geq e_{(II)} \geq e_{(III)}$, the different possibilities are summarized in the following scheme

• $k \geq \lambda + 2\mu:$ $e_{(II)} \leq (e_{(I)} + e_{(III)})/2 \Rightarrow \mathcal{E}_a \geq \mathcal{E}_{c_1} \geq \mathcal{E}_{c_3} \geq \mathcal{E}_{c_2} \geq \mathcal{E}_b$

                  $e_{(II)} \geq (e_{(I)} + e_{(III)})/2 \Rightarrow \mathcal{E}_a \geq \mathcal{E}_{c_3} \geq \mathcal{E}_{c_1} \geq \mathcal{E}_{c_2} \geq \mathcal{E}_b \quad (3.10a)$

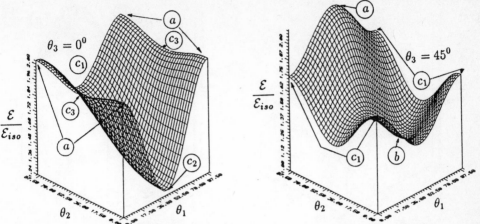

Figure 3.1. *Specific strain energy vs. Euler's angles for* $k > \lambda + 2\mu$ *(*$k = 40$, $\lambda = 6$, $\mu = 7$; $e_{(I)} = 10$, $e_{(II)} = 5$, $e_{(III)} = -5$).

- $k \leq \lambda + 2\mu$: $e_{(II)} \leq (e_{(I)} + e_{(III)})/2 \Rightarrow \mathcal{E}_b \geq \mathcal{E}_{c_2} \geq \mathcal{E}_{c_3} \geq \mathcal{E}_{c_1} \geq \mathcal{E}_a$

    $e_{(II)} \geq (e_{(I)} + e_{(III)})/2 \Rightarrow \mathcal{E}_b \geq \mathcal{E}_{c_2} \geq \mathcal{E}_{c_1} \geq \mathcal{E}_{c_3} \geq \mathcal{E}_a.$ (3.10b)

It must be noted that c–type solutions never correspond to absolute maxima or minima. In Fig. 3.1 some plots of the specific strain energy (in dimensionless form) as a function of the Euler's angles are shown for a material with low shear stiffness. Stationarity points are also indicated. Fig. 3.2 has the same meaning, but refers to a material with high shear stiffness.

4. CONCLUDING REMARKS

The problem of finding the mutual orientations of the material symmetry axes and the axes of principal strains that maximize (or minimize) the specific strain energy, has been dealt with for generally orthotropic materials and cubic materials. In any case, optimal solutions require that principal stresses be collinear with principal strains (Eq. (2.10)). Even though the complete solution seems to be too involved to be analytically derived for generally orthotropic materials (Sec. 2), *trivial* and *almost trivial* solutions were obtained (i.e., solutions where at least one of the axes of orthotropy is collinear with one of the principal strains) – Eq.s (2.11) and (2.12); these solutions recover the results already obtained by other authors in the simpler 2–D case.

For cubic materials (Sec. 3) the complete solution is found and it shown that, if the material has low shear stiffness, the most flexible solution is obtained if principal axes of strain and orthotropy are collinear, whereas the stiffest solution is characterized by equal axial strains along the axes of orthotropy (and *vice–versa* for materials with high shear stiffness) – Eq.s (3.10).

The results here obtained are of particular interest for structures reinforced by three–dimensional arrays of fibers of equal properties embedded in a matrix (3–D fiber composites). When the global stiffness (or flexibility) of the structure has to be maximized,

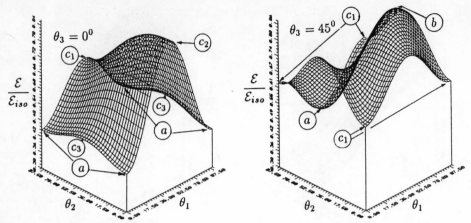

Figure 3.2. *Specific strain energy vs. Euler's angles for $k < \lambda + 2\mu$ ($k = 10$, $\lambda = 6$, $\mu = 7$; $e_{(I)} = 10$, $e_{(II)} = 5$, $e_{(III)} = -5$).*

the present analytical solution allows to optimally design the structure by properly orienting the fibers throughout the body.

REFERENCES

[1] N.V. BANICHUK, *Optimization Problems for Elastic Anisotropic Bodies*, Arch. Mech., Vol. 33, N. 6, pp. 347–363 (1981).

[2] P. PEDERSEN, *On Optimal Orientation of Orthotropic Materials*, Structural Optimization, Vol. 1, N. 2, pp. 101–106 (1989).

[3] P. PEDERSEN, *Bounds on Elastic Energy in Solids of Orthotropic Materials*, Struct. Opt., Vol. 2, N. 1, pp. 55–62 (1990).

[4] G. SACCHI LANDRIANI, M. ROVATI, *Optimal Design for 2–D Structures made of Composite Materials*, to appear on ASME J. of Engng. Materials and Technology.

[5] N. OLHOFF, J. THOMSEN, *Optimization of Fiber Orientation and Concentration in Composites*, Proc. CEEC Course on *Analysis and Design of Structures made of Composite Material*, Milano 28 May–1 June 1990.

ACKNOWLEDGEMENTS

The present work has been made possible by financial supports of Italian Ministry of Education (M.P.I.), which is here gratefully acknowledged. The authors are also indebted to Prof. N. Olhoff (University of Aalborg, DK) and Prof. P. Pedersen (The Technical University of Denmark) for the useful discussions on the subject.

Strategies for Interactive Structural Optimization and Composite Material Selection

Christoph Seeßelberg, Gunter Helwig, Horst Baier
Dornier GmbH
D-7990 Friedrichshafen, Postfach 1420

0   Introduction

Standard structural optimization often treats a "cleanly" defined mathematical problem which not always can really match the underlying practical design problem. This for example relates to the only vaguely possible mapping of the real problem to a strict mathematical problem formulation. An often proposed relief is supposed in using the designers experience and insight to be introduced especially via interactive approaches. But practical experience for larger scale problems shows, that one could be easily lost in a tremendous bulk of data to be evaluated. So proper data condensation and data preparation techniques a required. In the following two examples with different applications are presented.

1   Composite Material Selection

Composite materials have due to their excellent specific stiffness and strength an increasing field of applications not only for air- and spacecrafts. They are different from "conventional" isotropic materials, because they can be tailored by the selection of fibers- and matrix-systems and especially by the geometric variables of fiber orientations and ply thicknesses . For many kinds of requirements today composite optimization programmes are available, which determine for minimum weight the orientations and ply thicknesses under constraints like strength, stiffness, and thermal expansion,but with preselected material system.

The optimization problem gets more complicated, when the material system becomes an additional variable to be determined optimally from a material data bank. A formal treatment via discrete optimization is out of scope because of unreliable algorithms and computational effort. So the automatic material selection is algorithmically supported by first constructing continuous interpolation functions over the material data base and then using a "pointer variable" going through this interpolations in parallel to the geometric design variables, leading to a continuous nonlinear optimization problem. After solving this problem, the material with an actual "pointer variable" being most closely to the determined one is selected. Then, the optimization problem for minimum weight of a laminate reads :

$$\min\{ \ f(t_i,\rho_i) = \sum_{i=1}^{n} \rho_i \cdot t_i \ \mid \ g_j(t_i,\rho_i,\alpha_i) \le 0 \ ; \ j=1,m\}$$

1-1

(t_i = i-th ply thickness; ρ_i = i-th density; α_i = i-th ply orientation)

ρ is the "pointer variable" and t and α are the geometric variables for the m layers. Fig.1-1 shows the section-wise linearized strength of the data base in table 1-1. All other properties have to be arranged in the same manner. It is worthwhile to mention, that the here choosen "pointer variable" must not necessarily be the density, but could be any other property, for instance Youngs modulus E. But for weight optimization the density as variable makes results directly interpretable.

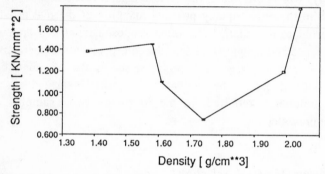

Fig. 1-1 Linearized strength over density as "pointer variable"

		Aramid		CFRP		Glass	
		Kevlar49	T300	M-40	GY-70	E-Glas	S-Glas
ρ	[g/cm**3]	1.38	1.58	1.61	1.74	1.99	2.04
σ_{1z}	[N/mm**2]	1380	1450	1100	750	1200	1780
σ_{1D}	[N/mm**2]	275	1400	1100	700	700	700
σ_{2z}	[N/mm**2]	27	55	50	40	65	65
σ_{2D}	[N/mm**2]	138	170	150	130	150	192
τ_{12}	[N/mm**2]	44	90	75	70	62	62
E_1	[KN/mm**2]	75.8	135	220	290	83	53.8
E_2	[KN/mm**2]	5.5	10	7	5	46	13.4
G_{12}	[KN/mm**2]	2.0	5	5	5	13	4.46
ν	[-]	0.34	0.27	0.35	0.41	4.36	0.29
α_1	[1.e-6/K]	-4	-0.6	-0.8	-1	6	6
α_2	[1.e-6/K]	50	30	30	30	26	26

Table 1-1 Databank of 6 different materials

In the following a simple example is given for a 2 layer composite with 2 load sets and the requirement : Find optimal thicknesses, angles and related materials out of databank, such that
- the weight is minimum
- Stiffness E ≥ 400000 N/mm
- strength criteria is satisfied under applied loads

Fig 1-3 shows the input (weight: 16.0 kg/m²) and results (weight: 5.23 kg/m²).

It is noticable that layer 1 with ρ=1.73 g/cm³ is close to the material with the highest Youngs modulus and so contributes to the stiffness requirement. Layer 2 is in the neighbourhood of a material with a much smaller Youngs modulus but significant higher strength and therefore makes the composite stronger. The right discrete material decision only the user can make, is to use GY-70 for layer 1 and M-40 for layer 2 with the corresponding optimal orientations and thicknesses.

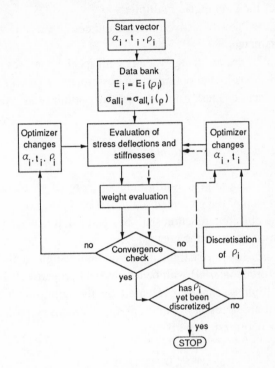

Fig 1-2 Flow diagram of optimization with material selection

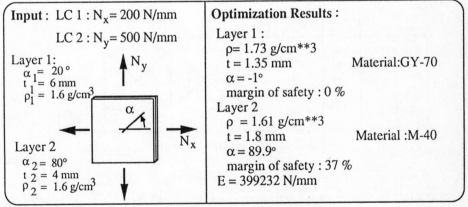

Fig 1-3 Two layer composite

2 *Postprocessing of the optimization results*

Special emphasis should also be given to a kind of postprocessing after any kind of structural optimization, which then serves for further understanding of the underlying design problem and for guiding a probably necessary mathematical reformulation. This postprocessing is based on different steps:

- Evaluation of the Lagrangian multipliers λ_i.
- Estimation of the new objective function and design variable values for changed problem parameters.
- Establishment of design change vectors with special purposes.

Those tasks can all be done by the optimization postprocessor INFO /2/, which needs for those calculations - except one task - only the present optimization results and gradients.

2.1 *Identifying the "costs" of constraints*

The evaluation of the Lagrangian multipliers λ_i in order to identify the "costs" of constraints is the first step of the presented optimization - postprocessing.

If a better value of the objective function shall be obtained, it is most efficient to increase the level of those constraints with the greatest λ_i.

Example: Ten bar truss, stress constraints; The optimal weight shall be decreased. For which bars should be used materials with better physical properties ?

The Lagrangian multipliers can be evaluated by the solution of the Kuhn-Tucker conditions 2-1, which are necessary conditions for non-convex problems and adequate conditions for convex optimization problems.

$$\nabla f(\underline{x}^{opt}) - \underline{A}(\underline{x}^{opt}) \cdot \underline{\lambda} = 0 \quad ; \quad \underline{A} = [\nabla g_1, \nabla g_2, ..., \nabla g_{m_a}] \qquad \text{2-1a}$$

$$g_j(\underline{x}^{opt}) \leq 0 \quad ; \quad j = 1, m \qquad \text{2-1b}$$

$$\lambda_j \cdot g_j(\underline{x}^{opt}) = 0 \quad ; \quad \lambda_j \leq 0 \; ; \quad j = 1, m \qquad \text{2-1c}$$

stress constraint	λ_i	weight difference for 10% increase of stress allowable bar i [%]
bar 8	-1231	-2.01
bar 4	-1150	-2.03
bar 2	-963	-1.74
bar 3	-783	-1.34
bar 1	-737	-1.21
bar 6	-616	-1.02
bar 5	-17	-0.02
bar 7	-0.2	-0.42

Tab 2-1 λ_i (ten bar truss example)

Tab 2-1 shows the λ_i of the ten bar truss example. The second column of the table contains the rate of change of the objective function caused by a 10 % increase of the stress allowable belonging to the bar. It is obvious that the increase of the stress allowables of those bars (bar 4, bar 8) with the highest λ_i causes the greatest improvements of the objective function (2 %).

In most cases the number of optimization variables is not equal to the number of active constraints m_a, so that the coefficient matrix $\underline{\underline{A}}$ in 2-1a is not quadratic.

For ($m_a > n$) n of m_a linear independent gradients of constraints have to be choosen. In the case of ($m_a < n$) the λ_i can be computed by several methods :

• Choosing m_a of n equations in 2-1, so that $\underline{\underline{A}}$ becomes quadratic.

• Activating non active constraints, so that $\underline{\underline{A}}$ becomes quadratic :

$$g_j = \frac{S}{S_{all}} - 1 \leq 0 \; ; \; S_{all} \overset{!}{=} S$$

$$\tag{2-2}$$

• The compensation method :

$$\min \|R\| = \| \nabla f - \underline{\underline{A}} \cdot \underline{\lambda} \|$$

$$\tag{2-3}$$

The accuracy of the results provided by methods 1 and 2 depends highly on the set of choosen equations respectively activated constraints, while the compensation method is the most reliable method.

Experience shows, that the Langrangian multipliers are very sensitive according to the obtained accuracy of the computed optimal design.

	x^*	Optimum x^{opt}	Difference
weight [KN]	5.476	5.4615	-0.27 %
cross sectional area bar 5 [mm**2]	944	1003	+6.3 %
cross sectional area bar 7 [mm**2]	130	126	-3.2 %
cross sectional area bar 8 [mm**2]	3370	3374	+0.12 %
λ (σ bar 5)	20.8	-16.8	-181 %
λ (σ bar 7)	-2.56	-0.192	+93 %
λ (σ bar 8)	-1230	-1231	+0.08 %

Tab 2-2 λ_i for two different design vectors near the optimum

E.g. the Lagrangian multipliers of the ten bar truss have been evaluated for two different design points : the optimum $\underline{x}^{opt}$ and a point $\underline{x}^*$ lying very close to the optimum (difference of the objective function values : 0,27 %). Tab 2-2 shows the results : The differences of the λ_i are much higher (up to 181 %). The positive value of $\lambda_5(\underline{x}^*)$ indicates a non optimal design vector, although $\underline{x}^*$ is a satisfactory solution of the optimization problem.

2.2 Problem Parameter Sensitivity Analysis

After evaluation of the Lagrangian multipliers it has been decided to relax the constraint with the highest λ_i to obtain a great improvement of the objective function. (Ten bar truss: increase of stress allowable bar 8, see Tab. 2-1).

We want to estimate the new optimal objective function and design variable values for the increased value of the stress allowable bar 8.

In general: We refer to the estimation of the current design vector $\underline{x}$ due to changes of parameters δP like constraint levels (stress allowables, deflection-, frequency bounds etc.), structure loads and other fixed parameters of the formulation.

The standard approach is simply to change the parameter and then reoptimize the problem. This method provides solutions with the highest quality, but of course it is computationally expensive. So we have to find a more efficient method to estimate the new design vector. The Expanded Design Space Method (EDSM) which has been proposed by Vanderplaats /1/ is such a method. It has got the advantage, that maximally one additional structure analysis is needed.

The method is to introduce one additional design variable $x_{n+1} = P$ according to the problem parameter P and then perform a single step of the feasible directions algorithm. The actual steplength depends on the value of δP. For the calculation of the search direction we need except the existing derivatives $\partial f/\partial x_i$ and $\partial g_j/\partial x_i$ only the derivatives of the objective functions and constraints with respect to the problem parameters: $\partial f/\partial P$ and $\partial g_j/\partial P$. In many cases this derivatives can be determined very easily by simple considerations :

For example the structure loadvector $\underline{F}$ is the problem parameter : $\underline{F}_1 = P \cdot \underline{F}_0$; then for the objective function (structure weight) can be said : $\partial f/\partial P = 0$. For the derivatives of the active stress constraints we get

$$g_j = \frac{P \cdot \sigma_j}{\sigma_{j,all}} - 1 \leq 0 \quad \rightarrow \quad \frac{\partial g_j}{\partial P} = 1$$

$$2\text{-}4$$

If the determination of the derivatives by simple consideratons is not possible, they can be computed by the finite difference method with one additional structure analysis. For the ten bar truss the quality of the estimations can be seen in fig. 2-1. Fig.2-1a shows the optimal weight with respect to the change of the stress allowable bar 8 estimated by EDSM versus the exact optimal objective function.

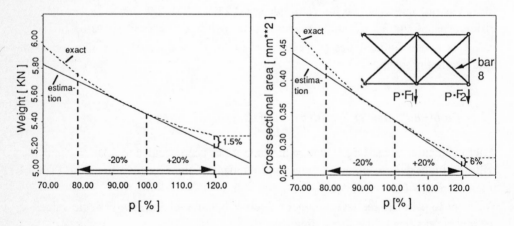

Fig 2-1 Estimation of the weight (a) and cross sectional area bar 8 (b)

As supposed the estimation forms a tangent line to the curve of the exact objective function. For changes up to 20% of the stress allowable bar 8 the error is lower than 1.5%. Fig.2-1b shows the estimated versus the exact cross sectional area of bar 8.

Investigations of further examples show, that for typical structures and changes of the problem parameters up to 20% the EDSM provides reasonable estimations for the new optimal design vector and related optimal objective function.

2.3   Design change vectors with special purposes

In many cases the mapping of the real design problem to a strict mathematical problem formulation is only vaguely possible. The results of those optimizations are often not satisfactory, because for example only the most important of some different objectives had been taken into account, while other objectives were put at a fixed value and handled as constraints.

In such cases it is a relief to use the designers experience and insight to perform - algorithmically derived - interactive design change vectors with special purposes. By this way the optimization results can be improved step by step.

Those design change vectors can have different purposes, for example :

a) *Leading from an unfeasible ($\underline{x}$) to a feasible design ($\underline{\tilde{x}}$) with a minimum increase of the objective functions value.*

The search direction $\underline{s}$ results from the solution of the linear optimization problem

$$\min_{\underline{s}} \{\nabla f^T \cdot \underline{s} \mid g_j + \nabla g_j{}^T \cdot \underline{s} \leq 0 \; ; \; j=1,m\} \qquad\qquad 2\text{-}5$$

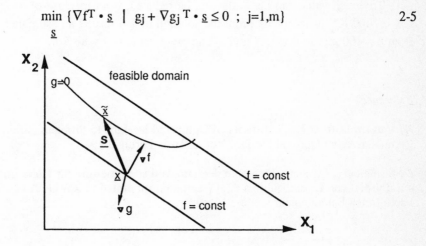

Fig. 2-2 Step from an unfeasible to a feasible design (2-dim. problem)

b) *Maximum improvement of a second-objective which has been treated as a constraint.*

The objective function's value is low enough but the second objective treated as constraint g_j isn't yet satisfactory. The problem is to achieve a maximal improvement of constraint g_j while accepting an exactly defined increase ε of the objective function value. The search direction $\underline{s}$ results from the solution of the linear optimization problem

$$\min_{\underline{s}} \ \{\nabla g_i^T \cdot \underline{s} \mid \nabla f^T \cdot \underline{s} \leq \varepsilon \ ; \ \nabla g_j^T \cdot \underline{s} \leq 0 \ ; \ j=1,m \ ; \ j \neq i \ \} \qquad 2\text{-}6$$

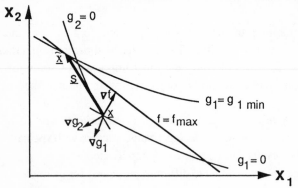

Fig. 2-3 Improvement of the constraint g_1 (2-dim. problem)

Example: Ten bar truss; 10 stress constraints and 1 deflection bound (node 6). The optimal weight is 6.34 KN ; the deflection constraint has the upper-bound value of 100 mm. To which value can the deflection be reduced, if an increase of weight from 6.34 KN to 7.00 KN is accepted? From the solution of 2-6 results an estimated improvement from u = 100 mm to u = 86.7 mm. (The exact value is u = 88.5).

References

/1/ Vanderplaats, G.N.; Sensitivity of Optimized Designs to Problem Parameters in: Computer Aided Optimal Design, Vol I, Troia Portugal 1986

/2/ Seeßelberg, C.; Neue Elemente der Tragwerksoptimierung für frühe Entwurfsphasen am Beispiel von Leichtbaubrücken, Dissertation submitted to RWTH Aachen, Fachbereich Bauingenieurwesen

OPTIMAL DESIGNS WITH SPECIAL STRUCTURAL AND MATERIAL BEHAVIOUR

SOME ASPECTS ON STRUCTURAL OPTIMIZATION
OF CERAMIC STRUCTURES

H.A. Eschenauer   Th. Vietor
Research Laboratory for Applied Structural Optimization
at the
Institute of Mechanics and Control Engineering
University of Siegen
FRG

Abstract: The use of advanced materials (ceramics, fibres, semiconductors) will be indispensable to future developments of numerous industrial branches. It is especially the material behaviour which for this reason has to be carefully considered in order to find optimal layouts of components and design variants, i.e. various kinds of failure and reliability criteria must be taken into account. Thus, the common deterministic optimization has to be extended by discrete and stochastic algorithms. For that reason, particular material values and their laws have to be described by means of distribution functions. In this connection it is important to establish appropriate stochastic optimization algorithms. Latest results are shown for a special component made of glass-ceramics as part of a large mirror.

1. INTRODUCTION

For a material to possess just one outstanding property is not generally sufficient in engineering applications; the overall behaviour with regard to a range of different properties is usually more important. The complexity of factors involved in finding "optimal" layouts and/or materials to use in different applications indicates a general increase in the importance of cooperation between material specialists, designers, and mechanicians. An essential improvement of the component behaviour at certain load combinations is achieved when not only the optimal design can be found but when corresponding optimal material properties or the optimal reliability of a component at given real material characteristics can be established as well. When carrying out reliability analyses for structures made of advanced materials, the material properties play an increasingly important role. A typical example of relevant material properties is the highest possible yield point.

During the last years a number of important papers on the probability and reliability theory in structural mechanics have been published, e.g. by E. HAUGEN [1], I. ELISHA-KOFF [2], G. SCHUELLER [3], P. THOFT-CHRISTENSEN and Y. MUROTSU [4], R.E. MELCHERS [5], D. FLADE [6]. The decisive breakthrough in applying stochastical procedures in structural optimization is still to come although some fundamental papers have been published by mathematicians, e.g. by MARTI [7], KALL [8] and PREKOPA [9]. FREUDENTHAL [10] built up models using approaches from probability theories in order to describe the breaking behaviour of brittle materials. Special stochastic optimization problems applied to the design of structures can be found in [11] to [15]. In the following, a special type of ceramic materials, the so-called glass-ceramics, are included into the analyses for a mirror structure and its optimal layout.

2. MECHANICAL PROPERTIES OF GLASS-CERAMICS

2.1 Theoretical and Practical Strength of Ceramics

In the case of ceramic materials there is an essential difference between the theoretical strength σ_{th} and the experimentally determined practical strength. GRIFFITH found the relation for the theoretical stress $\sigma_{th} = (2E\gamma/(\pi l))^{1/2}$, where l denotes the half length of an elliptical crack within a body, γ is a special energy expression, E is the YOUNG's modulus [16]. Using the average numerical values $E = 10^5$ N/mm^2, $\gamma = 10^{-3}$ N/mm and $l = 3 \cdot 10^{-7}$ mm, this approximately yields to $\sigma_{th} = 1.5 \cdot 10^4$ N/mm^2. The theoretical strength of a solid body approximately is 1/5 to 1/10 E. In practice dense ceramic materials as quartz-glass have tensile strengths up to 50 to 100 N/mm^2. These values are not constant, but depend on the special matrix of the material.

2.2 Statistical Definition of the Strength

The evaluation of the strength tests follows from statistical methods. Hereby, the stress of a component is determined by the weakest point inside the component, which usually can be approximated by a WEIBULL-distribution (Fig. 1). This belongs to the "extreme value distributions" or asymptotic distributions of the FISHER-TIPPETT-type. By means of the "weakest-link-theory", a WEIBULL-distribution can be given for the reliability $P_R = 1 - P_F$ (P_F = failure probability) for the one-dimensional stress-state as follows [27]

$$P_R = \exp\left[-V\left(\frac{\sigma-\sigma_u}{\sigma_0}\right)^k\right], \tag{1}$$

where σ denotes the stress and σ_0, σ_u and k material constants; σ_u gives the stress with the failure probability 'zero' and k is the so-called WEIBULL-modulus, a measure for the scattering. By extending to a multiple-axes stress-state, the following expression is valid:

$$P_R = \exp\left[-\left(\frac{1}{k!}\right)\left(\frac{1}{\sigma_c}\right)^k \frac{1}{V_c} \int_V (\sigma_1^k + \sigma_2^k + \sigma_3^k) dV\right] \tag{2}$$

with the material constants V_c, σ_c, k and the principal stresses σ_i (i = 1, 2, 3). The integration considers positive principal stresses only. This failure model does not cover the mutual influence of the principal stresses on the reliability which may lead to an overestimation of the reliability.

2.3 Normal Stress Hypothesis

In order to evaluate a multiple axes stress state in a component, a single axis reference stress σ_p is usually defined. There are several strength hypotheses for determining

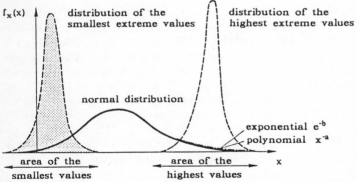

Fig. 1 Extreme-value distribution

the reference stress. Each of these hypotheses has a limited range of validity. For brittle materials, like the quartz-glass used here, the normal stress hypothesis according to RANCINE and LAMÉ is taken as a basis because these materials do not have a distinct yield point and the failure is caused by separation fracture. The normal stress hypothesis states that the largest occuring normal stress causes the fracture:

$$\sigma_p = \max \{\sigma_1, \sigma_2, \sigma_3\}. \tag{3}$$

2.4 Stochastic Material Parameter

An important characteristic of ceramics/glass-ceramics is the large scattering of material parameters. The scattering for each parameter can be considered by a distribution function. For the YOUNG's modulus E, the WEIBULL-distribution $D(E)$ reads as follows

$$D(E) = 1 - \exp\left[-\left(\frac{E}{\alpha}\right)^k\right], \tag{4}$$

with the parameters k and α.

3. SOLUTION CONCEPT FOR THE OPTIMIZATION

3.1 Mathematical Definitions

The present structural optimization task shall be considered as a Multicriteria-Optimization-Problem (MC-Problem). A continuous, deterministic MC-Problem can mathematically be defined by the following formulation [24, 26]:

$$\text{"Min"} \{f(x) \mid h(x) = 0, g(x) \geq 0\}, \tag{5}$$
$$x \in \mathbb{R}^n$$

with the symbols $\mathbb{R}^n$ set of real numbers,
              f vector of m objective functions,
              $x \in \mathbb{R}^n$ vector of n design variables,
              g vector of p inequality constraints,
              h vector of q equality constraints

and $X := \{x \in \mathbb{R}^n \mid h(x) = 0, g(x) \geq 0\}$
        "feasible domain" where $\leq$ is to be interpreted for each single component.

3.2 Problem Definition

a) Design Variables

Fig. 2 illustrates the structure of a circular mirror plate with the following continuous (c) and discrete (d) design parameters:
- mirror shape (circle, rectangle, hexagon),
- core cell structure (quadratic, triangular, hexagonal),
- cell size or rib distance (c,d),
- height of cell structure (c,d),
- thickness of layers (c,d),
- thickness of boundary stiffening (c,d),
- arrangement of the supports (d).

b) Objective Functions

The following objectives are chosen:
1) Weight of the mirror plate

$$f_1(x) := W_M. \tag{6}$$

Maximal cell size and minimal thickness of the cell walls and layers (Fig. 2) are the essential design parameters in order to reduce the weight of the mirror plate.

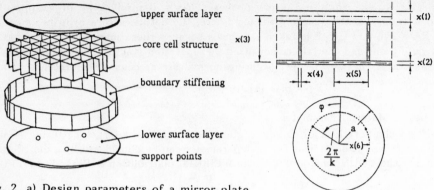

Fig. 2 a) Design parameters of a mirror plate
        b) Point-supported mirror plate

2) Surface accuracy

The accuracy of the mirror surface can generally be described by the following standard deviation as a criterion [23]:

$$f_2(\mathbf{x}) := rms = \left[\frac{\mathbf{v}^T \mathbf{v}}{n-q}\right]^{\frac{1}{2}} \tag{7}$$

with $\mathbf{v}$ vector of the deformations from ideal surface,
      n number of nodal points,
      q number of degrees of freedom.

c) Constraints

For the optimal layout, the following constraints have to be taken into account:

1) Quilting-effect [23]

The so-called quilting-effect plays an important role as far as the surface accuracy is concerned. The surface of each core cell is slightly deformed (pillow shape) by the polishing load. This pillow-shaped deformation leads to a periodical deformation of the total mirror surface. The mean square deviation of the surface deformation can be estimated by the bending of an quadratic plate clamped at all edges:

$$g_1(\mathbf{x}) := q_{rms} = c\,\frac{p\,a_s^4}{K_f} \le \frac{\lambda}{40} \qquad (\lambda = 633 \text{ nm}) \tag{8}$$

with $K_f = \dfrac{E t_f^3}{1-\nu^2}$ plate stiffness of the front plate,

      p polishing load ($p \approx 0,5$ N/cm²),
      a_s inner lateral length of the quadratic cell core,
      c constant factor from rms-calculation ($c \approx 0,004$ to 0,006),
      t_f thickness of a front plate.

2) Failure criteria

The normal stress hypothesis according to (3)

$$g_2(\mathbf{x}) = 1 - \frac{\sigma_P}{\sigma_{Pfeas}} \tag{9}$$

and the failure probability according to (2) with $P_F = 1 - P_R$

$$g_3(\mathbf{x}) = 1 - \frac{P_F}{P_{Ffeas}} \qquad (10)$$

are chosen.

3) Constraints of design parameters (bounds)
The terms for the most important quantities are given in Fig. 2a:
- cell size (quadratic) or rib distance (open structure),
- thickness of the upper surface layer,
- thickness of the lower surface layer,
- rib thickness,
- thickness of boundary stiffening,
- core height.

3.3 Structural Analysis

Because of the requirements mentioned above, we chose a porous, orthotropic linear-elastic plate as a basis for the structural analysis calculations. In addition to calculations by the FE-methods SAPV or ANSYS, an analytical approach for a point-supported circular plate was used. Fig. 2 shows the arrangement of point supports for a circular plate with constant plate thickness.
The differential equation of a point-supported plate is given by [25]

$$\Delta\Delta w = \frac{F}{K_o \pi a^2} - \frac{F}{K_o k} \sum_{j=1}^{k} \frac{\delta(r-b)}{b^2} \, \delta\left(\varphi - \frac{2j\pi}{k}\right), \qquad (11)$$

with δ = Delta-functions, K_o = plate stiffness for rectangular stiffened circular plate, a the radius of the plate, b the radius of the support circle and the load $F = p\pi a^2$ with the uniform pressure p. Development by Fourier-series

$$\sum_{j=1}^{k} \delta\left(\varphi - \frac{2j\pi}{k}\right) = \frac{a_o}{2} + \sum_{m=1}^{\infty} a_m \cos(km\varphi) \qquad (12)$$

with

$$a_m = \frac{1}{\pi} \sum_{j=1}^{k} \int_0^{2\pi} \delta\left(\varphi - \frac{2j\pi}{k}\right) \cos(km\varphi)d\varphi = \frac{k}{\pi} \qquad (13)$$

leads to

$$\Delta\Delta w = \frac{F}{K_o \pi a^2} - \frac{F}{K_o 2\pi b^2} \, \delta(r-b) - \frac{F}{K_o \pi} \frac{\delta(r-b)}{b^2} \sum_{m=1}^{\infty} \cos(km\varphi). \qquad (14)$$

The solution is carried out using the following approach [25]

$$w(r, \varphi) = \sum_{m=0}^{\infty} w_m(r) \cos(km\varphi). \qquad (15)$$

The examination of the calculated results by means of the Finite-Element-Method shows a good accord for the relation $\frac{h}{a} \leq \frac{1}{5}$ with h as the height and the radius a of the plate.

3.4 Treatment as a Discrete Optimization Problem

Because of different discrete variables (e.g. number of supports) the given problem shall be treated as a discrete optimization problem. The discrete optimization problem differs from the continuous one in so far as the design variables may take on values only from a given discrete set of values. This yields an incoherent design space with a finite number of points (Fig. 3).

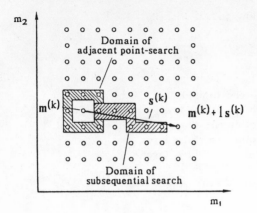

Fig. 3   k-th iteration in the integer design space $\mathbf{M} \in \mathbb{N}^2$

In analogy to the continuous problem, the discrete optimization problem is defined as follows [18, 26]:

$$\text{"Min"} \{f(\mathbf{x})\}, \ \mathbf{X}_d := \{\mathbf{x} \in \mathbb{R}^n \mid x_i \in \mathbf{X}_i; \ i = 1, \ldots, N; \ \mathbf{g}(\mathbf{x}) \geq 0 \ ; \ \mathbf{h}(\mathbf{x}) = 0\}, \qquad (16)$$
$$\mathbf{x} \in \mathbf{X}_d$$
$$\mathbf{X}_i := \{x_i^{(1)}, x_i^{(2)}, \ldots x_i^{(ni)}\}, \ \mathbf{X}_i \subset \mathbb{R} \ \forall \ i = 1, \ldots, N,$$

where $\mathbf{X}_i$ denotes the set of all n_i discrete values of the i-th design variable. Since the discrete, n-dimensional design space

$$\mathbf{X}_d := \{\mathbf{x} \in \mathbb{R}^n \mid x_i \in \mathbf{X}_i; \ i = 1, \ldots, N\} \qquad (17)$$

contains a finite number of points only, every discrete optimization problem can be transformed into an integer optimization problem with $x_i \triangleq m_i$, $m_i \in M_i := \{1, 2, \ldots, n_i\}$ and with the integer vector $\mathbf{m} \in \mathbf{M}$.

The integer gradient procedure (IG-Procedure) according to CHANARATNA et. al. [17] can be used for solving integer and, by this, discrete constrained optimization problems. The basic idea of integer problems is to gain an appropriate search direction at the point $\mathbf{m}^{(k)}$ in the form of an integer gradient by forming differences with integer gradient points. The standardization and rounding of the components yield the integer search direction shown in Fig. 3, in which one-dimensional minimization steps are carried out

$$\mathbf{m}^{k+1} = \mathbf{m}^k + \alpha \mathbf{s}^k, \quad (\alpha \in \mathbb{N}^1). \qquad (18)$$

This basic algorithm can further be improved by combining it with other search procedures (subsequential-search, adjacent-point-search [18]). Since this procedure does not consider the feasibility of the point $\mathbf{m}^k$, it is advisable to introduce a substitute objective function for constrained problems by means of an augmented penalty function [18].

3.5 Control of the Failure Probability

The inclusion of the failure probability into the optimization process means a considerate increase of the computation effort. Therefore, an approximation method of first order shall be demonstrated to show how failure probabilities of any density function can be calculated with a justifiable computation effort.

The failure probability is defined by the relation:

$$P_f = \int_{D_F} d(\mathbf{x})d\mathbf{x} \qquad \text{with} \qquad D_F = \{ \mathbf{x}\,|\,g(\mathbf{x}) < 0 \} \tag{19}$$

as the failure range in the space $\mathbf{X} \in \mathbb{R}^n$. The continuous variables $\mathbf{x}$ with random stochastic properties and the known density function $d(\mathbf{x})$ as well as correlated and uncorrelated variables are known.

From (19) one can see that the essential task of the probability theory is the calculation of multi-dimensional integrals. These calculations show some characteristic difficulties. This leads to the fact that the standard integration methods, i.e. the numerical integration and the Monte-Carlo technique, cannot be applied without modifications. A method has been developed basing upon the theory of asymptotic LAPLACE-Integrals. Since this theory is used in our investigations, it shall shortly be considered in the following.

By the relation

$$\mathbf{y} = \mathbf{T}(\mathbf{x}), \tag{20}$$

a transformation to the standardized normally-distributed and uncorrelated variables $\mathbf{y}$ is achieved. If this transformation is fulfilling special conditions [19] the following expression is valid

$$P_f = \int_{D_F=\{\mathbf{x}|g(\mathbf{x})<0\}} d(\mathbf{x})d\mathbf{x} \quad = \quad \int_{\Delta_F=\{\mathbf{y}|h(\mathbf{y})<0\}} \prod_i \varphi(y_i)\,dy_1 dy_2 \ldots dy_n, \quad h(\mathbf{y}) = g[\mathbf{T}^{-1}(\mathbf{y})] \tag{21}$$

with $\varphi(y_i)$ for the standard normal density function. This transformation is advantageous in so far as the variables y_i are uncorrelated and the density function is the multi-dimensional standard normal distribution . It can be seen as a disadvantage, however, that because of the transformation the integration range Δ_F is considerably "more complicated" than the range D_F. The transformation is given generally by the ROSEN-BLATT-Transformation, but for uncorrelated variables it is simplified to

$$y_i = \Phi^{-1}(D_i(x_i)), \tag{22}$$

where D_i denotes the distribution function of the variables x_i and Φ the standard normal-distribution function.

According to [20], for the integral

$$I(\lambda) = \int_{D_F} \exp(\lambda\,\Phi(\mathbf{x}))\,g_0(\mathbf{x})\,d\mathbf{x} \tag{23}$$

the approximative solution for $\lambda > 1$ leads to

$$I(\lambda) = \frac{(2\pi)^{(n-1)/2}\,g_0(\mathbf{x}_0)\,\exp(\lambda\,\Phi(\mathbf{x}_0))}{\lambda^{(n+1)/2}}\,|J|^{-1/2}. \tag{24}$$

Hereby J is a matrix containing the principal curvatures of the failure surface [20, 21]. Using this equation one can state the following failure probability [21]:

$$P_f \approx \Phi(-\beta) \prod_{r=1}^{R} (1 - x_r)^{-1/2}, \tag{25}$$

where

$$\beta = |\mathbf{y}^*| = \underset{\mathbf{y}}{\text{Min}}\{(\mathbf{y}^T\mathbf{y})^{1/2}\,|\,h(\mathbf{y}) = 0\} \tag{26}$$

is the minimal distance between the failure area $h(\mathbf{y}) = 0$ and the origin of the standard normally-distributed variables and x_r are the principal curvatures of the failure area in the point $\mathbf{y}^*$. By neglecting the influence of second order terms, the approach

$$P_f \approx \Phi(-\beta) \tag{27}$$

is valid which corresponds to a linearization of the failure area in the point $\mathbf{y}^*$ and to the succeeding integration over the above described linear failure area. With more than one failure criterion given, a failure probability can be stated for each criterion. The total failure probability can be approximated by various procedures (see [22]).

4. NUMERICAL RESULTS OF A MIRROR PLATE

4.1 Comparison of Continuous and Discrete Optimization

Some results of optimization calculations for the continuous, deterministic MC-Problem and for the discrete, deterministic MC-Problem are shown in Fig. 4 as functional-efficient boundaries. Quartz-glass with the material parameters $E = 72500 \text{ N/mm}^2$, $\nu = 0.17$ is chosen as material. Every point on a functional-efficient boundary corresponds to an optimal design. The number of support points is chosen as $k = 3$, 6 and 8. The distance between the functional-efficient boundaries clearly shows the improvement by increasing the number of support points. An larger number of support points ≥ 6 does not lead to an improvement. Therefore, the maximal feasible value is always assumed for the optimal design. Further calculations are to consider more than one support circle with the total number unchanged. The optimization shall then find the best possible distribution of points. The areas of the functional-efficient boundaries from 0 to 5 nm and from 40 to 60 nm are weakly-efficient solutions and therefore only theoretically important. The comparison of the designs of the continuous and of the discrete optimization model shows that unfavourable results are achieved in the case of discrete variables due to the limited design space. A functional-efficient boundary in the form of a continuous and monotonic decreasing function cannot be defined. By comparing the influence of single variables, it becomes obvious that the rib thickness of the examples shown in Fig. 4 always takes on the lower limit of 1.0 mm. The value increases above 1.0 mm in the weakly-efficient area from 0 to 5 nm only. The optimal radius of the support circle for all designs with 6 support points is found with a value of about 100 mm. In the first place this value only depends on the number of support points and on the material used.

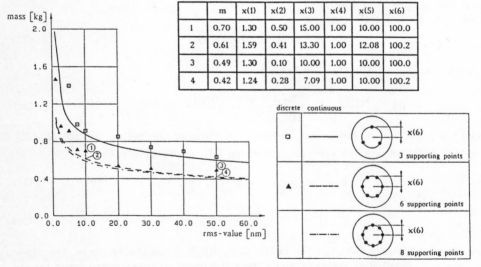

	m	x(1)	x(2)	x(3)	x(4)	x(5)	x(6)
1	0.70	1.30	0.50	15.00	1.00	10.00	100.0
2	0.61	1.59	0.41	13.30	1.00	12.08	100.2
3	0.49	1.30	0.10	10.00	1.00	10.00	100.0
4	0.42	1.24	0.28	7.09	1.00	10.00	100.2

Fig. 4 Functional-efficient boundaries for the point supported mirror

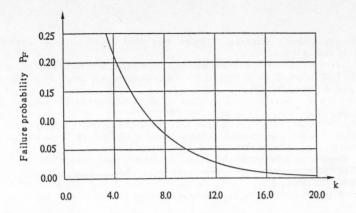

Fig. 5   Dependence of P_F = P(rms > 10 nm) on the WEIBULL-parameter k of the distribution of the YOUNG's modulus.

4.2 Calculations of the Reliability of a Chosen Design

In order to apply the reliability calculation presented in 3.5, the following failure probability is calculated for the design no. 1 shown in Fig. 4

$$P_F = P(\text{rms} > 10.0 \text{ nm}). \tag{28}$$

The following stochastic model is used:
- thickness of the upper layer x(1), normally-distributed, with the expected value E[x(1)] = 1.3 mm and the variance V[x(1)] = 0.13 mm,
- thickness of the lower layer x(2), normally-distributed, with the expected value E[x(2)] = 0.5 mm and the variance V[x(1)] = 0.05 mm,
- YOUNG's modulus, WEIBULL-distributed, with 1 < k ≤ 20 and the expected value of 72500 N/mm².

Fig. 5 shows the dependence of the failure probability on the parameter k of the WEIBULL-distribution. If one demands P_F < 0.01, k > 12 has to be demanded of the material. By reducing the variances of x(1) and x(2), a reduction of P_F is possible. This, however, increases the production effort.

5.   CONCLUSION

In order to improve the properties of high-efficient ceramics, the modelling of the structure plays a key role. As such, by optimizing the structure, it is possible to increase the absolute values of the properties, to decrease their scattering and to gain defined property values for special load cases. It is of equal importance to optimally model the component in view of "appropriate-for-ceramics" designs for special demands. This means that the scatterings of the properties have to be considered in the design process even if they are decreased by means of different measures.

The solution concept for the optimization is represented by a point-supported mirror plate as a special component with multiple objectives. Completing the results, a calculation of the reliability of the chosen designs is carried out.

REFERENCES

[1] Haugen, E.B.: Probabilistic Mechanical Design. New York, Chichester, Brisbane, Toronto:John Wiley & Sons 1979, 626 p.

[2] Elishakoff, I.: Probabilistic Methods in the Theory of Structures. New York: John Wiley & Sons 1979.

[3] Schuëller, G.: Einführung in die Sicherheit und Zuverlässigkeit von Tragwerken. Berlin, München: Wilhelm Ernst & Sohn 1987, 256 p.

[4] Thoft-Christensen, P. ; Murotsu, Y.: Application of Structural Systems Reliability Theory. Berlin, Heidelberg, New York, Tokyo: Springer-Verlag 1986.

[5] Melchers, R.E.: Structural Reliability Analysis and Prediction. Chichester: Ellis Horwood 1987, 400 p.

[6] Flade, D.: Methoden der wahrscheinlichkeitstheoretischen Analyse der Zuverlässigkeit von Tragwerken. Dissertation TH Darmstadt 1981.

[7] Marti, K.: Approximation stochastischer Optimierungsprobleme. Meisenheim am Glau: Verlag A. Hain 1979.

[8] Kall, P.: Stochastic Linear Programming. Berlin, Heidelberg, New York: Springer-Verlag 1976.

[9] Prekopa, A.: A Class of Stochastic Programming Decision Problems. Math. Oper. Stat. 3 (1972) 5, 349 - 354.

[10] Freudenthal, A.M.: Statistical Approach to Brittle Fracture. In: Liebowitz, H.(ed.): Fracture. An Advanced Treatise. New York, London: Academic Press 1968, 592 - 618.

[11] Stenvers, K.H.: Stochastische Vektoroptimierung zur Auslegung von Tragwerkstrukturen. Dissertation Universität GH Siegen 1985.

[12] Rackwitz, R.; Cuntze, R.: Formulations of Reliability-Oriented Optimization. Eng. Optimization 11, 1987, 69 - 72.

[13] Frangopol, D.M.: Multicriteria Reliability-Based Structural Optimization. Structural Safety, Vol. 3, No. 1, 1985, 23 - 26.

[14] Koski, J.; Silvenoinen, R.: Multicriteria Design of Ceramic Components. In [24], 447-462.

[15] Melchers, R.E.: On probabilistic absolute optimum design. Structural Optimization 1 (1989), 107 - 112.

[16] Salmang, H.; Scholze, H.: Keramik - Teil I: Allgemeine Grundlagen und wichtige Eigenschaften - Teil 2: Keramische Werkstoffe. Berlin, Heidelberg, New York: Springer-Verlag. 6. Auflage, 1982.

[17] Liebman, I.S.; Khachaturian, N.; Chanaratna, V.: Discrete Structural Optimization. ASCE, Journal of the Structural Division 107 (1987), ST II, 2177 - 2195.

[18] Schäfer, E.: Interaktive Strategien zur Bauteiloptimierung bei mehrfacher Zielsetzung und Diskretheitsforderungen. Dissertation Universität - GH Siegen 1990.

[19] Hohenbichler, M.; Rackwitz, R.: Non-Normal Dependent Vectors in Structural Safety . Journal of the Engineering Mechanics Division. ASCE, December 1981, 1227 - 1239.

[20] Bleistein, N.; Handelsman, R.A.: Asymptotic Expansions of Integrals. New York, Chicago: Holt, Rinehart and Winston 1975, 425 p.

[21] Breitung, K.; Hohenbichler, M.: Some Asymptotic Results in Second Order Reliability. Berichte zur Zuverlässigkeitstheorie der Bauwerke, Sonderforschungsbereich 96, Heft 69/ 1984, TU München.

[22] Ditlevsen, O.: Narrow Reliability Bounds for Structural Systems. J. Struct. Mechanics, 7 (4), 453 - 472 (1979).

[23] Eschenauer, H.A.: Telescope Designs - Structural Optimization of Mirror-Components. Proceedings of the 1. Convention of the IES "Engineering Achievements and Future Challenges", Singapore, 17-19 May 1990.

[24] Eschenauer, H.A.; Koski, J.; Osyczka, A.: Multicriteria Design Optimization. Berlin, Heidelberg, New York, London, Paris, Tokyo, Hong Kong: Springer-Verlag, 1990, 481 p.

[25] Budianski, M.P., Nelson, I.E.: Analysis of Mirror Supports for the University of California Ten Meter Telescope. Proceedings of the International Society for Optical Engineering. Berkeley, Ca., 1983.

[26] Eschenauer, H.A.: Structural Optimization - a Need in Design Processes. In: Eschenauer, H.A.; Mattheck, C.; Olhoff, N. (eds.): Lecture Notes in Engineering. Berlin : Springer-Verlag, (1991), 1-13.

[27] Munz, D.; Fett, T.: Mechanisches Verhalten keramischer Werkstoffe. WFT: Werkstoff-Forschung und -Technik. Herausgegeben von B. Ilschner, Band 8. Berlin, Heidelberg: Springer-Verlag, 1989, 244 p.

SHAPE OPTIMIZATION OF FRP DOME CLOSURES UNDER
BUCKLING CONSTRAINTS

J. Blachut
Department of Mechanical Engineering,
University of Liverpool,
P.O. Box 147,
Liverpool

1. Introduction

Recent references [1-4] provide exhaustive documentation on shape optimization of pressure vessel end closures under either static internal or external pressure.

Two shape optimization trends are apparent within internally pressurized dome ends but, with little emphasis on buckling constraints. In the first group, meridional shape of a dome closure is assumed to be fairly arbitrary as is illustrated in Refs. [5-7]. For example, optimal shape of a head, being approximated by two cubic segments, is sought via finite element approach in [5], under the von Mises stress constraints. The meridional shape of a CFRP filament wound end closure is approximated by B-splines in [6]. The membrane stress analysis shows how the Tsai-Wu failure index levels-out for an optimally shaped head. Buckle free meridional shapes are discussed in [7] for elastic cases. In the second group, the meridional shape is restricted to commonly used single or double knuckle torispherical or ellipsoidal shapes. This is mainly due to practical and safety requirements. An example of minimizing the maximum shearing stress in an elastic single and double knuckle torisphere is given in [8]. Limiting shape again to the single or double knuckle torisphere, the maximum limit load is searched for in [9] using sequential unconstrained minimisation techniques.

Optimal shape design of externally pressurized dome closures under buckling constraints has received even less attention than the internally pressurized heads. Their inherit sensitivity to initial geometric imperfections and large disparity between experimental and theoretical predictions make the analysis a still more important task than optimization [1,10].

It seems however, that some practically relevant shapes can be obtained through optimization of commonly used single knuckle steel heads. Elastic and elastic-plastic torispheres which withstand maximum or minimum buckling pressures are obtained numerically in [12] and experimentally verified in [13]. A similar problem is considered in [14] where the wall thickness varies in a step-wise manner.

Some recent numerical and experimental studies into application of composites in externally pressurized vessel end closures have been reported in [15,17]. Only commonly used shapes, i.e. hemispherical, torispherical and ellipsoidal, are considered. The three ways of manufacturing such components, i.e. vacuum bag/autoclave, filament winding and resin transfer moulding, are discussed. Whilst feasibility of these three routes is still under investigation, it seems that the meridional shape of an axisymmetric pressure vessel end closure is no longer a practical obstacle. Whichever way of vessel manufacturing is adopted one has to manufacture first a mandrel for vacuum bagging/filament winding or an appropriate moulding tool for Resin Transfer Mode (RTM). Accurate machining of the meridional shape in a male or female mould in a variety of materials is no longer a great difficulty. The shape of a composite dome closure will follow the previously machined shape simply by virtue of a replica cast (see [15,16] for details).

This ease of mandrel manufacturing makes optimal shaping of a meridian quite attractive from a practical point of view.

We assume in this paper that the meridional shape is composed of a number of circular segments which are convex to external pressure and that there is no shape discontinuity, up to the first derivative, at the adjacent segments.

The objective is to maximize the buckling pressure for a quasi-isotropic composite lay-up. The optimization technique is the complex method of Box using BOSOR4 code as the re-analysis tool.

2.   Preliminaries and Problem Statement

Let us consider an axisymmetric pressure vessel of constant thickness t and diameter D subjected to static external pressure p (Fig. 1). Its meridional shape is formed from N-1 individual segments y(i) which are joined at knots 2,3,...,N-1 with a continuous first derivative. These segments are assumed to be circular and convex to the applied pressure. The above requirements on meridional shape can formally be stated as follows:

- continuity at internal knots

$$y(i) = y(i-1), \qquad i = 2,3,\ldots,N-1 \tag{1}$$

- continuity of the first derivative at these knots

$$y'(i) = y'(i-1), \qquad i = 2,3,\ldots,N-1 \tag{2}$$

- individual segments convex to external pressure

$$y''(i) \leq 0, \qquad i = 1,2,\ldots,N-1 \tag{3}$$

- boundary conditions

$$x_1 = D/2; \qquad y_1 = 0, \tag{4}$$

$$y'_1 = -\infty, \tag{5}$$

$$x_N = 0; \qquad y'_N = 0. \tag{6}$$

Let us start construction of the multisegment meridian from the clamped edge (knot 1) and move towards the apex (knot N, see Fig. 1).

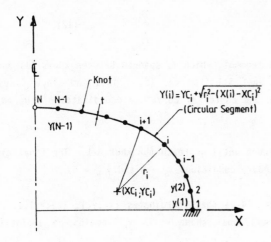

Figure 1   Geometry of a dome closure

The circular segment passing through knots i and i+1, with the first derivative at the knot i, is written as:

$$(x_i - xc_i)^2 + (y_i - yc_i)^2 = r_i^2, \tag{7}$$

$$(x_{i+1} - xc_i)^2 + (y_{i+1} - yc_i)^2 = r_i^2, \tag{8}$$

$$\frac{x_i - xc_i}{\sqrt{r_i^2 - (x_i - xc_i)^2}} = f_i. \qquad (9)$$

Assuming that coordinates (x_i, y_i), (x_{i+1}, y_{i+1}) of the first two knots and the derivative f_i are given then the eqs. (7-9) can be explicitly solved for r_i, the radius of the segment:

$$r_i = - [(x_{i+1} - x_i)^2 + (y_{i+1} - y_i)^2]/$$

$$2 \left[(x_{i+1} - x_i) \sqrt{\frac{f_i^2}{1 + f_i^2}} + (y_{i+1} - y_i) \sqrt{\frac{1}{1 + f_i^2}} \right], \qquad (10)$$

and for coordinates of the curvature center (xc_i, yc_i):

$$xc_i = x_i - r_i \sqrt{\frac{f_i^2}{1 + f_i^2}}, \qquad (11)$$

$$yc_i = y_i - r_i \sqrt{\frac{1}{1 + f_i^2}}. \qquad (12)$$

Before moving to the next segment, which is spanned between knots i+1 and i+2, we calculate the slope at the knot i+1 from eq. (8). This end slope is used in setting up equations for radius and center of curvature coordinates of the adjacent circular segment y(i+1).

The above procedure continues until we reach the knot N-1. The final circular arc is drawn using the apex boundary conditions $x_N = 0$ and $y'_N = 0$.

It becomes clear from eqs. (4-9) that coordinates $(x_2; y_2)$, $(x_3; y_3), \ldots, (x_{N-1}; y_{N-1})$ explicitly define the meridional shape while y_N remains an implicit one, depending on the vector $R = (x_2, y_2, \ldots, x_{N-1}, y_{N-1})$.

Our aim is to find such location of these N-2 knots, $(x_2, y_2), \ldots, (x_{N-1}, y_{N-1})$, in a multisegmental axisymmetric head, which maximizes the lowest dimensionless buckling pressure $p_m = \bar{p}_m / p_H$, i.e.

$$\max_{R \subset \Omega} (\min_{m=1,2,3} p_m) \qquad (13)$$

where,

$\bar{p}_1$ is bifurcation buckling pressure,

$\bar{p}_2$ is axisymmetric collapse pressure,

$\bar{p}_3$ is First Ply Failure (FPF) pressure, based on the Tsai-Wu failure criterion,

$\bar{p}_H$ is bifurcation, collapse or FPF pressure for a hemisphere (whichever is smaller).

Domain Ω contains $2(N-2)$ admissible coordinates of internal knots:

$$0 = x_N < x_{N-1} < \ldots < x_i < \ldots < x_1 = D/2, \tag{14}$$

$$y_N > y_{N-1} > \ldots > y_i > \ldots > y_1 = 0, \tag{15}$$

satisfying constraints (1), (2) and (3).

3. Single Knuckle Dome (N=3, D/t=500)

Analysis carried out in [17] shows that buckling (m=1) or axisymmetric collapse (m=2) are the controlling failure modes in a pre-preg Carbon Fibre Reinforced Plastic dome (CFRP) having D/t = 500. This eliminates the FPF mode (m=3) from eq. (13). However, all optimal solutions are checked against the First Ply Failure pressure based on the Tsai-Wu criterion to ascertain their FPF safety.

We consider here a symmetric $[0/60/-60]_s^{\circ}$ lay-up of CFRP material having $E_1 = E_2 = 70$ kN/mm^2, $G_{12} = 5$ kN/mm^2 and $\upsilon_{12} = \upsilon_{21} = 0.1$. The analysis methodology used in ref. [17] is applied here.

Different bounds are assumed for the domain Ω in the examples which follow.

In the first instance the domain Ω is restricted to (see Fig. 2a):

$$\Omega = (x_2, y_2) : \begin{bmatrix} y_2 > -x_2 + D/2 \\ x_2^2 + y_2^2 \leq (D/2)^2 \end{bmatrix} \tag{16}$$

with the constraints (1), (2) and (3) for continuity of shape, slope continuity and negative curvature.

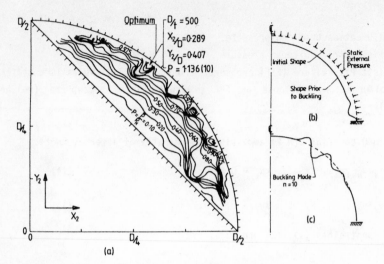

(a)

Figure 2   Contours of the objective within the feasible region (Fig. 2a).
            Prebuckling shape (Fig. 2b), and bifurcation mode (Fig. 2c) at the
            optimum

This corresponds to single knuckle torispheres which have $0 \leq r/D \leq 0.5$ and $0.5 < R_s/D < \infty$ where r and R_s are the knuckle and the spherical cap radii, respectively. Fig. 2a shows variation of the objective p inside the domain Ω. The optimal geometry is provided in Fig. 2b together with the deformed state prior to buckling. The bifurcation mode corresponding to $n = 10$ circumferential waves is given in Fig. 2c.

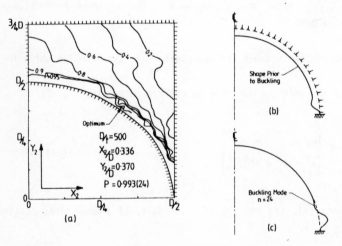

(a)

Figure 3   Location of the optimum (3a) together with the optimal shape (Fig. 3a &
            3b)

Buckling pressure of the optimal torisphere is 13.6% higher than that corresponding to a hemispherical dome of the same thickness and diameter.

In the second instance knots were allowed to be placed only outside the hemispherical profile (see Fig. 3a).

Contours of the objective are shown in Fig. 3a. The deformed shape and bifurcation mode of the optimal closure are given in Figs. 3b and 3c.

In both examples the optimum corresponds to the bifurcation mode (m=1).

4.   Multi Knuckle Dome (D/t=500)

In this paragraph the number of circular segments is increased from 2 to 4 and 8. Fig. 4a shows initial shape of a closure being made from 4 circles. Unknown coordinates of knots 2, 3 and 4 constitute a set of design variables (six in all).

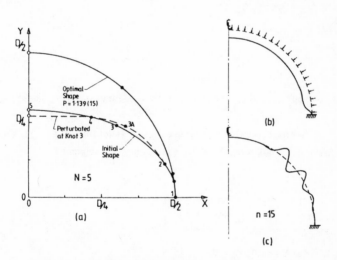

Figure 4   Initial and optimal shape (Fig. 4a). Prebuckling and buckling modes (Fig. 4b & c) at the optimum [(0.0,0.5) ; (0.497,0.048); (0.492,0.081); (0.320,0.369); (0.0,0.484)].

The broken line shows a shape once the knot number 3 is moved to a new position 3A. The slope at the knot 2 is preserved but, all three remaining segments are affected by this local perturbation. The optimal shape is also added in Fig.

4a. Its deformed shape is displayed at Fig. 4b and the bifurcation mode with n=15 circumferential waves is shown at Fig. 4c. The height of the optimal dome is about 4% less than that of a hemisphere. Its buckling pressure however, is 13.9% higher than the hemispheres having the same thickness. Details about a 8-segmental optimal dome are shown in Fig. 5.

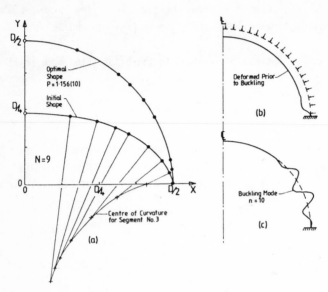

Figure 5   Initial and optimal shape together with the corresponding buckling mode

Table 1 contains maximal buckling pressures as a function of N. It seems that maximum buckling pressure is not sensitive to the number of segments.

N	3	5	9	11
P	1.136	1.139	1.156	1.159

Table 1   Buckling pressure vs. number of knots N.

5.   Conclusions

The optimal meridional shapes of axisymmetric FRP pressure vessel closure are found using the non-gradient search method of BOX. Thin domes of constant wall thickness are considered under two buckling constraints, i.e. bifurcation and axisymmetric collapse. The final designs are found to be safe in terms of the Tsai-Wu First-Ply-Failure criterion. It is expected however, that for thicker domes the First-Ply-Failure mode has to be included into stability constraints in order to achieve a feasible solution. Only circular segments are investigated in this paper. Other convex segments, like parabolic or cubic, need to be assessed. The optimization as well as the analysis method used in this paper are very flexible and can easily accommodate such shapes.

Finally it is worth mentioning that only perfect shells are discussed here despite known sensitivity of the buckling pressure to initial imperfections.

6. References

1. J. Krużelecki, M. Życzkowski, "Optimal design of shells - a survey", SM Archives, 10, (1985), 101-170.

2. Y. Ding, "Shape optimization of structures: a literature survey", Comp. Struct., 24, (1986), 985-1004.

3. R.T. Haftka, R.V. Grandhi, "Structural shape optimization - a survey", Computer Methods in Appl. Mech. and Eng., 57, (1986), 91-106.

4. A. Gajewski, M. Życzkowski, "Optimal structural design under stability constraints", Kluwer, The Netherlands, 1988.

5. C.V. Ramakrishnan, A. Francavilla, "Structural shape optimization using penalty functions", J. Struct. Mech., 3, 4, (1974-75), 403-422.

6. P.M. Martin, "Optimal design of filament wound composite pressure vessels", Proc. of the seventh OMAE Conference, M.M. Salama (ed.), ASME, (1988), 975-981.

7. W. Szyszkowski, P.G. Glockner, "Design for buckle free shapes in pressure vessels", Trans. of the ASME, J. of Pressure Vess. Technology, 107, (1985), 387-393.

8. J. Middleton, "Optimal design of torispherical pressure vessel end closures", Eng. Optimization, 4, (1979), 129-138.

9. J. Middleton, J. Petruska, "Optimal pressure vessel shape design to maximize limit load", Eng. Computations, 3, (1986), 287-294.

10. G.D. Galletly, J. Blachut, J. Krużelecki, "Plastic buckling of imperfect spherical shells subjected to external pressure", Proc. Instn. Mech. Engrs., C3, 201, (1987), 153-170.

11. G.D. Galletly, J. Krużelecki, D.G. Moffat, B. Warrington, "Buckling of shallow torispherical domes subjected to external pressure - a comparison study of experiment, theory and design codes", J. Strain Anal., 22, (1987), 163-175.

12. J. Blachut, "Search for optimal torispherical end closures under buckling constraints", Int. J. Mech. Sci., 31, (1989), 623-633.

13. J. Blachut, G.D. Galletly, D.N. Moreton, "Buckling of near perfect steel torispherical and hemispherical shells subjected to external pressure", AIAA J. (in press).

14. J. Blachut, "Optimally shaped torispheres with respect to buckling and their sensitivity to axisymmetric imperfections", Comp. Struct., 29, (1988), 975-981.

15. J. Blachut, G.D. Galletly, A.G. Gibson, "CFRP domes subjected to external pressure", J. Marine Struct., 3, 1990, 149-173.

16. F. Levy, G.D. Galletly, J. Mistry, "Buckling of composite torispherical and hemispherical domes", in Proc. of 'Composite Materials Design and Analysis, CADCOMP-90, Brussels', (eds.) W.P. de Wilde and W.R. Blain, Springer Verlag, 1990, 375-393.

17. J. Blachut, G.D. Galletly, "A numerical investigation of buckling/material failure modes in CFRP dome closures", in Proc. of 'Composite Materials Design and Analysis, CADCOMP-90, Brussels', (eds.) W.P. de Wilde and W.R. Blain, Springer Verlag, 1990, 395-411.

SHAPE OPTIMIZATION USING BOUNDARY ELEMENTS

Vladimir Kobelev

Institute for Problems in Mechanics, Academy of Sciences USSR

Av. Vernadskogo . 101, Moscow, SU-117526, USSR

1. INTRODUCTION.

The method of derivation of shape sensitivity analysis formula for the systems described by boundary integral equations is proposed. The boundary element method was used not only to solve the state equations, but also to provide sensitivity analysis. The analogous formulation of the problem, but treated using the different method, was studied in [1].

2. INTEGRAL EQUATION FOR LAPLACIAN

Consider the plane domain Ω with the smooth boundary $\Gamma+\gamma$ (Fig.1) On the boundary are given the mixed conditions for the Laplace equation:

$$\Delta w = 0 \text{ in } \Omega, \quad w = W(s) \text{ on } \gamma, \quad \frac{\partial w}{\partial n} \equiv u = U(s) \text{ on } \Gamma \quad (1)$$

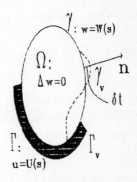

Figure 1.

We shall apply the boundary equation method to this problem. The derivation of boundary equations is a well known procedure and can be performed by several methods (See [2,3]). The common procedure for Laplace equation is based on the third Green identity and for a smooth boundary leads to the following *boundary integral equation for mixed boundary value problem*:

$$P \in \Gamma : \qquad \pi \, w_P = \int_\Gamma w_Q \, \frac{\partial}{\partial n} \ln R_{PQ} \, d\Gamma_Q - \int_\Gamma U_Q \ln R_{PQ} d\Gamma_Q +$$

$$+ \int_\gamma W_Q \, \frac{\partial}{\partial n} \ln R_{PQ} \, d\gamma_Q - \int_\gamma u_Q \ln R_{PQ} d\gamma_Q$$

$$(2.A)$$

$$P \in \gamma : \qquad \pi \, W_P \;=\; \int_\Gamma W_Q \, \frac{\partial}{\partial n} \ln R_{PQ} \, d\Gamma_Q - \int_\Gamma U_Q \ln R_{PQ} d\Gamma_Q +$$

$$+ \int_\gamma W_Q \, \frac{\partial}{\partial n} \ln R_{PQ} \, d\gamma_Q - \int_\gamma u_Q \ln R_{PQ} d\gamma_Q \qquad (2.B)$$

where P denotes an observation point and Q -an arbitrary source point on the boundary, R_{PQ}- the distance between points P and Q. The boundary equation is written here in two different forms, depending upon on what part of the boundary lies the observation point P. This division, not commonly provided, is significant for the application of boundary equations to problems with a free boundary.

3. FORMULATION OF SHAPE OPTIMIZATION PROBLEM

Our aim is to apply the boundary equation method to the following shape optimization problem:

Minimize the area I_o *of the domain* Ω (*objective functional*)

$$I_o = \int_\Omega d\Omega \to \min_{\Gamma_v \; \gamma_v} \qquad (3)$$

while the certain integral functionals I_ν *defined on the boundary values of unknown function u and its normal derivative* ($u = \partial w / \partial n$) *exceed the prescribed constants* c_ν ($\nu = 1, , M$)

$$I_\nu = \int_\Gamma F_\nu (w, u) \, d\Gamma \leqslant c_\nu \qquad (4)$$

The general definition of the shape optimization problem requires also the dependence of the functions U and W upon the current location of the unknown boundary, i.e. $W(s) = W(s, \gamma_v)$, $U(s) = U(s, \Gamma_v)$.

4. DOMAIN VARIATIONS, ADJOINT EQUATIONS AND OPTIMALITY CONDITIONS.

Let the normal variation of the domain be given by the function $\delta t(s)$, (s- arc length). This means, that after the variation the arbitrary boundary point P moves in the direction of outer normal on the distance δt. We shall associate the boundary values of functions with the moving point, i.e. w_P and u_P will denote the boundary values of an unknown function and its normal derivative before normal variation and $(w + \delta w)_P$ and $(u + \delta u)_P$ - corresponding values after variation. Note that variations of functions U on Γ_v and, respectively, function W on γ_v are explicitly prescribed by the rule which determines the variations of boundary conditions. Variations δu on γ and δw on Γ depend upon shape variations implicitly and should be treated as arbitrary. Mentioning this, write *the equations in variations* corresponding the equations (2)

$$P \in \Gamma: \quad \pi \, \delta w_P = \int_\Gamma \delta w_Q \frac{\partial}{\partial n} \ln R_{PQ} \, d\Gamma_Q + \int_\Gamma w_Q \delta\left(\frac{\partial}{\partial n} \ln R_{PQ} \, d\Gamma_Q \right) -$$

$$- \int_{\Gamma_V} \delta\left(U_Q \ln R_{PQ} d\Gamma_Q \right) + \int_{\gamma_V} \delta\left(W_Q \frac{\partial}{\partial n} \ln R_{PQ} \, d\gamma_Q \right) \qquad (5.A)$$

$$- \int_\gamma \delta u_Q \ln R_{PQ} d\gamma_Q - \int_{\gamma_V} u_Q \delta\left(\ln R_{PQ} d\gamma_Q \right)$$

$$P \in \gamma: \quad \pi \, \delta W_P = \int_\Gamma \delta w_Q \frac{\partial}{\partial n} \ln R_{PQ} \, d\Gamma_Q + \int_{\Gamma_V} w_Q \delta\left(\frac{\partial}{\partial n} \ln R_{PQ} \, d\Gamma_Q \right) -$$

$$- \int_{\Gamma_V} \delta\left(U_Q \ln R_{PQ} d\Gamma_Q \right) + \int_{\gamma_V} \delta\left(W_Q \frac{\partial}{\partial n} \ln R_{PQ} \, d\gamma_Q \right) \qquad (5.B)$$

$$- \int_\gamma \delta u_Q \ln R_{PQ} d\gamma_Q - \int_{\gamma_V} u_Q \delta\left(\ln R_{PQ} d\gamma_Q \right)$$

Equations (5.A)-(5.B) can be obviously written in the form

$$L_{11} \, \delta w + L_{12} \, \delta u = \delta M_1, \quad L_{21} \, \delta w + L_{22} \, \delta u = \delta M_2$$

where L_{ij} - integral operators, δM_i - integrals.

Now we prepared enough to derive adjoint equations and sensitivity analysis formula. For this purpose introduce Lagrange multipliers λ_ν and write the augmented Lagrange functional:

$$I = I_o + \sum_{\nu=1}^{N} \lambda_\nu \, (I_\nu - c_\nu) \qquad (6)$$

The Lagrange multipliers are defined such that

$$\lambda_\nu > 0 \qquad \text{if } I_\nu = c_\nu \qquad (7)$$

$$\lambda_\nu = 0 \qquad \text{if } I_\nu > c_\nu$$

It is easily show that the variation of I is equal to

$$\delta I = \int_{\gamma_V + \Gamma_V} \left(1 + k \, F \right) \delta t \, d\Gamma + \int_\Gamma \frac{\partial F}{\partial w} \delta w \, d\Gamma + \int_\gamma \frac{\partial F}{\partial u} \delta u \, d\gamma + \int_{\gamma_V} \frac{\partial F}{\partial w} \delta W \, d\Gamma$$

$$(8)$$

$$+ \int_{\Gamma_V} \frac{\partial F}{\partial u} \delta U \, d\gamma \, , \text{ where } F = \sum_{\nu=1}^{N} \lambda_\nu F_\nu \, , \ k \text{ - boundary curvature.}$$

Multiply (5.A) and (5.B) on the adjoint functions $\varphi(s)$ and $\psi(s)$ respectively, integrate along the corresponding parts of the boundary and the result add to the variation of augmented Lagrange functional δI:

$$\delta I^* = \delta I + \int_\gamma \varphi\left[L_{11}\delta w + L_{12}\delta u - \delta M_1\right]d\gamma + \int_\Gamma \psi\left[L_{21}\delta w + L_{22}\delta u - \delta M_2\right]d\Gamma \quad (9)$$

Changing the integration order and elimination of variations of the functions δu and δw leads to the adjoint equations in operator form

$$L_{11}\ \psi + L_{21}\ \varphi = \frac{\partial F}{\partial w}\ , \qquad L_{12}\ \psi + L_{22}\ \varphi = \frac{\partial F}{\partial u}$$

or, equivalently, in integral equation form:

$$P \in \Gamma : \ \pi\ \psi_P\ = \int_\Gamma \psi_Q\ \frac{\partial}{\partial n}\ \ln R_{PQ}\ d\Gamma_Q + \int_\gamma \varphi_Q\ \frac{\partial}{\partial n}\ \ln R_{PQ}\ d\gamma_Q + \frac{\partial F}{\partial w} \quad (10.A)$$

$$P \in \gamma : \qquad \int_\Gamma \psi_Q\ \ln R_{PQ}\ d\Gamma_Q\ + \int_\gamma \varphi_Q\ \ln R_{PQ}\ d\gamma_Q = \frac{\partial F}{\partial u} \quad (10.B)$$

Remark. Equations (2) and (10) are the equations with weak singularity (the kernel in (10.B) is, moreover, of Fredholm type). Changing the order of integration in (9) is possible for arbitrary functions from L_P on any Liapunov contour $\Gamma + \gamma$ [5].

The rest terms in the expression (9) give the *sensitivity analysis formula*:

$$\delta I = \delta I_\Gamma + \delta I_\gamma\ , \quad\quad (11)$$

$$\delta I_\Gamma\ = \int_{\Gamma_V}\left\{\ 1 + k\ F_P + \ \delta U_P \frac{\partial F}{\partial u} + \right.$$

$$+\int_\Gamma\left[\ \psi_Q\ w_P\ \delta\left(\frac{\partial}{\partial n}\ \ln R_{PQ}\ d\Gamma_Q\right)\right] +\int_\gamma\left[\ \varphi_Q\ w_P\ \delta\left(\frac{\partial}{\partial n}\ \ln R_{PQ}\ d\Gamma_Q\right)\right]-$$

$$\left.-\int_\Gamma\left[\ \psi_Q\ \delta\left(U_P\ \ln R_{PQ}\ d\Gamma_Q\right)\right] -\int_\gamma\left[\ \varphi_Q\ \delta\left(U_P\ \ln R_{PQ}\ d\Gamma_Q\right)\right]\right\}\ d\Gamma_P$$

$$\delta I_\gamma = \int_{\gamma_v} \left\{ 1 + k\, F_P + \delta W_P \frac{\partial F}{\partial w} + \pi\, \varphi_P \delta W_P + \right.$$

$$-\int_\Gamma \left[\varphi_Q\, u_P\, \delta\left[\ln R_{PQ}\, d\Gamma_Q \right] \right] - \int_\gamma \left[\varphi_Q\, u_P\, \delta\left[\ln R_{PQ}\, d\Gamma_Q \right] \right] +$$

$$+\int_\Gamma \left[\varphi_Q\, \delta\left[W_P \frac{\partial}{\partial n} \ln R_{PQ}\, d\Gamma_Q \right] \right] + \int_\gamma \left[\left. \varphi_Q\, \delta\left[W_P \frac{\partial}{\partial n} \ln R_{PQ}\, d\Gamma_Q \right] \right] \right\} d\Gamma_P$$

The symbols δI_Γ, δI_γ denote the variations of the objective functional due to variation of the parts Γ_v, γ_v of the unknown boundary. The variations of the kernels in (11) are calculated using the expressions:

$$\delta(\, d\Gamma\,) = k\, \delta t\, d\Gamma,$$

$$\vec{\delta R} = \vec{n}\, \delta t, \qquad \delta\, R = \frac{(\vec{n}, \vec{R})}{R} \delta t\,, \qquad \delta\, n_1 = n_2 \frac{\partial \delta t}{\partial s}\,, \qquad \delta\, n_2 = -n_1 \frac{\partial \delta t}{\partial s}$$

$$\delta \ln R = \frac{(\vec{n}, \vec{R})}{R^2} \delta t\,, \qquad \delta\, \frac{\partial}{\partial n} \ln R = \frac{(\vec{\delta n}, \vec{R})}{R^2} + \frac{\delta t}{R^2}\left[1 - \frac{(\vec{n}, \vec{R})^2}{R^2} \right]$$

5. DISCRETIZATION AND NUMERICAL EXAMPLES.

Equations (2),(10) and (11) form the basis of the gradient numerical procedure of shape optimization. On each iteration are solved the direct and adjoint problems and with the aid of sensitivity analysis formula performs the computation of the new boundary. As the design parameters were taken the radius-vectors of the nodal points. The discretization procedure for the equations (2) and (10) is common (See [2,3]).It is worth while mentioning, that the corresponding matrices for discrete analogs of the equations (2) and (10) are transposed. This circumstance allows to significantly reduce the computational expenses; for example, to solve both direct and adjoint linear equations it is enough to provide LU-factorization only once on each gradient iteration.

As an example of application of the derived expressions the following optimization problem was considered. On the inner and outer boundaries of a two-fold domain were given the conditions $W(\gamma)=1$, $W(\gamma_v)=0$ (Fig.2). The L_P-norm [4] of the normal derivative on the outer boundary was minimized. Due to the symmetry only a quarter of the domain was studied, and on the axes of symmetry were given the conditions $u=0$. The boundary was divided in 24 straight segments, and the linear interpolation was assumed. The program was written in Turbo-Pascal.

For the typical example each gradient iteration required 62 sec for 4.77 MHz XT without coprocessor, and about 4 sec for 16 MHz NEAT with coprocessor. Total optimization time was equal 324 sec and 46 sec respectively. Four

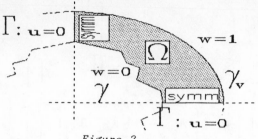

Figure 2.

-six gradient iterations was required to satisfy the optimality conditions with the precision 10^{-3}. Results are shown on Fig. 3-5. On the left pictures the initial, and on the right - the final shapes are drawn. The height of the rectangles along the free boundary is proportional to the value of the gradient of functional F at the corresponding node. Radius-vector on each iteration changes proportionally to the deviation of the gradient from its mean value.

6. CONCLUSIONS.

The numerical algorithm based on simultaneous solution of direct and adjoint integral equations and sensitivity analysis formula was developed. The algorithm was tested on the solution of 2-dimensional mixed boundary value problem for Laplace equation. The proposed method of derivation of gradient formulas is easily extended to 3-dimensional and vector problems.

This study was performed at the Institute for Mechanics and Control Engineering (University Siegen, FRG) and supported by Humboldt Foundation.

REFERENCES.

1. Choi J.H., Kwak B.M. Shape Design Sensitivity Analysis of Elliptic Problems in Boundary Integral Equation Formulation, *Mech. Struct. & Mach.*, 16(2),147-165 (1988)
2. Hartmann F. *Introduction to Boundary Elements. Theory and Applications*. Springer, Berlin, Heidelberg, 1989.
3. Gipson G.S. *Boundary Element Fundamentals - Basic Concepts and Recent Developments in the Poisson Equation*, Computational Mechanics Publications, Southampton,Boston,1987.
4. Banichuk N.V. *Introduction to Optimal Design*. Springer, Berlin, Heidelberg, 1990.
5. Zabreiko P.P., Koshelev A.I., Krasnosel'sky M.A., Mikhlin S.G., Rakovshchik L.S., Stet'senko V.Ya. *Integral Equations - a Reference Text*. Noordhoff, Leyden, 1975.

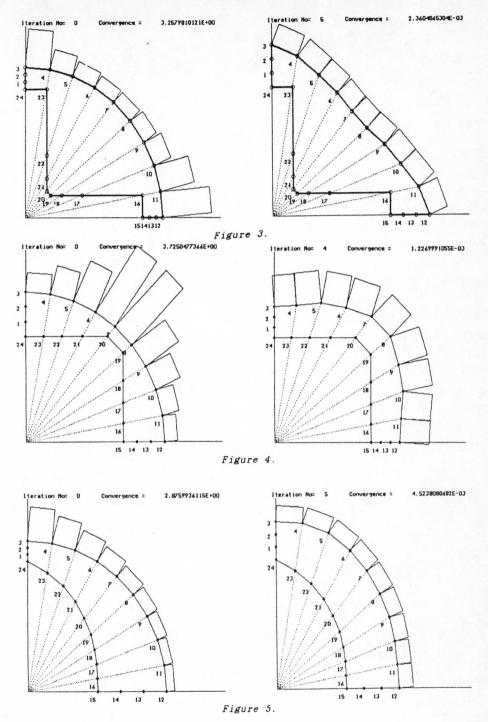

Figure 3.

Figure 4.

Figure 5.

3D-SHAPE OPTIMIZATION: DIFFERENT WAYS TO AN OPTIMIZED DESIGN

L.Harzheim and C.Mattheck
Nuclear Research Center Karlsruhe GmbH
Institute for Material and Solid State Research IV
Postfach 3640
7500 Karlsruhe, West Germany

Abstract

Shape optimization by the CAO-method (CAO: Computer Aided Optimization) [1] which is based on a computer simulation of biological growth, is a simple and effective tool in order to obtain optimized 2D- and 3D-components with homogenized surface stresses. Copying the growth of the trees the design proposal of the component to be optimized is coated with a thin layer which is allowed to grow according to an empirical volumetric swelling law to reach the desired homogeneous stress state. Nevertheless, it is not necessary to coat the whole surface with the layer, but it is also possible to let only parts of the surface grow. As a representative example of a 3D-component, the start proposal of a kinked bending bar is optimized in different ways by letting different parts of the surface grow. The different optimized shapes are of similar quality according to the homogeneity of the stress state, but they exhibit substantial differences in their shapes. In practice, this is of great advantage because you can select the design which is most easily manufactured or take into account design limitations with respect to the dimensions. Consequently, the variety offered by the CAO-method opens up the possibility of constructing components which comply with the restrictions in an effective way. The paper furthermore shows the influence of the history of optimization with respect to the ready-optimized shape

1. Introduction

The CAO-method which was presented in [1] is a powerful tool for optimizing engineering components. However, to utilize fully the advantages of this method and to get the most suitable optimized design some practice in handling the CAO-method is required. This is due to the fact that the optimized design is not unequivocally determined but depends on the path chosen to optimize the design. This property opens up the possibility for an experienced user to get an optimized design which fulfils the restrictions one needs or components which can be most easily manufactured. Therefore - as an example - a kinked bar loaded by bending is optimized in three different ways to show the different optimized shapes resulting from the different paths. This will give the reader a feeling of what happens by using different optimization histories and will enable him to generate the best suited design for his requirements.

The kinked bar to be optimized together with the boundary conditions are shown in fig. 1. The bar has a constant width with the curves in the knees designed as circles. In addition, the Mises stress along the right border is shown in the figure. It is visible that the upper and lower straight parts of the bar act only as a simple bending bar with a constant maximum bending stress $\sigma_{appl.}$. To be independent of the special value of the acting moment the Mises stress shown in the figure is normalized to $\sigma_{appl.}$. In the region of the knee the convex part is underloaded whereas in the concave part a notch stress appears.

The goal is to optimize the kinked bar with respect to the following two points:

1. The stress along the left and the right borders should be homogenized.

2. The stress along the left and the right borders should be diminished by a factor of nearly 1/3 relative to $\sigma_{appl.}$.

In applying the CAO-method it is possible to let grow three surfaces F1 and the opposite surface F2 and the upper surface F3 (see fig. 1). In this publication we will say "in-plane" growth if only F1 and F2 will be allowed to grow and "out-of-plane" growth if only F3 will be allowed to grow. In the following chapter we will explain three different paths of optimization:

1. In the first step only "in-plane" growth and in the second step only "out-of-plane" growth are allowed.

2. In the first step only "out-of-plane" growth and in the second step only "in-plane" growth are allowed.

3. Both "in-plane" and "out-of-plane" growths are allowed simultaneously in the same step.

For the calculation a Young's modulus $E = 210000 \ N/mm^2$ and a Poisson ratio $v = 0.3$ were chosen as material constants. The calculations were done using the FEM code ABAQUS [2] with HE8 elements.

2. Three Optimization Paths for the Kinked Bar

In this chapter the results of the three optimization paths will be presented. The shapes of the initial, the intermediate and the final designs will be shown with the corresponding stress distribution along the right-hand border. In some cases, the cross-sections of the upper straight part and the upper knee are shown to demonstrate the variation in the shape of the design. The corresponding cuts through the lower straight part and the lower knee lead to the same but mirrored cross-sections and are not shown for that reason. For all optimization paths both growth as shrinkage were allowed.

2.1 First Optimization Path

The chosen order of optimization steps in this path and the resulting designs are shown in fig. 2. In the first step only "in-plane" growth was allowed and $\sigma_{ref}^{(1)}/\sigma_{appl.} = 1.0$ was chosen. This means that only homogenization of the stress occurs but not reduction. This leads to a shrinkage of the convex parts of the knees due to underloading there and to a growth of the concave parts due to the notch stresses acting. The result is the design 2 shown in fig. 2. The cross-section is not changed in this first step and it is rectangular like in the starting design. To save computer time this optimization step was calculated in a 2D-model.

In the second step only "out-of-plane" growth was allowed and $\sigma_{ref}^{(2)}/\sigma_{appl.} = 1/3$ was chosen. This leads to an increasing growth from the center line to the borders reflecting the bending stresses in the bar. The resulting design 3 is shown in fig. 2 with the cross-section of the upper straight part and the region of the knee, respectively. The cross-section in the upper straight part is symmetric, whereas the cross-section in the region of the knee is enhanced towards the left side. This asymmetry arises from the fact that the stress has not been fully homogenized in the first step. The remaining higher stress in the concave part leads to slightly stronger growth than in the convex part where the stress is slightly less than 1. Nevertheless, in a good approximation, one has an optimized shape with a constant cross-section which can be manufactured easily.

2.2 Second Optimization Path

The order of the second path and the corresponding results are shown in fig. 3. Here in the first step only "out-of-plane" growth was allowed and a reference stress $\sigma^{(1)}_{ref}/\sigma_{appl.} = 1/3$ was chosen. This leads to a design with the Mises stress homogenized and reduced in just one step. Nevertheless, it is not possible to get a fully homogenized stress distribution in the first step because of the circular shapes in the concave parts of the bar which will not change by "out-of-plane" growth. Consequently, the stress is reduced but a notch stress still remains which, however, is reduced. The cross-section of the upper straight part of design 2 after step one is symmetric with decreasing thickness from the center to the borders reflecting the bending stresses. In contrast to the above the cross-section in the region of the knee shows a larger thickness at the concave side where the notch stresses appeared and a reduced thickness at the convex side which had been underloaded.

In the second step only "in-plane" growth was allowed with the same reference stress like in step one. This leads to a design for which the upper and the lower straight parts of the bar remain unchanged, whereas the concave parts of the knees grow and the convex parts shrink. The result is the design 3 with a homogenized stress distribution. Here the difference between the thickness of the two borders in the region of the knee is reduced by "in-plane" growth. However, the design is difficult to manufacture compared to the optimized design in the previous section.

2.3 Third Optimization Path

This optimization path needs only one step because "in-plane" and "out-of-plane" growths take place simultaneously with the desired maximum bending stress used as reference stress. This leads to an increase in the width and - in contrast to the sharp edges appearing in the optimized designs of the first and the second optimization paths - to a rounded cross-section (fig. 4). The cross-section of the optimized bar is not the same along the bar. It is symmetric in the straight parts of the bar, whereas the cross-section in the region of the knees is asymmetric. The thickness at the concave part is higher than at the convex side as a result of the notch stresses and the underloading part of the bar, respectively. It is interesting that similar cross-sections can be seen in nature. The photo shows a saw cut through a lateral root of a spruce tree indicating the high degree of shape optimization in nature. (The photo was kindly handed over to us by Dr. Wood, Oxford.)

Summarizing it turns out that the different optimization paths lead to very different shapes but of comparable fatigue resistances. The formation of the different shapes can be understood from the histories of the optimization procedures and enables the CAO-user to choose the most suitable optimization strategy to get that optimized design which complies best with the design restrictions.

3. Summary

By a specific example of three optimization paths the use and the resulting effects are demonstrated to show the reader the freedom he has in optimizing a design. The shapes of the three resulting optimized designs show substantial differences but they exhibit the same quality regarding stress distribution. This variety of the CAO-method offers the possibility of constructing the best suited design under a functional or manufacturing point of view.

It should be mentioned here that the three paths shown in this publication are not the only possibilities. For example, in the first optimization path one can choose $\sigma_{ref}^{(1)}/\sigma_{appl.} = 1/3$ for the first step instead of $\sigma_{ref}/\sigma_{appl.} = 1$. In this case one optimizes the design in one step and a second step is not necessary. The optimized shape in this example would have a rectangular cross-section and the shape of the left and the right borders would be the same as in the design 2 in fig. 2. But the width would be increased to reach the desired $\sigma_{ref}/\sigma_{appl.} = 1/3$. It is left to the reader to think about other possibilities and to imagine roughly how the optimized design would look like.

References

[1] C. Mattheck
    Engineering Components Grow like Trees
    Mat.-wiss. u. Werkstofftech. 21, 143-168, 1990.

[2] Hibbitt, Karlsson, Sorensen
    ABAQUS, User-Manual
    Providence, Rode Island, 1985.

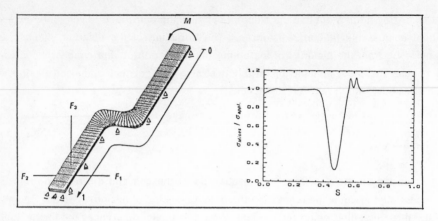

Fig. 1: FEM-model of the kinked bar to be optimized with the boundary conditions and the stress distribution along the right border. The surfaces which may grow are marked in addition

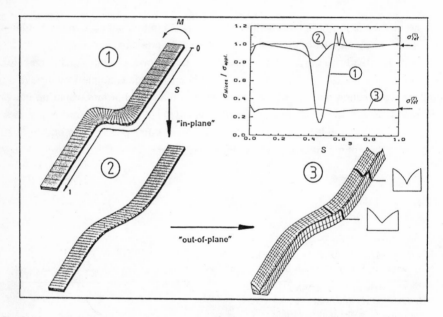

Fig. 2: The initial, the intermediate and the final designs resulting from the first optimization path with the corresponding stress distribution along the right border. (Optimization history: First growth "in-plane", second growth "out-of-plane")

MULTIPARAMETER DESIGN OPTIMISATION IN RESPECT OF STRESS CONCENTRATIONS

D. Radaj, S. Zhang

Daimler-Benz AG
Stuttgart, Germany

Abstract: The dependence of stress concentrations (structural stress, notch stress and crack stress intensity) on the dimension ratios in the structure under review, can be presented within delineated dimension areas as a simple approximation formula. The dimension ratios occur multiplicatively in this formula, with initially unknown exponents (smaller than one). The exponents can be determined from a small number of supporting point values of the stress concentration. They indicate to the designer the dimension ratios to which the stress concentration reacts particularly "sensitively", the direction in which this occurs and how large the possible reduction is. On the other hand, it becomes clear which dimension ratios have no or little influence on the stress concentration. The designer is thus provided with a valuable aid for structural optimisation. The method is demonstrated by way of example for the notch stress concentrations of welded cruciform joints.

1 Introduction

The method presented below for multiparameter structural optimisation in respect of stress concentrations is based on the following basic idea. The elastic stress concentration factors K can be represented within sufficiently narrowly defined parameter areas by a simple formula in which the (ν = 1, 2, ..., n) mutually independent dimension ratios λ_ν of the problem under review (it is the ratios that matter and not the dimensions themselves) are given different exponents n_ν, multiplied with each other, and combined with a coefficient k placed in front of the product [1, 2]:

$$K = k \, \lambda_1^{n_1} \, \lambda_2^{n_2} \, \lambda_3^{n_3} \dots \lambda_n^{n_n} \tag{1}$$

In a log-log plotting of the two-parameter dependence of the notch stress concentration factor K_t for instance, what then appears is a rhombic-shaped field of straight and parallel intersecting lines, Fig. 1 (two rhombic-shaped fields are shown one above the other here).

If this simplified representation is sufficiently accurate (verified in [1] for numerous notch cases), it should suffice for determining the entire line field from a minimum number of actually calculated dimension combinations (n+1 combinations are sufficient). This offers a considerable saving in terms of computing effort. Especially when

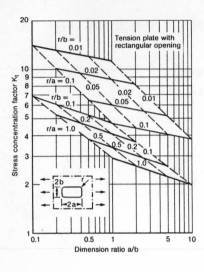

Fig. 1: Notch stress concentration factor K_t for rectangular opening in tension plate as a function of dimension ratios a/b and r/b; results of balancing calculation in two rhombic-shaped fields; logarithmically linearised approxi‑mation formula; after Radaj [1].

dealing with problems with more than two dimension ratios, the number of dimension combinations which requires to be analysed quickly becomes exceedingly large even if a narrow area of ratios is to be covered uniformly (i. e. not only with the minimum number of ratios).

The simplified formula in equation (1) provides a further advantage in that it is possible to immediately detect the direction in which the various parameters influence the stress concentration factor (positive or negative exponent), and to what extent this occurs (small or large exponent). In other words, it is immediately recognised to which dimension ratios the stress concentration reacts particularly "sensitively" and by which ratios it is more or less unaffected.

The simplified formula may relate to the concentration of structural stress, notch stress or the crack stress intensity. Formulae of this kind are known for the structural stress concentration at welded tube joints [3] and included in design specifications. Corresponding formulae apply to the notch stress concentration, for instance, at openings and inclusions [4] or at concave fillets of crankshafts [5]. In fracture mechanics, such formulae are remotely comparable in which a stress intensity factor determined for an infinite plate or solid is combined with correction factors in which the finite-dimension ratios occur. Preference can be given quite generally to the multiplicative form over the corresponding additive form [1]. This does not rule out most accurate results occasionally being obtained in individual cases with mixed forms.

2 Approximation formula for tensile plate with rectangular opening

The simplified formula for the two-parameter dependence of the notch stress concentration in the plate with rectangular opening and rectangular core (rectangles with rounded corners) has been examined in detail for 7 load cases each. The balancing calculation based on the least-squares fit was performed in the area under review (corresponding to one of the two curve fields in Fig. 1) for 16 available initial values each, i. e. significantly more than the three required as a minimum. This does not mean, though, that the approximation formula was significantly improved by the large number of initial values. On the one hand, the initial values exist with only limited accuracy, on the other hand, the actual curve pattern may deviate greatly from the simple approximation.

For comparison purposes, the curve field in question is now fixed with a small number of initial or supporting values. According to equation (1), three supporting points suffice for the two-parameter dependence in order to determine the quantities k, n_1 and n_2 definitely without a balancing calculation. As the approximation can be improved by including a small number of additional supporting points in combination with a balancing calculation, this latter procedure is adopted.

The question which then remains is how to best select the distribution of the supporting points. Closely adjoining supporting points in the middle of the area worsen the approximation in the more distant area because the limited accuracy of the initial values has a more intensified effect there and because the non-linearity of the actual curve pattern is not balanced in that area. The argument referring to the intensification of the inaccuracy does not apply to the choice of supporting points exclusively at the outer edge of the area although the argument relating to the unbalanced non-linearity remains valid. A further point to consider is that the initial values at the outer edge are generally less accurate than in the middle of the area as a consequence of the more extreme dimension ratios.

What has been stated above is confirmed by the balancing calculations performed for the stress concentration factors of the rectangular opening in the tension plate. The supporting points were varied as shown in Fig. 2 (the rhombic-shaped field is replaced by a square field). The mean and maximum deviation of the initial values from the balanced values related to the balanced values, Δ_m and Δ_{max}, are

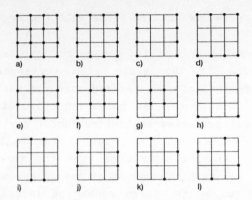

Fig. 2: Variation of number and position of supporting points for balancing calculation; rhombic-shaped field replaced by square field.

Table 1: Accuracy of results after balancing calculation; notch stress concentration factor for rectangular opening in tension plate; small corner rounding radii (upper field in Fig. 1); $K_t = k(r/b)^{n_1}(a/b)^{n_2}$.

Supporting points (Fig. 2)		Parameters in eq.(1)			Deviation	
Position	Number	k	n_1	n_2	Δ_m	Δ_{max}
a	16	1.988	-0.376	-0.098	0.025	0.051
b	12	2.036	-0.371	-0.099	0.026	0.063
c	8	2.098	-0.362	-0.097	0.026	0.055
d	8	2.106	-0.364	-0.108	0.029	0.059
e	8	1.929	-0.386	-0.094	0.026	0.051
f	8	2.048	-0.366	-0.101	0.024	0.052
g	4	1.633	-0.425	-0.080	0.037	0.098
h	4	2.179	-0.355	-0.105	0.033	0.083
i	4	2.150	-0.358	-0.127	0.032	0.084
j	4	2.050	-0.364	-0.089	0.024	0.075
k	4	1.911	-0.388	-0.105	0.028	0.078
l	4	1.948	-0.383	-0.084	0.028	0.054

presented in Table 1 together with the quantities k, n_1 and n_2. The results with only 4 supporting points approximately at the middle of the edges of the field represent the optimum. They are equivalent to those with all 16 supporting points. The corresponding investigation for a 5 x 5 field reveals the combination of the edge points mentioned with the centre point of the field to be the optimum choice.

3 Approximation formula for welded cruciform joint

The procedure described, verified for the two-parameter notch stress problem of the tension plate with rectangular opening is now applied to a multiparameter problem, namely to the contour model of a welded cruciform joint with fillet welds. The characteristic dimensions of this model are the tensile plate thickness t_1, the transverse plate

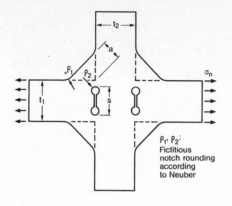

Fig. 3: Dimension parameters of investigated cruciform joint contour model; variation of the dimension ratios a/t_1, t_2/t_1, s/t_1, ρ_1/t_1 and ρ_2/t_1.

ρ_1, ρ_2:
Fictitious
notch rounding
according
to Neuber

thickness, t_2, the throat thickness, a, the slit length, s, the radii of curvature, ρ_1 and ρ_2, at the fillet transition and weld root, Fig. 3. The (fictitious) rounding of the fillet transition and weld root corresponds to Neuber's approach for the microsupporting effect at the sharp notch root when subjected to fatigue loading [6, 7]. The mutually independent dimension ratios have to be introduced into the approximation formula, equation (1). The ratios selected are the plate thickness ratio, t_2/t_1, the weld throat thickness ratio, a/t_1, the slit length ratio, s/t_1, and the radius ratios, ρ_1/t_1 and ρ_2/t_1. As the dimension ratios should be independent of each other, the throat thickness, a, should be measured from the ideal weld root point, irrespective of the slit length, s. As the notch stress maximum is dependent only on the curvature radius of the notch in question, not on that of the more distant notches, the influence of ρ_2/t_1 is ignored in the equation for K_{t1} and that of ρ_1/t_1 likewise in the equation for K_{t2}. The two approximation equations then read:

$$K_{t1} = k_1 \left(\frac{a}{t_1} \right)^{n_1} \left(\frac{t_2}{t_1} \right)^{n_2} \left(\frac{s}{t_1} \right)^{n_3} \left(\frac{\rho_1}{t_1} \right)^{n_4} \tag{2}$$

$$K_{t2} = k_2 \left(\frac{a}{t_1} \right)^{m_1} \left(\frac{t_2}{t_1} \right)^{m_2} \left(\frac{s}{t_1} \right)^{m_3} \left(\frac{\rho_2}{t_1} \right)^{m_4} \tag{3}$$

In other words, these are two four-parameter notch stress problems. To determine the 5 unknown quantities in each of the above equations, 9 supporting points were selected in each case, 4 more than required as a minimum: the basic model with a/t_1 = 1, t_2/t_1 = 1, s/t_1 = 1, ρ_1/t_1 = 0.1, ρ_2/t_1 = 0.1 and 8 further models, each with a variation of only one dimension ratio with the other ratios remaining unchanged, in concrete terms a/t_1 = 0.25 and 4.0, t_2/t_1 = 0.25 and 4.0, s/t_1 = 0.5 and 1.5, ρ_1/t_1 = 0.25 and 0.5, ρ_2/t_2 = 0.05 and 0.2. The result of the

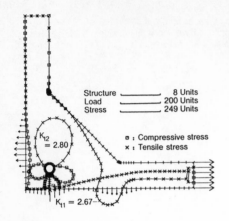

Fig. 4: Notch stress analysis using boundary element method for symmetry quarter of cross tension joint model under tensile load (right-hand edge); reaction forces (left-hand and bottom edge); boundary stress distribution plotted to the inside.

Table 2: Notch stress concentration factors of fillet transition (K_{t_1}) and of weld root (K_{t_2}) of cross joint contour model for 9 dimension variants (supporting points)

Model No.	Dimension ratios a/t_1	t_2/t_1	s/t_1	ρ_1/t_1	ρ_2/t_1	Stress concentration K_{t1}	K_{t2}
1	1.0	1.0	1.0	0.1	0.1	2.67	2.80
2	1.0	1.0	1.0	0.25	0.05	1.99	3.66
3	1.0	1.0	1.0	0.5	0.2	1.59	2.19
4	1.0	1.0	0.5	0.1	0.1	2.54	2.11
5	1.0	1.0	1.5	0.1	0.1	2.82	3.49
6	4.0	1.0	1.0	0.1	0.1	2.41	0.88
7	1.0	4.0	1.0	0.1	0.1	2.65	2.59
8	0.25	1.0	1.0	0.1	0.1	5.71	6.46
9	1.0	0.25	1.0	0.1	0.1	2.68	2.95

balancing calculation is considered to be approximatively valid even if all the dimension ratios are varied simultaneously provided the following region is not exceeded: $0.2 \leq a/t_1 \leq 5.0$, $0.2 \leq t_2/t_1 \leq 5.0$, $0.3 \leq s/t_1 \leq 1.6$, $0.08 \leq \rho_1/t_1 \leq 0.6$ and $0.04 \leq \rho_2/t_1 \leq 0.25$. Moreover, ρ_1/a and ρ_2/a are limited similar to ρ_1/t_1 and ρ_2/t_1.

The initial values K_t at the 9 supporting points were calculated using the boundary element method, Fig. 4 (a symmetry quarter of the contour model is sufficient). The parameters evaluated were the notch stress concentration at the fillet transition for K_{t_1} and the same at the end of the slit for K_{t_2}, Table 2. The result of the balancing calculation reads:

$$K_{t1} = 1.192 \left(\frac{a}{t_1} \right)^{-0.311} \left(\frac{t_2}{t_1} \right)^{-0.004} \left(\frac{s}{t_1} \right)^{0.130} \left(\frac{\rho_1}{t_1} \right)^{-0.392} \tag{4}$$

$$K_{t2} = 1.155 \left(\frac{a}{t_1} \right)^{-0.720} \left(\frac{t_2}{t_1} \right)^{-0.047} \left(\frac{s}{t_1} \right)^{0.433} \left(\frac{\rho_2}{t_1} \right)^{-0.371} \tag{5}$$

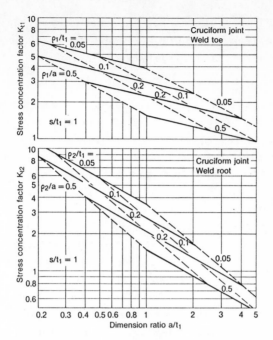

Fig. 5: Notch stress concentration factors (K_{t1} and K_{t2}) for the cruciform joint model according to equations (4) and (5) with s/t_1 = 1.0: fillet transition (on top) and weld root (at bottom).

The stress concentration factor K_{t1} increases with t_1/a and t_1/ρ_1, and weaker with s/t_1, whereas the influence of t_2/t_1 is negligible. The stress concentration factor K_{t2} increases in a similar way with t_1/a, t_1/ρ_2 and s/t_1. The result of the calculation is displayed as a graph in Fig. 5 for s/t_1 = 1.0.

Mean and maximum deviation of the initial values from the balanced values related to the balanced values, are in the case of equation (4), Δ_m = 0.104 and Δ_{max} = -0.207, in the case of equation (5), Δ_m = 0.060 and Δ_{max} = 0.141. The approximation is thus not so good in the first case as in the second one.

Random-sample comparison calculations with dimension ratios within the stated ranges, reveal deviations less than Δ_m. The deviations become significantly larger when exceeding the range limits, for instance for s/t_1 = 0.2 and 2.0 (the former with ρ_2/t_1 = 0.2 corresponding to a circular hole without a slit); they were close to the stated Δ_{max} values.

A sign of the high stability of the balancing calculation is the fact that both when the term with t_2/t_1 is removed from the above equations and also when the term with ρ_2/t_1 is added to equation (2) or the term with ρ_1/t_1 to equation (3), largely identical n_ν, m_ν and k values were determined with the remaining terms.

The stress concentration factors calculated with the boundary element method for the basic model, $K_{t1} = 2.67$ and $K_{t2} = 2.80$, were compared with data from Radaj [6, 7] (in [6] Fig. 161b for a/b = 1.0), which had been determined using the same method. The correspondence within the limits of the possible read-off and extrapolation accuracy is rather good. The values of $K_{t1} = 4.05$ and $K_{t2} = 5.73$, determinable according to Rainer [8] for the cruciform joint under consideration (complex approximation formulae based on finite element calculations with the addition of the solution for the shoulder bar according to Neuber) are, by contrast, too high. The simpler approximation formulae of Yung and Lawrence [9] lead to the acceptable values $K_{t1} = 2.33$ and $K_{t2} = 3.16$, the formula of Turmov [10], on the other hand, to the value $K_{t1} = 1.48$ which is too small. What is remarkable with the approximation formulae according to [9, 10] is that they are based on the simple $\sqrt{t/\rho}$ dependence of the stress concentration factor, however in the form of the "one-plus-formula" discussed in [1] and not found to be better. The above mentioned dependence with the square root is not confirmed by the equations (4) and (5), for which there are plausible arguments.

4 Concluding remarks on the optimisation task

The task of structural optimisation which is set to the designer is effectively supported by the simple approximation formula presented for stress concentration. First of all, it is necessary to specify which dimension ratios mainly determine the respective stress concentration. Only they are included in the approximation formula. Their initially unknown exponents are then determined on the basis of a small number of variation calculations. As soon as this has been done, it is possible to ascertain without any further variant or optimisation calculations, (i. e. without the need to call in an analysis engineer), what effects possible design measures have and to which dimension ratios the stress concentration factor reacts particularly "sensitively". For instance, it was in no way predictable in the case of the cruciform joint that the ratio t_2/t_1 has no effect and that the influence of the ratio a/t_1 is particularly pronounced. It is once again stressed that the approximation approach elucidated above applies equally well to notch stress concentration, structural stress concentration and crack stress intensity.

References

[1] Radaj, D.: Zur vereinfachten Darstellung der mehrparametrigen Formzahlabhängig keit. Konstruktion 38 (1986) No. 5, p 193-197.

[2] Radaj, D.; Zhang, S.: Mehrparametrige Strukturoptimierung hinsichtlich Spannungserhöhungen. Konstruktion 42 (1990) No. 9, p. 289 - 292.

[3] Wardenier, J.: Hollow section joints. Delft University Press, Delft 1982.

[4] Radaj, D.; Schilberth, G.: Kerbspannungen an Ausschnitten und Einschlüssen. DVS-Verlag, Düsseldorf 1977.

[5] Eberhard, A.: Einfluß der Formgebung auf die Spannungsverteilung von Kurbelkröpfungen mit Längsbohrungen. MTZ Motortechn. Zeitschr. 34 (1973) No. 7, p. 205-210 and No. 9,, p. 303-307.

[6] Radaj, D.: Gestaltung und Berechnung von Schweißkonstruktionen - Ermüdungsfestigkeit. DVS-Verlag, Düsseldorf 1985.

[7] Radaj, D.: Design and analysis of fatigue resistant welded structures. Abington Publishing, Cambridge 1990.

[8] Rainer, G.: Parameterstudien mit Finiten Elementen - Berechnung der Bauteilfestigkeit von Schweißverbindungen unter äußeren Beanspruchungen. Konstruktion 37 (1985) No. 2, p. 45-52.

[9] Yung, J.-Y.; Lawrence, F. V.: Analytical and graphical aids for the fatigue design of weldments. Fatigue Fract. Engng. Mater. Struct. 8 (1985) No. 3, p. 223-241.

[10] Turmov, G. P.: Determining the coefficient of concentration of stresses in welded joints. Avt. Svarka 10, p. 14-16.

SENSITIVITY ANALYSIS
– PROGRAMME SYSTEMS

METHOD OF ERROR ELIMINATION FOR A CLASS OF SEMI-ANALYTICAL SENSITIVITY ANALYSIS PROBLEMS

NIELS OLHOFF and JOHN RASMUSSEN
Institute of Mechanical Engineering
Aalborg University, DK-9220 Aalborg, Denmark

Abstract: The semi-analytical method of sensitivity analysis [1-3] of finite element discretized structures is indispensable in a computer aided engineering environment for interactive design and optimization. However, it has been shown [3-10] that the method may exhibit serious inaccuracies when applied to structures modeled by beam, plate, shell, and Hermite elements.

The inaccuracy of primary concern is associated with the dependence of design sensitivity error on finite element mesh refinement [3-10], but also errors subject to the pertubation of design variables may manifest themselves. Truncation errors due to conditioning of algebra and limited computer precision will not be considered here.

In this paper we present a new method developed in [10] for elimination of inaccuracy in semi-analytical sensitivity analysis for a class of problems. The method is advantageous from the point of view that problem dependent, exact error analysis is not required, and that it both eliminates the dependence of the error of the sensitivity on finite element mesh refinement and on design variable pertubation. Also, the method is computationally inexpensive because the differentation of the stiffness components can be exclucively carried out via a forward difference scheme, provided that a set of simple correction factors has been computed. The correction factors may be determined once and for all for a given type of finite element, or as an initial step of the procedure.

1. INTRODUCTION

Among the different methods available for sensitivity analysis of a finite element discretized structure, i.e., the overall finite difference technique, the analytical technique, and the semi-analytical technique [1-4], the latter is preferable for a broad class of problems.

The method is based on the global equilibrium equations for a finite element discretized problem

$$[S]\{D\} = \{F\} \ , \tag{1}$$

where $\{F\}$ is the vector of external loading, $[S]$ the stiffness matrix, and $\{D\}$ the resulting displacement vector. In a design problem, $[S]$ and $\{D\}$ depend on a vector $\{a\}$ of design variables a_j, $j = 1,..,J$. We shall assume that the external loads are independent of design so that $\partial\{F\}/\partial a_j = \{0\}$, $j = 1,..,J$.

The primary goal of design sensitivity analysis is to determine the sensitivities $\partial\{D\}/\partial a_j$ of the nodal displacements with respect to design. To this end, (1) is differentiated with respect to a_j, $j = 1,..,J$, and with design independent external loads, we obtain

$$[S(\{a\})] \ \frac{\partial\{D\}}{\partial a_j} = \{\bar{F}\}_j \qquad j = 1, \ldots, J \ , \tag{2}$$

where

$$\{\bar{F}\}_j = - \ \frac{\partial[S(\{a\})]}{\partial a_j} \{D\} \qquad j = 1, \ldots, J \ , \tag{3}$$

is the so-called *pseudo load vector* associated with the design variable a_j.

The sensitivities $\partial\{D\}/\partial a_j$ can now be solved from (2) using the same factorization of the global stiffness matrix $[S]$ as is employed in the initial solution of the finite element equilibrium equations (1) for the nodal displacements $\{D\}$ in a given step of redesign.

With {D} obtained from (1), the determination of $\partial\{D\}/\partial a_j$ from (2) only requires knowledge of the pseudo loads $\{\bar{F}\}_j$ from (3), where the design sensitivities $\partial[S]/\partial a_j$ of the stiffness matrix must be available. If the latter sensitivities are determined analytically, the above approach is called the method of *Analytical* sensitivity analysis, and if they are determined by numerical differentation, the term *Semi-analytical* sensitivity analysis is used.

In the recent papers [3–10] it has been demonstrated that the method of semi-analytical sensitivity analysis may suffer serious accuracy drawbacks when applied to finite element discretized structures modeled by beam, plate, and Hermite elements. Thus, in the papers [8,10], error analyses were carried out for a model problem of a finite element discretized beam, whose length was taken as a design variable, i.e., a simplified type of shape optimization problem was studied. Along with the expected and acceptable feature that the sensitivity error is proportional to the relative pertubation of the design variable, it was also found in [8,10] that, unfortunately, the sensitivity error is at the same time proportional to the square of the number of finite elements used to model the beam.

The source of the latter severe inaccuracy problem was found to be two–fold in [10]. Firstly, the components of the stiffness matrix of the finite element used, depend on the design variable in three different powers because the element both possesses translational and rotational degrees of freedom. Secondly, given this fact, the order of approximation behind a standard forward finite difference scheme (or, for that matter, a central finite difference scheme) for numerical differentiation of the stiffness components, is insufficient to make associated stiffness errors equal (preferably to zero), which was found to be a requirement for elimination of error dependence on mesh refinement in [10].

We shall now consider a method [10] for an extended class of problems, that does not only eliminate the accuracy problem pertaining to the number of finite elements used in the discretization, but also removes the error subject to the pertubation of design variables.

2. CLASS OF PROBLEMS AND METHOD OF ERROR ELIMINATION

Suppose that a global or local finite element stiffness matrix $S(\{a\})$ is to be differentiated numerically with respect to a design variable a_j, $j = 1,..,$ J, and assume that the typical a_j–dependent stiffness components s_r, $r = 1,..,$ R, contain a_j in different negative integer powers and have the form

$$s_r(l) = p_r + q_r l^{-r} \quad , \quad r = 1,..,R, \tag{4}$$

where $l = a_j$, and $R \geq 1$. Eq. (4) implies that l will typically be a characteristic element length or dimension in the plane of the finite element. The terms p_r and coefficients q_r, $r = 1,..,$ R, depend, in general, on the remaining design variables, i.e.,

$$p_r = p_r(a_1,.., a_{j-1}, a_{j+1},.., a_J) \ , \quad q_r = q_r(a_1,.., a_{j-1}, a_{j+1},.., a_J) \ , \quad r = 1,..,R \ . \tag{5}$$

If s_r, $r = 1,..,$ R, have the form (4), we introduce a substitution of variable such that a_j is represented by the *reciprocal* variable z,

$$z = l^{-1} \ , \tag{6}$$

whereby (4) can be written in the form $s_r^*(z) = p_r + q_r z^r$, $r = 1,..R$, such that we have $s_r^*(z) = s_r^*(l^{-1}) = s_r(l) = s_r(z^{-1})$. As is obvious from the latter relationships, the introduction of $a_j = z$ rather than $a_j = l$ as a design variable presents no barring for practical application, and can be easily implemented. Let us refrain from applying asterisks as indicators of stiffness components given as functions of z, and just write

$$s_r(z) = p_r + q_r z^r \quad , \quad r = 1,..,R \ . \tag{7}$$

We shall now assume that the numerical differentiation with respect to z of a given stiffness component s_r given by (7) is performed by means of a standard finite difference operator $d_z^{(m)}$, where m designates the order of the polynominal approximation of $s_r(z)$ that constitutes the basis for computation of the finite difference approximation $d_z^{(m)}s_r$ to the exact first derivative $\partial s_r/\partial z$.

For, e.g., first through fourth order approximation of s_r, we have the following well–known formulas for computation of the first derivative:

$$(m = 1) \qquad d_z^{(1)} s_r = \frac{\Delta s_r}{\Delta z} = \frac{1}{\Delta z} \left[s_r(z + \Delta z) - s_r(z) \right] \tag{8a}$$

$$(m = 2) \qquad d_z^{(2)} s_r = \frac{1}{2\Delta z} \left[s_r(z + \Delta z) - s_r(z - \Delta z) \right] \tag{8b}$$

$$(m = 3) \qquad d_z^{(3)} s_r = \frac{1}{6\Delta z} \left[-s_r(z + 2\Delta z) + 6s_r(z + \Delta z) - 3s_r(z) - 2s_r(z - \Delta z) \right] \tag{8c}$$

$$(m = 4) \qquad d_z^{(4)} s_r = \frac{1}{12\Delta z} \left[-s_r(z + 2\Delta z) + 8s_r(z + \Delta z) - 8s_r(z - \Delta z) + s_r(z - 2\Delta z) \right] \tag{8d}$$

The fact that the computational cost increases with the number of incremented values of z at which the stiffness components s_r have to be evaluated, must naturally be taken into account in the computational procedure.

Let us denote by η_z the relative increment (pertubation) of the design variable z, i.e.,

$$\eta_z = \frac{\Delta z}{z} . \tag{9}$$

For a given order m of approximation, we now express the *finite difference approximations* $d_z^{(m)} s_r$ in terms of the *exact first derivatives* $\partial s_r / \partial z$ and the *relative error factors* $\alpha_r^{(m)}$:

$$d_z^{(m)} s_r = \frac{\partial s_r}{\partial z} \left(1 + \alpha_r^{(m)} \right) \quad , r = 1,.., R . \tag{10}$$

It is easily verified that, due to the form of Eqs. (7), (8) and (10), *the relative error factors* $\alpha_r^{(m)}$ *will be independent of the actual value of the design variable z. Thus, if a particular* $\alpha_r^{(m)}$ *is non-vanishing, it will only depend on the relative pertubation* η_z *as defined by (9).*

The crucial point is that *when written in the form of (7), the stiffness components* s_r, $r = 1,.., R$, *depend on the design variable* $a_j = z$ *in non-negative integer powers* $r = 1,.., R$. *This implies that numerical differentiation of these stiffness components by means of a formula from among Eqs. (8), all of which are based on standard polynomial approximations, will furnish exact derivatives of all the stiffnesses* s_r, $r = 1,.., R$, *provided that the order m of the polynomial approximation behind the applied formula is equal to or larger than R, i.e., m ≥ R.*

Similar advantage is not achieved if the stiffness components s_r are considered functions of a design variable appearing in *negative powers* as in (4), because such a form of the stiffness components *cannot* be represented exactly by *any* standard polynomium of the design variable in question.

Thus, only when using $a_j = z$ as a design variable, can we make the finite difference based first derivatives coincide with the exact first derivatives of the stiffness components. This requires that we take m ≥ R, and can be expressed as

$$d_z^{(m)} s_r = \frac{\partial s_r}{\partial z} \quad \Leftrightarrow \quad \alpha_r^{(m)} = 0 , \quad r = 1,.., R, \quad \text{if } m \geq R . \tag{11}$$

The fact that the relative error factors $\alpha_r^{(m)}$ of the derivatives of the a_j–dependent stiffnesses s_r, $r = 1,.., R$, become equal for m ≥ R, implies that *the contribution from the design variable* $a_j = z$ *to the sensitivity error associated with finite element mesh refinement is eliminated if we take* m ≥ R. Moreover, the vanishing of the error factors associated with m ≥ R implies that they have become independent of the value of the relative pertubation η_z of the design variable $a_j = z$. This means that also the numerically computed derivatives of the stiffness components become independent of η_z if m ≥ R. From this we may conclude that *even the contribution to the sensitivity error from the pertubation of the design variable* $a_j = z$ *is eliminated if we take* m ≥ R.

We shall now implement the above results in an efficient computational procedure of low cost. This means that use of higher order formulas in (8) is to be limited as much as possible, because, as we have already discussed, the computational cost increases with m.

Write now (11) for the smallest order m of approximation which yields the *exact* derivatives $\partial s_r / \partial z$ of all the z–dependent stiffness derivatives, i.e., m = R:

$$d_z^{(R)} s_r = \frac{\partial s_r}{\partial z} , \quad r = 1,.., R . \tag{12}$$

Since the derivative at the right hand side is independent of η_z (but depends on z), the same holds true for the derivative at the left hand side.

Next, consider (10) in the case of first order approximation, m = 1, where $d_z^{(1)} s_r$ is the simple first order derivative $\Delta s_r / \Delta z$, cf. (8a), so that we may write

$$c_r \frac{\Delta s_r}{\Delta z} = \frac{\partial s_r}{\partial z} ; \quad c_r = \frac{1}{1 + \alpha_r^{(1)}} , \quad r = 1,..,R , \tag{13}$$

where the dimensionless factors c_r, r = 1,.., R, will be termed *correction factors* in the sequel. The correction factors c_r must follow the error factors $\alpha_r^{(1)}$ in being independent of the actual value of the design variable z and only dependent on the value of the relative increment $\eta_z = \Delta z/z$. It is characteristic, though, that we will always have $c_1 = 1$, cf. (13) with r = 1, since the first order formula (8a) yields the exact result for the linear stiffness component s_1.

Obtain now the following expression for the correction factors c_r by combination of (12) and (13):

$$c_1 = 1; \quad c_r = \frac{d_z^{(R)} s_r}{\dfrac{\Delta s_r}{\Delta z}} , \quad r = 2,.., R. \tag{14}$$

Define then the corrected first derivatives of the stiffness components $(\Delta s_r / \Delta z)_{corr}$, r = 1,..,R, as

$$\left(\frac{\Delta s_r}{\Delta z} \right)_{corr} = c_r \frac{\Delta s_r}{\Delta z} , \quad r = 1,.., R. \tag{15}$$

It is the primary objective of the new computational procedure to determine these *corrected* derivatives, and we see by means of (13) that they correspond to the *exact* derivatives,

$$\left(\frac{\Delta s_r}{\Delta z} \right)_{corr} = \frac{\partial s_r}{\partial z} , \quad r = 1,.., R. \tag{16}$$

Let us now set up the *computational procedure* pertaining to a given design variable $a_j = z$:

(0)   Choose an appropriate value of the relative increment $\eta_z = \Delta z/z$ to be used throughout.

(I)   As an initial step, determine for all subsequent computations, the correction factors c_r, r = 1,.., R, by means of Eq. (14) with the denominator given by (8a) and the numerator given by (8b) if R = 2, (8c) if R = 3, etc. The values of s_r to be used are obtained from (7) on the basis of an arbitrarily chosen value of z and the above value of η_z, which furnishes $\Delta z = \eta_z z$.

(II)   In all subsequent computations where it is required to determine the derivatives of the stiffness components subject to a specified value of the design variable z, apply Eq. (15) with the values of c_r, r = 1,.., R, obtained in step (I). The determination of $\Delta s_r / \Delta z$, r = 1,.., R, on the right hand side of (15) implies a simple, first order forward difference approach based on formula (8a), and approximate derivatives are then corrected by means of the factors c_r such that we obtain $(\Delta s_r / \Delta z)_{corr}$, r = 1,.., R, which correspond to the exact derivatives, cf. (16).

(III)   The corrected values of the first derivatives of the stiffness components with respect to the design variable $a_j = z$ can now be assembled in $\partial[S(\{a\})]/\partial a_j$ and substituted into (3) along with the displacement vector $\{D\}$ obtained from the analysis problem (1). The pseudo load vector $\{\bar{F}\}_j$ determined from (3) is then substituted into (2), which we can solve for the desired displacement derivatives $\partial\{D\}/\partial a_j$.

Since the result $(\Delta s_r/\Delta z)_{corr}$, $r = 1,..$, R, obtained in step (II) subject to any specified value of z, is *independent* of the relative pertubation η_z, *it is not required to assign* η_z *a small value* in step (0). In fact, values of η_z taken in the range between 10^{-1} and 1 have a beneficial effect on reducing truncation errors due to the conditioning of the algebra and the computational accuracy of the computer.

It should be observed that the initial steps (0) and (I) may actually be executed once and for all for the type of element and the element design variable a_j in question. Thus, *precomputed values of the correction factors* c_r *for a given value of* η_z *will be applicable for all future sensitivity analysis problems involving the given type of element and design variable* a_j, *provided that the original value of* η_z *is used.*

It is a notable feature of the computational procedure that, *after the initial steps* (0) *and* (I), *it is only required to perform the numerical differentation of the stiffness components by means of simple, computationally inexpensive, first order finite differences, and yet exact stiffness derivatives are obtained.*

The above results are valid, and the computational procedure is applicable, for any of the design variables a_j, and for any set of values assigned to them. Attention should also be drawn to the fact that the correction factors c_r, cf. (14), and the computational procedure as such, are *independent* of actual values of the coefficients p_r and q_r of the typical stiffness components, s_r, $r = 1,..,R$, in Eq.(7). This implies that *although the finite element mesh alterations inherent in shape optimization problems imply changes of the coefficients* p_r *and* q_r *through changes of the values of the design variables, see* (5), *these changes will not affect the computational procedure and the applicability of the initially determined values of the correction factors* c_r.

Thus, within the class of stiffness matrices considered, our development may be said to represent a general, efficient, and cost competitive method for elimination of inaccuracy subject to both finite element mesh refinement and pertubation of design variables in semi–analytical sensitivity analyses. Here, we tacitly assume that an accurate and efficient solution procedure for linear equations, is available for solution of the pure analysis problems (1) and (2).

3. EXAMPLE

We consider an example that has been adopted for display and numerical investigation of the semi–analytical sensitivity inaccuracy problem in [5,6] and error analyses in [8,10]. The example pertains to a finite element modeled uniform Bernoulli–Euler beam of constant bending stiffness EI and variable length L, see Fig. 1. The beam is loaded by a given, concentrated bending moment M at the free end, i.e., the external nodal load vector $\{F\}$ and the associated nodal displacement vector $\{D\}$ of the *analysis problem* (1) are

$$\{F\}^T = \{\ 0,\ 0,\ \ldots\ .\ ,\ M\}^T\ ,\qquad \{D\}^T = \{u_1,\ \theta_1,\ldots,\ u_n,\ \theta_n\}^T\ . \tag{17}$$

As in [5,6,8,10], the study will be devoted to the sensitivity of the transverse deflection $u_n(L) = ML^2/2EI$ at the free end with respect to a change of the length L of the beam, so the *exact* result within Bernoulli–Euler theory is

$$\frac{\partial u_n}{\partial L} = \frac{ML}{EI}\ . \tag{18}$$

The example only involves *one* design variable $a_j = a$ since only the total beam length L may vary. Let us choose to discretize the beam into a total number of n finite elements of equal length $l = L/n$, see Fig.1. In order to investigate the inaccuracy problem associated with finite element mesh refinement, the choice of design variable must both reflect the discretization and the beam length L, and two alternate choices of the design variable a will be considered, namely

$$a = l \quad and \quad a = z = l^{-1}\ ,\qquad where\ l = L/n\ . \tag{19}$$

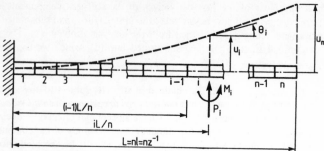

Fig. 1. Global finite element model.

We adopt a beam finite element which is exact within Bernoulli–Euler theory, whereby the element stiffness matrix is (see, e.g., [11]):

$$
\begin{bmatrix}
s_{11} & s_{12} & -s_{11} & s_{12} \\
s_{12} & s_{22} & -s_{12} & \frac{1}{2}s_{22} \\
-s_{11} & -s_{12} & s_{11} & -s_{12} \\
s_{12} & \frac{1}{2}s_{22} & -s_{12} & s_{22}
\end{bmatrix}
$$

with stiffness components given by

$$
s_{11} = 12\,\frac{EI}{l^3}\ , \quad s_{12} = 6\,\frac{EI}{l^2}\ , \quad s_{22} = 4\,\frac{EI}{l}\ . \tag{20}
$$

In the Table, we now present numerically computed values of the subject design sensitivity for a series of values of the number n of elements used in the finite element modeling of the beam, see Fig. 1. The unit values in the second column of the Table serve as the exact result to be compared with by sensitivities determined by the semi–analytical method.

Firstly, we present some results which we obtain [10], if we apply a standard approach of semi–analytical sensitivity analysis, where the current method for error elimination is *not implemented*.

Thus, the values of the semi–analytical sensitivity $(\Delta u_n/\Delta L)_l$ in the third column are based on a standard first order forward difference approximation of stiffness derivatives with the use of l as a design variable [10], and the relative increment is taken to be $\eta_l = \Delta l/l = 10^{-4}$. The results show that the semi–analytical sensitivities become increasingly inaccurate with increasing n, and that even the sign becomes wrong when we use more than 63 finite elements. The subsequent column displays the relative error ε_l, and clearly illustrates the n^2–dependence of ε_l. The results [10] for $(\Delta u_n/\Delta L)_z$ in the fifth column are based on application of the reciprocal design variable $z = l^{-1}$ and the value $\eta_z = \Delta z/z = 10^{-4}$ of the relative increment, i.e., the same value as was used for $\eta_l = \Delta l/l$, and again standard semi–analytical sensitivity analysis is carried out. Except for having the correct sign for all values of n, the sensitivities are seen to be no more accurate than when l is used as a design variable, and we notice again that the relative error ε_z exhibits severe n^2–dependence. The sources of these problems are revealed in [10].

Let us now adopt the present method of error elimination to help this unfortunate state of affairs. Then z is required to be the design variable, i.e., $a_i = a = z$, and from (7) and (20) we identify R = 3 and the typical z–dependent stiffness components to be

$$
s_1 = s_{22} = 4EIz\ , \quad s_2 = s_{12} = 6EIz^2\ , \quad s_3 = s_{11} = 12EIz^3\ , \tag{21}
$$

where z = n/L, cf. (19). Since we seek to determine $\Delta u_n/\Delta L$ and know that ML/EI factors out from the result, we assign M, L, and EI unit values in the subsequent computations, whereby we simply have z = n.

TABLE

Computed displacement design sensitivities vs. number n of finite elements used in beam model

n	$\dfrac{\partial u_n}{\partial L}$	$\left(\dfrac{\Delta u_n}{\Delta L}\right)_1$	ε_1	$\left(\dfrac{\Delta u_n}{\Delta L}\right)_z$	ε_z	$\left(\dfrac{\Delta u_n}{\Delta L}\right)_{corr}$	ε
1	1	1.000	.000	1.000	.000	1.000	0.3E-14
2	1	.999	-.001	1.001	.001	1.000	0.9E-14
3	1	.998	-.002	1.002	.002	1.000	0.3E-13
4	1	.996	-.004	1.004	.004	1.000	0.3E-13
5	1	.994	-.006	1.006	.006	1.000	-0.1E-13
6	1	.991	-.009	1.009	.009	1.000	0.2E-12
7	1	.988	-.012	1.012	.012	1.000	-0.2E-12
8	1	.984	-.016	1.016	.016	1.000	0.1E-12
9	1	.980	-.020	1.020	.020	1.000	0.6E-13
10	1	.975	-.025	1.025	.025	1.000	0.1E-13
12	1	.964	-.036	1.036	.036	1.000	0.3E-12
14	1	.951	-.049	1.049	.049	1.000	-0.1E-11
16	1	.936	-.064	1.064	.064	1.000	0.7E-12
18	1	.919	-.081	1.081	.081	1.000	0.4E-12
20	1	.900	-.100	1.100	.100	1.000	0.2E-12
22	1	.879	-.121	1.121	.121	1.000	-0.8E-12
24	1	.856	-.144	1.144	.144	1.000	0.2E-11
26	1	.831	-.169	1.169	.169	1.000	0.2E-11
28	1	.804	-.196	1.196	.196	1.000	-0.4E-11
30	1	.775	-.225	1.225	.225	1.000	0.1E-11
32	1	.744	-.256	1.256	.256	1.000	0.3E-11
34	1	.711	-.289	1.289	.289	1.000	-0.3E-11
36	1	.676	-.324	1.324	.324	1.000	0.8E-12
38	1	.639	-.361	1.361	.361	1.000	-0.4E-11
40	1	.600	-.400	1.400	.400	1.000	-0.1E-11
42	1	.559	-.441	1.441	.441	1.000	0.9E-11
44	1	.516	-.484	1.484	.484	1.000	0.3E-12
46	1	.471	-.529	1.529	.529	1.000	0.4E-11
48	1	.424	-.576	1.576	.576	1.000	0.5E-11
50	1	.375	-.625	1.625	.625	1.000	0.1E-11
52	1	.324	-.676	1.676	.676	1.000	0.9E-11
54	1	.271	-.729	1.729	.729	1.000	0.7E-11
56	1	.216	-.784	1.784	.784	1.000	-0.1E-10
58	1	.159	-.841	1.841	.841	1.000	0.7E-11
60	1	.100	-.900	1.900	.900	1.000	-0.4E-11
62	1	.039	-.961	1.961	.961	1.000	-0.1E-10
64	1	-.024	-1.024	2.024	1.024	1.000	0.2E-10
66	1	-.089	-1.089	2.089	1.089	1.000	-0.5E-11
68	1	-.156	-1.156	2.156	1.156	1.000	-0.2E-10
70	1	-.225	-1.225	2.225	1.225	1.000	-0.3E-11
72	1	-.296	-1.296	2.296	1.296	1.000	0.1E-10
74	1	-.369	-1.369	2.369	1.369	1.000	0.4E-11
76	1	-.444	-1.444	2.444	1.444	1.000	-0.1E-10
78	1	-.521	-1.521	2.521	1.521	1.000	0.3E-11
80	1	-.600	-1.600	2.600	1.600	1.000	0.4E-11
82	1	-.681	-1.681	2.681	1.681	1.000	0.6E-11
84	1	-.764	-1.764	2.764	1.764	1.000	0.4E-10
86	1	-.849	-1.849	2.849	1.849	1.000	0.6E-10
88	1	-.936	-1.936	2.936	1.936	1.000	0.4E-11
90	1	-1.024	-2.024	3.025	2.025	1.000	0.8E-12
92	1	-1.115	-2.115	3.116	2.116	1.000	0.4E-11
94	1	-1.208	-2.208	3.209	2.209	1.000	0.2E-10
96	1	-1.303	-2.303	3.304	2.304	1.000	0.2E-10
98	1	-1.400	-2.400	3.401	2.401	1.000	-0.4E-10
100	1	-1.499	-2.499	3.500	2.500	1.000	0.8E-11

Multiplier : ML/EI

In step (0) of the computational procedure, we select η_z as $\eta_z = 10^{-1}$. In the initial step (I), based on the arbitrarily chosen value $z = 200$ and thus $\Delta z = 20$, from the expressions

$$c_1 = 1 \quad ; \quad c_r = \frac{1}{6} \frac{-s_r(z+2\Delta z) + 6s_r(z+\Delta z) - 3s_r(z) - 2s_r(z-\Delta z)}{s_r(z+\Delta z) - s_r(z)} \quad , \quad r = 2,3, \qquad (22)$$

we determine the correction factors as $c_1 = 1$, $c_2 = 0.952381..$, $c_3 = 0.906344...$.

Based on these values, we now in step (II) apply the formula

$$\left(\frac{\Delta s_r}{\Delta z}\right)_{corr} = c_r \frac{1}{\Delta z}\left[s_r(z+\Delta z) - s_r(z)\right] \quad , \quad r = 1,2,3 \quad , \qquad (23)$$

cf. (15) and (8a), for computation of the corrected derivatives of the stiffness components $(\Delta s_r/\Delta z)_{corr}$, $r = 1, 2, 3$, for all the values of $n = z$ required in the Table. For each of these values of $n = z$, we then follow the scheme of step (III) of the computational procedure and obtain the desired displacement design sensitivity. The results are denoted by $(\Delta u_n/\Delta L)_{corr}$ and are listed in the second–last column of the Table, with associated errors ε given in the last column. The errors ε are simply computed as the difference between $(\Delta u_n/\Delta L)$ and unity.

The results illustrate most convincingly that the types of error considered in this paper can be completely eliminated by means of the proposed method. The errors are clearly seen to be small truncation errors due to the conditioning of the algebra and limited precision of the computer. Except for this type of error, the values of the semi–analytical sensitivities cannot be distinguished from the exact ones.

ACKNOWLEDGMENTS – Authors gratefully acknowledge stimulating discussions with prof. Gengdong Cheng. This work received support from the Danish Technical Research Council (Programme of Research on Computer Aided Design) and from the Research Council (Konsistoriums forskningsudvalg) of the University of Aalborg.

REFERENCES

[1] O.C. Zienkiewicz & J.S. Campbell: Shape Optimization and Sequential Linear Programming, in: Optimum Structural Design, Theory and Applications (Eds. R.H. Gallagher and O.C.Zienkiewicz), Wiley and Sons, London, 109–126 (1973).

[2] B.J.D. Esping: Minimum Weight Design of Membrane Structures, Ph.D. Thesis, Dept. Aeronautical Structures and Materials, The Royal Institute of Technology, Stockholm, Report 83–1 (1983)

[3] G. Cheng & Y. Liu: A New Computation Scheme for Sensitivity Analysis, Eng. Opt. 12, 219–235 (1987)

[4] R.T. Haftka & H.M. Adelman: Recent Developments in Structural Sensitivity Analysis, Structural Optimization 1, 137–151 (1989).

[5] B. Barthelemy, C.T. Chon & R.T. Haftka: Sensitivity Approximation of Static Structural Response, Finite Elements in Analysis and Design 4, 249–265 (1988)

[6] B. Barthelemy & R.T. Haftka: Accuracy Analysis of the Semi–Analytical Method for Shape Sensitivity Calculation, AIAA Paper 88–2284, Proc. AIAA/ASME/ASCE/ASC 29th Structures, Structural Dynamics and Materials Conf., (held in Williamsburg, Va., April 18–20, 1988), Part 1, 562–581 (1988)

[7] K.K. Choi & S.–L. Twu: On Equivalence of Continuum and Discrete Methods of Shape Sensitivity Analysis, AIAA J. (to appear)

[8] P. Pedersen, G. Cheng & J. Rasmussen: On Accuracy Problems for Semi–Analytical Sensitivity Analyses, Mech. Struct. Mach. 17, 373–384 (1989)

[9] G. Cheng, Y. Gu & Y. Zhou: Accuracy of Semi–Analytic Sensitivity Analysis, Finite Elements in Analysis and design 6, 113–128 (1989).

[10] N. Olhoff & J. Rasmussen: On Elimination of Inaccuracy in Semi–Analytical Sensitivity Analysis, Report No. 28, Institute of Mechanical Engineering, University of Aalborg, Denmark, October 1990. (Submitted to Structural Optimization).

[11] R.D. Cook, D.S. Malkus & M.E2. Plesha: Concepts and Applications of Finite Element Analysis, 3rd Edition Wiley & Sons, New York (1989)

DESIGN OPTIMIZATION WITH A COMMERCIAL FINITE ELEMENT PROGRAM

Guenter Mueller

Peter Tiefenthaler

CAD-FEM GmbH, D-8107 Ebersberg / Muenchen

1   Introduction

Engineering design is an iterative process that strives to obtain a best or optimum design. The optimum design is usually one that meets the design requirements with a minimum expense of certain factors such as cost and weight.

The usual path to the optimum design is a traditional one. The desired function and performance of the design are first defined. Then, primarily from experience, a trial configuration is developed with the intent of meeting the function and performance requirements. Next an analysis of the trial arrangement is performed and the results are evaluated against the design requirements. The design configuration is then usually altered in an attempt to better meet the design needs and the cycle begins again.

Design optimization is a mathematical technique that seeks to determine a best design based on criteria set up by the engineer. The technique generates, analyzes, evaluates, and regenerates series of variations until specified criteria are met. The engineer determines the criteria and bounds for the design problem, sets the problem up, but leaves the task of controlling and executing the design cycle to the design optimization routine.

This paper presents the design optimization module in the finite element program ANSYS. The optimization strategie is briefly described and examples are given to show the ease of use and the wide range of applicability.

2   The Optimization Approach in ANSYS

2.1 Definitions

Design Variable (DV) - design variables represent those aspects of the design that can be varied to obtain a minimum objective function. Design variables should be independent variables. DVs are often geometric parameters such as length, thickness, radius, material orientation or even node coordinates.

State Variable (SV) - state variables represent the response of the design to loadings and boundary conditions, and to changes in geometry. Each state variable is a function of one or more of the design variables. Limits placed on state variables act to limit design response and define design feasibility. Stresses, displacements, temperatures, and natural frequencies are typical state variables.

Objective Function (OBJ) - the objective function is a single variable that characterizes the aspect to be minimized. It is a function of one or more of the design variables. Typically weight or cost, it can be virtually any design characteristic desired.

Constrained/Unconstrained Problem - a design problem subject/not subject to limits

Feasible/Infeasible Design - a design that satisfies/violates the constraints (limits)

2.2. The ANSYS Design Optimization Cycle

The optimization module within ANSYS uses approximation techniques to characterize the analysis of a design with a set of quadratic functions at each design loop. These functions define an approximate subproblem to be minimized, yielding a better design vector for the next design loop. The approximate subproblems are updated at each design loop to account for the additional information, and the process continues until convergence criteria are met. This procedure attempts to gain maximum information from each finite element solution while preserving generality in the choice of design variables, constraints, and objective. Following determination of the approximations, the constrained approximate problem is converted into an unconstrained one by using penalty functions. The search for the minimum of the unconstrained problem is then performed using SUMT. As a result of this search, a new trial design is determined and then analyzed, creating a new design set. The changes in the objective function and the design variables between this design and the best design yet encountered are evaluated to determine if convergence has occurred and a possible minimum is reached. Too many sequential infeasible designs or too many design loops will cause termination. A design loop is one cycle through design optimization. If neither convergence nor termination occur, the approximations are updated to account for the new design set and the cycle is repeated. This process continues automatically until convergence or termination is indicated. These steps (Fig. 2.1) are discussed briefly in the following sections. More details can be taken from [1].

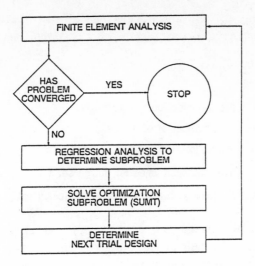

Fig. 2.1 The ANSYS Design Optimization Procedure

2.3   The Approximation of the Objective Function and State Variables

By default, the objective function curve in ANSYS is approximated in the form of a quadratic equation including cross product terms. The state variables are approximated in the same fashion as the objective function, but the default equation fit does not contain cross product terms. By using these approximations, highly nonlinear, arbitrary functions can be represented.

$$\tilde{H} = \underbrace{\underbrace{\underbrace{a_0 + \sum_{n=1}^{N} a_n X_n}_{\text{linear}} + \sum_{n=1}^{N} b_n(X_n)^2}_{\text{quadratic}} + \sum_{n=1}^{N-1} \sum_{\eta=n+1}^{N} c_{n\eta} X_n X_\eta}_{\text{quadratic with cross terms}}$$

where: H = approximation of objective function or state variable

X_n = design variable n

N = total number of design variables

a_n = coefficient for linear term

b_n = coefficient for squared term

$c_{n\eta}$= coefficient for cross term

The coefficients a_n, b_n, c_n are determined by minimizing the weighted least squares error of a set of trial designs.

$$E^2 = \sum_{s=1}^{S} W_s (H_s - \widetilde{H}_s)^2$$

where:

E^2	=	the weighted least squares error
S	=	total number of design sets
W_s	=	weight for design set s
H_s	=	exact function value for design set s
$\widetilde{H}_s$	=	approximated function value for design set s

The weighting can be based on the design variables, the objective function, feasibility or the product of all three. For each function, a multiple regression coefficient R is calculated. R measures how well the fitted equations match the actual function data and varies from zero to one. At each design loop, either a full or a partial equation fit is performed depending on the number of design sets s and the nature of the fit. For partial fits, terms are added or dropped each loop. The terms that are added to the current fit are those which reduce E. A term is dropped if its removal from the curve fit results in an insignificant decrease in R.

Once an objective function approximation is determined, the constrained problem is changed into an unconstrained one by using penalty functions to enforce design and state variable constraints. The unconstrained function, also termed the response surface, is the objective function approximation function plus penalties.

$$\phi_k = \widetilde{F}(1 + r_k)\left(\sum_{n=1}^{N} P_{nk} + \sum_{m=1}^{M} P_{mk} \right)$$

where:

ϕ_k	=	k^{th} response surface for a given design loop
$\widetilde{F}$	=	objective function approximation for a given design loop
P_{nk}	=	penalty function for design variable X_n and response surface ϕ_k
P_{mk}	=	penalty function for state variable G_m and response surface ϕ_k
r_k	= $3^{(k-3)}$ =	penalty multiplier
k	=	response surface number (varies from 1 to 5 within a design loop)
M	=	total number of state variables
N	=	total number of design variables

2.6 Looping until Convergence / Termination

With the new design variables another finite element analysis is done and an improved approximate subproblem is generated which allows to determine a new design vector. The looping continues until either convergence occurs or the problem terminates because the total number of design loops is reached. Convergence is defined to have occurred when all constraints on design variables and state variables are satisfied and the change of the objective function is within a given tolerance.

3   Application Examples

One of the strenghts of the ANSYS Optimization technique is that it can handle a wide variety of optimization problems. It has been applied to statics, dynamics, acoustics, heat transfer as well as electrostatics and magnetic field problems and it is even applicable to coupled problems e.g. thermal-stress or electro- magnetic-thermal stress analysis.
The following examples are taken from various publications of the program developer and from project reports of our company [3, 4, 5, 6, 7, 8, 9, 10, 11, 12, 13, 14]

- Optimization of a Connecting Piece under tensile load [3]
  The objective function to be minimized is weight. The shape of the model can be described by only three design variables, which define the width in the middle of the structure and the outer contour by using a spline fit. Only one state variable - the maximum allowable stress - is defined.

- Design of a Microstrip Transmission Line [4]
  Objective was to find the minimum strip width with the constraints that the impedence be 47.67 Ohms +/- 1 % and the strip width fall between 0.5 and 3.0 mm.

- Minimization of the Dimensions of a coaxial cable [4]
  It is decided that the electrical field intensity be limited to 40 % of the dielectric strength.

- Optimum Design for an Involute Spur Gear [5]
  The idea was to find the minimum stress in the fillet. Design variables have been number of fillets, radius and pressure angle.

- Axisymmetric curved membrane [6]
  The membrane is simply supported at the outside edge and loaded by a concentrated force in the center. The objective is to find the minimum force to displace the center by 4 mm. The shape of the membrane has been defined by cubic splines.

- Minimization of Costs for a Radial Cooling Fin Array [7]
  Objective was to minimize cost as function of fin thinkness, fin radius, distance of fins and fluid mean stream velocity. Side constraint was heat loss which was requested to be larger than 25000 Watts.

- Minimization of the Weight of a Truck Frame [8]
  Weight could be reduced by 25 % while limiting the maximum combined stress level and torsional stress level in any beam element. The problem was solved on a personalcomputer in 510 minutes. 40 loops have been needed for convergence.

- Frequency Tuning of a Bell [9]
  Given is the initial shape of the Minor Third Bell. The objective is to find the shape for a Major Third Bell. To get the tune for a Major Third Bell the designer has to make sure that given frequency ratios are met. Therefore the objective function results in a minimization of the sum of the squares of the deviation of the frequency ratios. Design variables are the radii at 9 points, the height and 2 thicknesses of the bell. As state variables the ratio of the frequency of the octave and the frequency of the major 3 is chosen. For this investigation axisymmetric elements allowing nonaxisymmetric loading are chosen. Thus the problem can be reduced by 1 dimension which is an important aspect for time consuming optimization runs.

- Minimization of the Stress Intensity in the Fillet of a Housing of a Potentiometer [10]
  The goal of this optimization problem is, to reduce the maximum stress intensity which occurs in the fillet of a housing. The shape of this fillet is described by four design variables. This optimization problem is only constrained in the design variables, a state variable was not defined. The maximum equivalent stresses between the initial and the final design have been reduced up to 37%. Besides, the stresses along the fillet path are much more homogeneous in the final design.

- Failure Optimization of a Thick-walled Pressure Vessel Unter Internal
  Pressure [11]
  For thick pressure vessels, the state of stress or strain is three-
  dimensional. A 3-dimensional quadratic failure criterion includes
  interaction among the stress or strain components. In this problem, the
  Tsai-Wu quadratic interaction failure criterion is used with layered
  solid elements to optimize the winding angle in a pressure vessel with
  cylindrical orthotropy (one of the axes of orthotropy is parallel to
  the longitudinal axis of the cylinder). The stress distribution across
  the thick-wall of the cylinder is also studied using the optimum angle
  configuration.

- An Acoustic Design Optimization Technique for Automobile Audio Systems
  [12]
  This Example demonstrates the application of an optimization technique
  to the acoustic design of an audio system. The air inside the cabin
  enclosure has been modeled using the acoustic fluid element of the
  ANSYS program, while the walls were assumed to be rigid surfaces. For
  a given location of the speakers in the automobile passenger cabin, the
  orientation of the speakers and the damping of enclosure walls have
  been varied in an attempt to obtain a flat sound level response over
  a frequency band in the audio range. The objective function of the
  design is the difference between the maximum and minimum response sound
  pressure level, (SPL), at a given location.

- Design Optimization of Ultrasonic Plastic Welding Equipment with the
  ANSYS Program [13]
  Both piezoelectric and design optimization capabilities available in
  the ANSYS program have been used to optimize the shape of a booster in
  an ultrasonic plastic welding equipment. The absolute value of the
  difference between the computed amplitude of the displacement at the
  bottom of the booster and the 2.2 expected value was taken as the
  objective function to be minimized. The first natural frequency of the
  booster was defined as the only state variable and was bounded between
  39.5 and 40.5 KHz. Three independant design variables define the shape
  of the structure.

- Weight Optimization of a lifting magnet [14]
  In a lifting magnet the solenoid coil sets up a magnetic field that
  passes through the core, across an air gap, and into the armature. The
  object of the optimization was, to minimize the weight of the whole
  magnet. The cross section of the keeper and the core as well as the

length of the air gap were defined as design variables. The only state
variable was the magnetic force, which should have a fixed value. Under
this conditions 25 % of weight could be saved.

4   Conclusion

A first step into application of optimization techniques in engineering
is done. The technique offered in the ANSYS program combining finite
element and optimization technology has been presented. The technique
first reduces the global problem to a set of relationships between the
objective functions, design variables, and state varibles. The reduced
problem is then minimized using the Sequential Unconstrained Minimization
Technique (SUMT). The advantage of this approach is that it is not
limited to any particular class of problem and that it does not require
derivative information. The optimization module is implemented in a way
engineers who have no in-depth knowledge of this technique may easily
apply it. Though the application is still limited to a small number of
parameters due to high computing times, the algorithms are available to
allow more parameters as more powerful computers come up. It is
emphasized that optimization requires features as database, solid
modeling and parametric language. Solid modeling and parametric language
are prerequisites for shape optimization. Even complicated 3-d models can
be described with a minimum of variables. Besides, the optimization
module from ANSYS can be used for other purposes like parameter studies
or to curve fit results from analysis or measurements. This method can
be used to correlate experimental results. Although the SUMT method,
which is implemented in the ANSYS optimization module, is very powerful,
because every optimization variables can be assigned to any physical
meaning, the results may not reflect the absolute optimum or the number
of necessary iterations may be quite high. It is acknowledged that more
powerful optimization strategies are available. These can be interfaced
with the ANSYS program.

References

[1]   Kohnke, P.C.: ANSYS Theoretical Manual
         Swanson Analysis Systems, Inc., Houston, PA, U.S.A., 1987
[2]   Vanderplaats, G.N.: Numerical Optimization Techniques for
         Engineering Design, McGraw Hill, NewYork, 1971
[3]   Tumbrink, H.-J., Weight Optimization of a Connecting Piece,
         internal report, Lucas Girling GmbH, Koblenz, 1986

[4]   Ostergaard, D.F.: Adapting Available Finite Element Heat
        Transfer Programs to Solve 2-D and 3-D Electrostatic Field
        Problems, IEEE-IAS Society Meeting, Toronto, 1985.
[5]   Beazley, P.P.: ANSYS Design Optimization Seminar Notes
        Swanson Analysis Systems, Inc., Houston PA, U.S.A., 1987
[6]   Mueller, G., Tiefenthaler, P.: Shape Optimization of an
        Axisymmetric Membrane, CAD-FEM GmbH internal report, Ebers-
        berg/Muenchen, 1988
[7]   Imgrund, M.C.: Applying a New Numerical Technique to the
        Solution of Optimum Convective Surfaces and Associated Thermal
        Design Problems, Swanson Analysis Systems, Inc.,
        Houston PA, U.S.A., 1985
[8]   Johnson, D.H.: Finite Element Optimization of the WABCO 170 Ton
        Haulpack Truck Frame, Earthmooving Industry Conference,
        Pretoria, Illinois, U.S.A., 1986
[9]   Schweizerhof, K., Mueller, G.: Frequency Tuning of a Bell
        CAD-FEM GmbH internal report, Ebersberg/Muenchen, 1986
[10]   Dr.R.Kammerer: Mechanisches Auslegen von Formteilen und Werk-
        zeugen - Wege zu optimalen Ergebnissen mit FEM, Sonderdruck aus
        "Kunststoffe", Heft 10/1988, Carl Hanser Verlag, München
[11]   Frank J. Marx and Prashant Ambe: ANSYS Revision 4.4 Tutorial
        Composite Structures, Swanson Analysis Inc., Houston/PA, USA,
        May 1989
[12]   M.A. Jamnia and C. Rajakumar: An Acoustic Design Optimization
        Technique for Automobile Audio Systems, SAE Technical Paper
        Series, Detroit, 1989
[13]   A. Schaller: Design Optimization of Ultrasonic Plastic Welding
        Equipment with the ANSYS Program, ANSYS 1989 Conference
        Proceedings, Pittsburgh/PA, May 1989
[14]   30 lb. Solenoid Valve Assembly, Moog, Inc. Space products
        Division, ANSYS Application Example, Swanson Analysis, Inc.,
        Houston/PA, USA 1988

ANSYS uses extended interior penalty functions in the unconstrained problem. Because this type of penalty function is defined and continuous over all of design space, it is possible to converge to a feasible minimum from outside the feasible region. The objective function approximation and the state variable approximations with their constraints are collectively known as the approximate subproblem. It is this subproblem that will be minimized (optimized).

2.4 Minimization of the Approximate Problem

The penalized approximate objective function is minimized using the sequential unconstrained minimization technique (SUMT). This technique searches for the minimum of the current approximation of the objective function [2].

For each design loop, five response surfaces are calculated and minimized, using the result from one response surface as the starting point for the next. The minimum of each response surface is found by a series of unidirectional searches in design space, starting at the previous best design. This sequence continues until the change in the minimum of the response surface is less than a small tolerance. The unidirectional searches that occur during this looping process are performed using an iterative algorithm.

2.5 New Trial Design

When the minimum has been determined for the approximate subproblem, a new design is computed. A partial step is taken between the best design encountered so far and the design predicted by minimizing the current approximate subproblem.

$$X_{n_{new}} = X_{n_{best}} + A\,(X_{n_{predict}} - X_{n_{best}}) \quad \text{(for } n = 1, N)$$

where:

$X_{n_{new}}$ = new value of design variable X_n

$X_{n_{best}}$ = value of design variable X_n that was used in the best design

$X_{n_{predict}}$ = predicted value of design variable X_n based on minimization of the current approximations.

A = value between 1 and 0 chosen by the program or by the user

Improvement of Semi-Analytic Sensitivity Analysis and MCADS

Cheng Gengdong, Gu Yuanxian, Wang Xicheng

Research Institute of Engineering Mechanics

Dalian University of Technology,

Dalian 116024, China

ABSTRACT

Sensitivity analysis is one of the cornerstones for the development of computer aided structural shape optimization softwares. The semi-analytic sensitivity analysis method proposed in recent years has provided an easy way to integrate the existing general purpose FEM and optimization packages [1,2]. Nevertheless, the method is suffered from the inaccuracy problem if the size of the perturbed bending elements becomes too small. An alternative forward/backward finite difference scheme has been studied to reduce the possible error and implemented in MCADS [3,4]. This paper further investigates the problem and suggests ways to improve the accuracy of the semi-analytic sensitivity analysis. It is shown in this paper that for annular plate and Euler beam elements the approximate sensitivity calculated with the semi-analytic method can be improved by adding second order terms. For a typical beam example the improvement is estimated by making use of computer algebra software "MUMATH" to manipulate the approximate sensitivity and its second order correction. Numerical results also show that for Timoshenko beam elements the accuracy of semi-analytic method is much less dependent upon the element size, which is a desirable feature.

The second part of the paper introduces MCADS system--a Micro-Computer Aided Design System. We have incorporated several newest techniques, such as semi-analytic method for design sensitivity analysis, optimization-analysis modeling for shape design, application oriented user interfaces and coupling of automatic optimization and user's intervention into the system. A practical example from industry involving coupling field structural shape optimization is given.

1. INTRODUCTION

There has been great interest in developing methods for sensitivity analysis due to the importance of sensitivity information for many areas, such as structural design optimization, system identification and statistical structural analysis. The implementation of sensitivity calculation is followed with particular interest. The semi-analytic sensitivity analysis method proposed in recent years has provided an easy way to integrate the existing general purpose FEM and optimization packages, and has been in wide use. Nevertheless, the method is suffered from the inaccuracy problems if the size of the perturbed bending elements becomes too small. This paper has further demonstrated the problem for annular plate elements. To reduce the possible error and maintain the easy-of-implementation of the semi-analytic method we have proposed an alternative forward/backward finite difference scheme in [1]. The present paper examines another two ways to alleviate the error problem. One way is to add second order terms to the approximate sensitivity. Numerical results for Euler beam and annular plate elements are presented. For a typical beam example, by manipulating the approximate sensitivity and its second order correction with the aid of computer algebra software "MUMATH", we show the error is reduced by one order if the second order term is added. Since the error may be related to the large shear deformation induced by the pseudo-load, we have tested Timoshenko beam elements and observed encouraging numerical results.

The second part of the paper introduces MCADS system--a Micro-Computer Aided Design System. We have incorporated several newest techniques such as semi-analytic method for sensitivity analysis, optimization-analysis modeling for shape design, application oriented user interfaces and coupling of automatic optimization and user's intervention into the system. And a practical example from industry involving coupling field structural shape optimization is presented.

2. SECOND ORDER TERM CORRECTION AND ERROR ESTIMATION

Consider a structure whose shape is to be optimized. Denote $d=(d_1, d_2, ...d_i, ...d_n)$ the shape design variables, K global stiffness matrix, P

external load. The nodal displacement vector u satisfies the equation

$$K(d)u(d) = P(d) \qquad (1)$$

Let us perturb one design variable d_i by Δd_i and define a new design $d'=(d_1,d_2,...d_i+\Delta d_i,...d_n)$. The displacement vector u' of the new design satisfies

$$K(d')u'(d') = P(d') \qquad (2)$$

For simplicity we assume the external load P(d') not changed, and we further define ΔK and Δu in

$$K(d') = K(d) + \Delta K \qquad (3\text{-}a)$$
$$u'(d') = u(d) + \Delta u \qquad (3\text{-}b)$$

Substituting Eq.(3) in (2) and applying Eq.(1), we have

$$K(d)\Delta u + \Delta Ku + \Delta K\Delta u = 0 \qquad (4\text{-}a)$$

The first order approximation of Δu, denoted by $\Delta u'$, is

$$K(d)\Delta u' = -\Delta Ku \qquad (4\text{-}b)$$

In the traditional semi-analytic method, the approximate sensitivity is given by

$$\frac{\partial u}{\partial d_i} \simeq \frac{\Delta u'}{\Delta d_i} \qquad (5)$$

To improve the accuracy of the approximate sensitivity (5), we suggest to add the second order correction $\Delta u''$ which satisfies

$$K\Delta u'' = -\Delta K\Delta u' \qquad (6)$$

and to replace the approximate sensitivity (5) by

$$\frac{\partial u}{\partial d_i} = \frac{(\Delta u' + \Delta u'')}{\Delta d_i} \qquad (7)$$

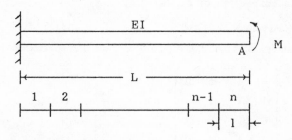

Fig. 1

In order to estimate the error of approximate sensitivity (5) and (7), let us consider a typical example in sensitivity study, i.e. a cantilever shown in Fig.1 [1]. The length L, uniform bending rigidity EI of the beam, and the moment M acting on its tip A are given. From beam theory, the deflection and rotation of the cantilever beam are

$$u(x) = \frac{Mx^2}{2EI} \ , \qquad \vartheta(X) = \frac{\partial u}{\partial x} = \frac{Mx}{EI} \qquad (8)$$

The tip displacement u_A and its sensitivity with respect to L are

$$u_A = \frac{ML^2}{2EI} \ , \qquad \frac{\partial u_A}{\partial L} = \frac{ML}{EI} \qquad (9)$$

In what follows we apply finite element method with cubic shape function to solve the problem. The beam is arbitrarily subdivided into n elements, and the length of the last element is l. We write down the displacement vector of the last element as

$$u_n = \frac{M}{2EI} \left\{ (L-l)^2 \quad 2(L-l), \quad L^2, \quad 2L \right\}^T \qquad (10)$$

The Eq. (10) is obtained from Eq. (8) as well as the results generated from FEM because the later provides exact solution in this problem.

Next, we consider to perturb the length L of the beam. We assume the first n-1 elements have fixed length and only the last element's length l is changed to be l+Δl. In such a case, the pseudo-load in Eq. (4) becomes

$$-\Delta Ku = - \sum_{i=1}^{n} \Delta k_i u_i = \Delta k_n u_n \qquad (11)$$

and the first order of Δu is

$$\Delta u' = -K^{-1} \Delta k_n u_n \qquad (12)$$

where Δk_n is the change of stiffness matrix of the last element. In the same way, the second order correction of the pseudo-load is

$$-\Delta K \Delta u' = -\Delta k_n \Delta u' \qquad (13)$$

and the second order correction of Δu is

$$\Delta u'' = -K^{-1} \Delta K \Delta u' = -K^{-1} \Delta k_n \Delta u' \qquad (14)$$

The stiffness matrix k_n of the last element before perturbation is

$$k_n(l) = \frac{2EI}{l^3} \begin{bmatrix} 6 & 3l & -6 & 3l \\ 3l & 2l^2 & -3l & l^2 \\ -6 & -3l & 6 & -3l \\ 3l & l^2 & -3l & 2l^2 \end{bmatrix} \qquad (15)$$

and the finite difference of stiffness matrix is

$$\Delta k_n(l) = -2EIB = -2EI \begin{bmatrix} 6\beta_3 & 3\beta_2 & -6\beta_3 & 3\beta_2 \\ 3\beta_2 & 2\beta_1 & -3\beta_2 & \beta_1 \\ -6\beta_3 & -3\beta_2 & 6\beta_3 & -3\beta_2 \\ 3\beta_2 & \beta_1 & -3\beta_2 & 2\beta_1 \end{bmatrix} \qquad (16)$$

where
$$\beta_1 = \frac{\eta}{l(1+\eta)}, \qquad \beta_2 = \frac{\eta^2+2\eta}{l^2(1+\eta)^2}$$

$$\beta_3 = \frac{\eta^3+3\eta^2+3\eta}{l^3(1+\eta)^3}, \qquad \eta = \frac{\Delta l}{l} \tag{17}$$

And we also need to mention the fact that k_n and Δk_n in Eq (11-14) have the dimension as same as the total degree of structural nodal freedoms, but we only write them in 4*4 matrix. The similar for u_n and Δu_n. This is well accepted in computational mechanics.

Since the pseudo-load and its correction in Eq. (11) and (13) are concentrated forces and moments acting on the finite element nodes, the displacements and rotations obtained from Eq (12) and (14) are equal to the analytic solution. Though we can not write down the expressions of K and K^{-1} because no detail of subdivision for the first n-1 elements is specified, the effect of K^{-1} upon Eq. (11) and (13) can be obtained by analytic solution. To do this, we consider the same cantilever with arbitrary load vector F acting on the two nodal points A and B of the last element,

$$F = \left\{ P_B, \ M_B, \ P_A, \ M_A \right\}^T$$

From beam theory, the displacement vector of the last element is given by

$$u = \left\{ \begin{array}{c} u_B \\ \vartheta_B \\ u_A \\ \vartheta_A \end{array} \right\} = \frac{1}{6EI}CF$$

$$= \frac{1}{6EI} \begin{bmatrix} 2(L-1)^3 & 3(L-1)^2 & (L-1)^2(2L+1) & 3(L-1)^2 \\ 3(L-1)^2 & 6(L-1) & 3(L^2-1^2) & 6(L-1) \\ (L-1)^2(2L+1) & 3(L^2-1^2) & 2L^3 & 3L^2 \\ 3(L-1)^2 & 6(L-1) & 3L^2 & 6L \end{bmatrix} \begin{bmatrix} P_B \\ M_B \\ P_A \\ M_A \end{bmatrix} \tag{18}$$

Based on Eq. (12), (16) and (18), we obtain

$$\frac{\Delta u'}{\Delta l} = \frac{-1}{\Delta l}\frac{C}{6EI}(-2EIB)u_n = \frac{1}{3\Delta l}Tu_n \tag{19}$$

$$\frac{\Delta u''}{\Delta l} = \frac{1}{9\Delta l} \, T^2 u_n \tag{20}$$

Note the fact that in Eq (19) and (20) both the vector $\Delta u'$ and $\Delta u''$ are 4*1 vector, the interesting quantity, i.e., approximate $\partial u_A/\partial l$ and its correction are the third component of $\Delta u'/\Delta l$ and $\Delta u''/\Delta l$. And further we have

$$T = CB \tag{21}$$

the elements of which are

$$T_{11} = -36\beta_2 Ll + 36\beta_3 Ll^2 - 18\beta_3 L^2 l + 18\beta_2 L^2 + 18\beta_2 l^2 - 18\beta_3 l^3$$

$$T_{12} = -18\beta_1 Ll + 18\beta_2 L l^2 - 9\beta_2 L^2 l + 9\beta_1 L^2 + 9\beta_1 l^2 - 9\beta_2 l^3$$

$$T_{13} = -T_{11}, \qquad T_{14} = T_{12}$$

$$T_{21} = -36\beta_3 Ll + 36\beta_2 L - 36\beta_2 l + 36\beta_3 l^2$$

$$T_{22} = -18\beta_2 Ll + 18\beta_1 L - 18\beta_1 l + 18\beta_2 l^2$$

$$T_{23} = -T_{21}, \qquad T_{24} = T_{22}$$

$$T_{31} = -18\beta_3 L^2 l + 18\beta_2 L^2 - 9\beta_2 l^2 + 6\beta_3 l^3$$

$$T_{32} = -9\beta_2 L^2 l + 9\beta_1 L^2 - 6\beta_1 l^2 + 3\beta_2 l^3 \tag{22}$$

$$T_{33} = -T_{31}$$

$$T_{34} = -9\beta_2 L^2 l + 9\beta_1 L^2 - 3\beta_1 l^2 + 3\beta_2 l^3$$

$$T_{41} = -36\beta_3 Ll + 36\beta_2 L - 18\beta_2 l + 18\beta_3 l^2$$

$$T_{42} = -18\beta_2 Ll + 18\beta_1 L - 12\beta_1 l + 9\beta_2 l^2$$

$$T_{43} = -T_{41}$$

$$T_{44} = -18\beta_2 Ll + 18\beta_1 L - 6\beta_1 l + 9\beta_2 l^2$$

To simplify the further calculation, we introduce the approximation of β_1, β_2, and β_3

$$\beta_1 \simeq \frac{\eta - \eta^2}{l} \qquad \beta_2 \simeq \frac{2\eta - 3\eta^2}{l^2} \qquad \beta_3 \simeq \frac{3\eta - 6\eta^2}{l^3} \tag{23}$$

Then we obtain

$$\sum_{i=1}^{4} T_{3i} u_{ni} = 6\eta lL - 36\eta^2 \frac{L^3}{l} + 18\eta^2 L^2 - 3\eta^2 l^2$$

and

$$\frac{\Delta u_A'}{\Delta l} = \frac{ML}{EI} - \frac{6\eta ML^3}{EI \, l^2} + \frac{3\eta ML^2}{EI \, l} - \frac{\eta Ml}{2EI} \tag{24}$$

in which ML/EI is the exact value of sensitivity $\partial u_A/\partial L$, and the remaining parts in Eq. (24) is the error linearly dependent upon η. The term $-6\eta ML^3/EIl^2$ is inversely proportional to l^2 and is the most troublesome term when the size l of the perturbed element tends toward zero.

With the aid of computer algebra software "MUMATH" we derive the expressions of T^2, and obtain

$$\sum_{i=1}^{4} (T^2)_{3i} u_{ni} = 108\eta^2 \frac{L^3}{l} -54\eta^2 L^2 +18\eta^2 l^2 -648\eta^3 \frac{L^3}{l} +54\eta^3 L^2 +270\eta^3 Ll$$

$$-108\eta^3 l^2 +1296\eta^4 \frac{L^4}{l^2} -648\frac{\eta^4 L^3}{l} +54\eta^4 L^2 -108\eta^4 Ll +63\eta^4 l^2 \qquad (25)$$

and the second order correction is

$$\frac{\Delta u'_A{}'}{\Delta l} = 6\eta \frac{ML^3}{EIl^2} -3\eta \frac{ML^2}{EIl} +\eta \frac{Ml}{EI} + \frac{M}{EI} \Bigg[-36\eta^2 \frac{L^3}{l^2} +3\eta^2 \frac{L^2}{l} +15\eta^2 L -6\eta^2 l$$

$$+72\eta^3 \frac{L^4}{l^3} -36\eta^3 \frac{L^3}{l^2} +3\eta^3 \frac{L^2}{l} -6\eta^3 L +3.5\eta^3 l \Bigg]$$

The improved approximate sensitivity with the second order correction is

$$\frac{\partial u_A}{\partial L} \simeq \frac{\Delta u'_A +\Delta u'_A{}'}{\Delta l} = \frac{ML}{EI} + \frac{\eta Ml}{2EI} + o(\eta^2) \qquad (26)$$

It can be seen from Eq. (26) that the part of the error linearly dependent upon η is no longer inversely proportional to l^2, but proportional to l. The approximate sensitivity is improved indeed. Though the part of the error proportional to η^2 may still have some troublesome terms, but they are well suppressed by η^2.

3. NUMERICAL RESULTS FOR EULER BEAM AND ANNULAR PLATE ELEMENTS

To demonstrate the improvement by adding the second order correction, we present three examples.

Example 1. Consider a simply supported beam shown in Fig.2. At the middle of the beam we have a concentrated force P. The other data are shown in Fig.2. The sensitivity of the deflection at the middle of the beam is studied, its analytic value is 0.025.

In Table 1, the figures under the column GFD, SAM and SAMI are results from the global finite difference method, semi-analytic method and semi-analytic method with the second order correction, respectively. The improvement is easily seen. GFD method also suffers from the conditional error when Δl is small and n is large.

The following two examples deal with annular plates under axisymmetric loads. The shape function within each element is cubic function $w(r)=ar^3 +br^2+cr+d$, w is the deflection of the plate. Both examples show the error problem of semi-analytic method when the size of perturbed element is too small. And the error is not serious as long as we choose reasonably small Δa. But the accuracy is improved by adding the second order term correction.

Example 2. Consider an annular plate whose inner edge is free and the outer edge is simply supported and subject to uniform moment M. The outer radius a is the design variable. The theoretical results are from plate theory. The data and numerical results are shown in Table 2 and the attached picture.

Example 3. Consider an annular plate whose outer edge is fixed and inner edge is free and subject to an uniformly distributed line loads Q. The outer radius a is the design variable. Other data and numerical results are given in Table 3 and the associated figure.

4. NUMERICAL EXAMPLE FOR TIMOSHENKO BEAM

Various explanations have been given to numerical error in semi-analytic method. One argument [5] is that sensitivity field is not a reasonable displacement field though it is obtained as a displacement field induced by the pseudo-load acting on the original structure. The pseudo-load is a special load and the sensitivity is dominated by shear deformation. To better model a beam structure with large shear deformation, Timoshenko beam is a proper choice. The same simply supported beam as in Fig.2 is treated but with Timoshenko beam elements. The results are presented in Table 4. By comparing Table 1 and 4, it is obvious that for approximate sensitivity by semi-analytic method the error problem is much less severe for Timoshenko beam elements than for Euler beam elements. Similar observation is expected for thick plate and thick shell elements.

5. MCADS AND APPLICATIONS

MCADS (Micro-Computer Aided Design System) is a general purpose structural shape and size optimization program implemented on IBM PC and

APOLLO. A general purpose FEM structural analysis program DDJ-W is adopted as a black box on which MCADS has been built. DDJ-W has been developed by Prof. Zhong Wanxie's group in Dalian University of Technology since 1980 and is well distributed inside China. It has many interesting features such as a rich element library, free format input, chain master-slave displacement transformation and flexible nodal specification. In this way, DDJ-W is capable of modeling complicated and practical structures. Semi-analytic method for sensitivity computation is applied to integrate DDJ-W and optimization packages. In principal, any structure DDJ-W can analyze can be optimized by MCADS. Preprocessor MESHG and postprocessor GRAPH have made MCADS even more user friendly. To meet the need of various design variables, an application oriented programming interface in the form of source file is left open. By modifying those source files, the user can describe structural shape and design elements at his/her will. For example, user can describe structural shape/size and design elements by engineering parameters appearing in engineering drawing.

A number of applications have been done successfully by using MCADS [6]. Fig. 5 shows a typical turbine engine disk with hub and rim. The present study is concerned with optimum design of the profile of the disk. The object is to lower the maximum stress and stresses at certain points, limit the radial displacement of the rim and minimize the structural weight. 18 engineering parameters are chosen as natural design variables. A special subroutine is written for describing the structural shape and design elements. The initial and final designs are shown in Fig.5. The reductions of stress and structural weight are 14.91% and 18.52%, respectively.

CONCLUDING REMARKS

With the alternative forward/backward finite difference scheme of the second order correction, the error in semi-analytic method can be reduced to great deal. Use of thick beam or plate element is also an alternative to avoid the accuracy problem. Nevertheless, in many cases, the semi-analytic method provides excellent accuracy as long as the perturbation of design is probably chosen.

REFERENCE

[1] Cheng, G.D. & Liu, Y.W. 1987: A new computational scheme for sensitivity analysis, J. Eng. Opt. 12, 219-235

[2] Haftka, R.T. & Adelman, H.M. 1989: Recent developments in structural sensitivity analysis, J. Stru. Opt. 1, 137-151

[3] Cheng, G.D.; Gu, Y.X. & Zhou, Y.Y. 1989: Accuracy of semi-analytic sensitivity analysis, J. Finite Elements in Analysis and Design, 6, 113-128

[4] Gu, Y.X. & Cheng, G.D. 1990: Structural shape optimization integrated with CAD environment, J. Stru. Opt. 2, 23-28

[5] Barthelemy, B. & Haftka, R.T. 1988: Accuracy analysis of the semi-analytical method for shape sensitivity calculation, AIAA paper 88-2284, Proc. AIAA/ASME/ASCE/AHS/ASC 29th Structures, Structural Dynamics and Materials Conf. (held in Williamsburg, Va., April 18-20), Part 1, pp572-581

[6] Gu, Y.X. & Cheng, G.D. 1990: Microcomputer-based system MCADS integrated with FEM, optimization and CAD, J. Computational Mechanics and Applications, 1, 71-81

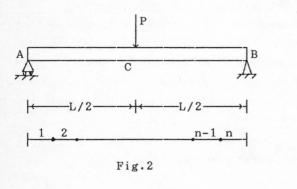

Fig.2

$L = 200.0 in$

$F = 10.0 in^2$

$J = 10.0 in^4$

$\Delta L = 10^{-RS} L$

$W_c = PL^3 / 48EJ$

$$\frac{\partial W_c}{\partial L} = 25000$$

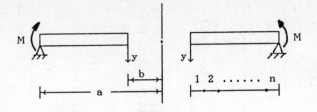

Fig.3

a=1, b=0.2, D=1, $\nu=1/6$, M=200

$$y = -\frac{M}{2D}\left[\frac{a^4-a^2b^2-2a^2b^2\ln(a/b)}{a^2(1+\nu)+b^2(1-\nu)}\right] = 69.27$$

$$\frac{\partial y}{\partial a} = 163.8$$

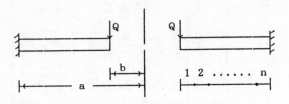

Fig.4

a=1, b=0.2, D=1, $\nu=1/6$, Q=20,

$$y_{MAX} = -\frac{Qb}{8D}\left[a^2-b^2+\frac{2b^2(a^2-b^2)-8a^2b^2\ln\frac{a}{b}+4a^2b^2(1+\nu)(\ln\frac{a}{b})^2}{a^2(1-\nu)+b^2(1+\nu)}\right] = -0.5057$$

$$\frac{\partial y_{max}}{\partial A} = -1.166$$

Table 1.

RS	GFD	n=2 SAM	n=2 SAMI	n=4 SAM	n=4 SAMI	n=8 SAM	n=8 SAMI	n=16 SAM	n=16 SAMI	n=32 SAM	n=32 SAMI	n=64 SAM	n=64 SAMI
1	2758	1885	2481	992	2094	-2621	2498	-17040	52730	-74740	673800	-305500	$1.1\cdot10^7$
2	2525	2427	2521	2311	2514	1846	2486	-17.7	2427	7472	3037	-37290	18970
3	2503	2493	2502	2481	2502	2433	2502	2241	2501	1476	2497	-1587	2494
4	2500	2499	2500	2498	2500	2493	2500	2474	2500	2397	2500	2090	2500
5								2497		2490		2459	
6										2499		2496	
10										2500		2502	2502

Table 2.

All entries in units of 10^{-3}.

RS	n=2 GFD	n=2 SAM	n=2 SAMI	n=4 GFD	n=4 SAM	n=4 SAMI	n=8 GFD	n=8 SAM	n=8 SAMI	n=16 GFD	n=16 SAM	n=16 SAMI	n=32 GFD	n=32 SAM	n=32 SAMI
1	1719	123	1299	1727	-4264	3557	1728			2521			1728		
2	1639	1441	1630	1646	890	1614	1647	-1333	1611	1728	1647	30272			
3	1631	1611	1631	1638	1560	1637	1639	1334	1638	1648	1639	-3230	1646		
4	1630	1628	1630	1637	1629	1637	1638	1608	1638	1639	1638	1635	1639		
5	1630	1637	1636	1637	1638	1635	1638	1639	1637	1639	1637	1639			
6	1637	1636	1637	1638	1638	1638	1639	1626	1638	1637	1595	1655			
7	1637	1637	1637	1639	1639	1639	1639	1637	1638	1638	830	1418			
8	1637	1637	1637	1632	1636	1636	1616	1643	1643	3122	1979				
9	1639	1637	1637	1715	1618	1618	1485	1512	1512	6952	-61883				
10	1639	1653	1653	1786	1267	1267	2010	1948	1948						

Table 3.

RS	n=2 (10^{-3}) GFD	SAM	SAMI	n=4 (10^{-3}) GFD	SAM	SAMI	n=8 (10^{-3}) GFD	SAM	SAMI	n=16 (10^{-3}) GFD	SAM	SAMI	n=32 (10^{-3}) GFD	SAM	SAMI
1	1217	629	1042	1218	-496	957	1219	-5021	5710	1219	-2313	1064	1219	-95585	18170
2	1169	1098	1167	1171	953	1161	1171	368	1143	1171	-197	1250	1171	-11342	3637
3	1165	1157	1165	1166	1144	1165	1166	1084	1166	1167	843	1165	1167	-119.6	1163
4	1164	1163	1164	1166	1166	1163	1166	1158	1166	1166	1134	1166	1166	1037	1166
5	1164	1164		1165	1165			1165	1165		1163		1166	1153	1166
6								1166	1166		1166	1166	1167	1166	1167
7													1195	1185	1185
8							1165	1165	1165	1168	1175	1175	1296	1039	1039
9							1164	1163	1163		1165	1175	3279	2166	2166
10				1171			1122	1161	1162	1205	1144	1144	10246	12446	12446

Table 4.

RS	n=2 GFD	SAM	SAMI	n=4 SAM	SAMI	n=8 SAM	SAMI	n=16 SAM	SAMI	n=32 SAM	SAMI
1	2858	1885	2481	1177	2189	1700	2524	1632	2531	1613	2511
2	2625	2427	2521	2347	2526	2501	2621	2494	2620	2492	2620
3	2600	2499	2500	2508	2510	2599	2600	2599	2600	2599	2600
4	2600	2499	2500	2508	2510	2599	2600	2599	2600	2599	2600
5					2510		2600		2600	2600	2600
6					2510		2600		2600		
7							2600				
8											

SIMULATION APPROACH TO STRUCTURAL OPTIMIZATION USING DESIGN SENSITIVITY ANALYSIS

Victor P. Malkov, Vassili V. Toropov
Department of Solid Mechanics, Gorky University,
Gagarin Ave. 23, 603600 Gorky, USSR

Let us consider a general structural optimization problem in the following form:
minimize

$$F_0(\underline{x}), \quad \underline{x} \in R^N \tag{1}$$

subject to

$$F_j(\underline{x}) \leq c_j \quad, \quad j = 1, \ldots, M \tag{2}$$

and

$$A_i \leq x_i \leq B_i \quad, \quad i = 1, \ldots, N \tag{3}$$

where $\underline{x} = (x_1, x_2, \ldots, x_N)$ is a vector of design variables, $F_0(\underline{x})$ is an objective function, $F_j(\underline{x}), j=1,\ldots,M$ are constraint functions, A_i and B_i are given lower and upper bounds on the design variables, which define a search region in the N-dimensional space of design variables.

The essence of the simulation approach (Malkov et. al. 1982, Toropov 1989) consists of the iterative replacement of the detailed model of a structure (i.e. the implicit functions $F_j(\underline{x}), j=0,\ldots,M$ with the simulation model (i.e. the explicit functions $\widetilde{F}_j(\underline{x})$), valid in a subregion of the original region $[A_i, B_i], i=1,\ldots, N$. The simulation model can then conveniently be used in a particular nonlinear mathematical programming problem of a step.

In every k-th step of the iterative procedure the following mathematical programming problem is formulated and solved:

find the vector $\underline{x}_*^{(k)}$ that minimizes the objective function

$$\widetilde{F}_0^{(k)}(\underline{x}) \tag{4}$$

subject to

$$\widetilde{F}_j^{(k)}(\underline{x}) \leq c_j \quad , \quad j = 1, \ldots, M \tag{5}$$

and

$$A_i^{(k)} \leq x_i \leq B_i^{(k)} \tag{6}$$

$$A_i^{(k)} \geq A_i \quad , \quad B_i^{(k)} \leq B_i \quad , \quad i = 1, \ldots, N$$

where $\widetilde{F}_j^{(k)}(\underline{x}), j=0,\ldots,M$ are the functions, which approximate the functions of the initial optimization problem (1) - (3).

Let us consider the main problem i.e. the problem of the simulation model formulation. To simplify notation, we will suppress the indices k and j on the functions $\widetilde{F}_j^{(k)}(\underline{x})$.

Assume that the designed structure is decomposed into S individual substructures and that the vector of design variables $\underline{x}$ can be divided into S+1 subvectors $\underline{x}^{(s)}, s=1,\ldots,S+1$. Each subvector $\underline{x}^{(s)}$ describes the s-th substructure and consists of n_s components. The last subvector $\underline{x}^{(S+1)}$ contains the design variables, which describe the structure as the whole (the global variables).

Assume now that the behaviour of an individual s-th substructure can be described by the function

$$f_s = f_s(\underline{x}^{(s)}) \quad , \quad s = 1, \ldots, S \tag{7}$$

and that the behaviour of the structure as the whole can be expressed in the following general form:

$$\widetilde{F} = \widetilde{F}\left(f_1, \ldots, f_S, \underline{x}^{(S+1)}, \underline{a}\right) \tag{8}$$

Let us consider the problem of constructing the function (8) assuming that all functions (7) are known (Toropov 1989). The vector $\underline{a} = (a_0, a_1, \ldots, a_L)^T$ in expression (8) consists of L+1 parameters, which are defined on the basis of results of numerical experiments (response analyses and sensitivity analyses) with the detailed model for points located in the design variable space R^N in accordance with some design (plan) of experiments. Then the weighted least-squares method leads to the following problem:

find the vector $\underline{a}$ that minimizes the function

$$G(\underline{a}) = \sum_{p=1}^{P} \left\{ w_p^{(0)}(F(\underline{x}_p) - \widetilde{F}_p)^2 + \sum_{i=1}^{N} \left[w_p^{(i)}(F(\underline{x}_p)_{,i} - \widetilde{F}_{p,i})^2 \right] \right\} \tag{9}$$

where

p is the number of the current point of the plan of experiments,

P is the total number of such points,

$\underline{x}_P$ is the vector of design variables that defines the currrent point,

$F(\underline{x}_P)$ is the value of the implicit function $F(\underline{x})$ in (1),(2) as the result of the response analysis at the point $\underline{x}_P$,

$\widetilde{F}_P = \widetilde{F}_P(f_1,...,f_S, \underline{x}^{(S+1)}, \underline{a})$ is the value of the explicit function (4), (5) at the point $\underline{x}_P$,

$$F(\underline{x}_P),_i = \frac{\partial F(\underline{x})}{\partial x_i}$$ is the value of the derivative of the im-

plicit function (1),(2) with respect to design variable x_i as the result of the sensitivity analysis at the point $\underline{x}_P$ (the first order sensitivity),

$\widetilde{F}_{P,i}$ is the value of the derivative of the explicit function (4),(5) with respect to x_i at the point $\underline{x}_P$,

$w_P^{(o)}$ and $w_P^{(i)}$ are the weight coefficients, which correspond to values of the functions (1),(2) and their sensitivities at the point $\underline{x}_P$.

To solve the identification problem it is necessary to choose the structure of the function (8), i.e. to define $\widetilde{F}$ as a function of parameters $\underline{a}$ only. The simplest case is a linear function of parameters $\underline{a}$:

$$\widetilde{F}(\underline{a}) = a_o + a_1 f_1 + ... + a_S f_S +$$
$$+ a_{S+1} x_1^{(S+1)} + ... + a_L x_{n(S+1)}^{(S+1)}$$

(10)

where the functions f_s, $s = 1,...,S$ describe the behaviour of individual substructures. To simplify notation, rewrite (10) in the following form:

$$\widetilde{F}(\underline{a}) = a_o + \sum_{l=1}^{L} a_l \varphi_l$$

(11)

The problem of estimation of the parameters $\underline{a}$ leads to the linear system of normal equations:

$$[\varphi]^{\tau}[w][\varphi]\,\underline{a} = [\varphi]^{\tau}[w]\,\underline{F} \qquad (12)$$

where

$[\varphi]$ – is the rectangular matrix consisting of $P(N+1)$ rows and $L+1$ columns of elements φ_{ℓ} and $\varphi_{\ell,i}$, which are defined by expression (11) for every p-th point of the plan of experiments,

$[w]$ – is the diagonal matrix $P(N+1) \times P(N+1)$ consisting of weights for all the plan points,

$\underline{F}$ – is the vector containing the values $F(\underline{x}_P)$ and $F(\underline{x}_P)_{,i}$ for p= 1,...,P.

In the case of linear function

$$\underline{F} = \left\{ \begin{array}{c} F(\underline{x}_1) \\ F(\underline{x}_1)_{,1} \\ \cdots \\ F(\underline{x}_1)_{,N} \\ \cdots \\ \cdots \\ \cdots \\ F(\underline{x}_P) \\ F(\underline{x}_P)_{,1} \\ \cdots \\ F(\underline{x}_P)_{,N} \end{array} \right\}, \qquad \underline{a} = \left\{ \begin{array}{c} a_o \\ a_1 \\ \vdots \\ \vdots \\ a_L \end{array} \right\},$$

$$[\varphi] = \begin{bmatrix} 1 & \varphi_{11} & \cdots & \varphi_{1L} \\ 0 & \varphi_{11,1} & \cdots & \varphi_{1L,1} \\ \cdot & \cdot & & \cdot \\ 0 & \varphi_{11,N} & \cdot\cdot & \varphi_{1L,N} \\ \cdot & \cdot & \cdot & \cdot \\ \cdot & \cdot & \cdot & \cdot \\ \cdot & \cdot & \cdot & \cdot \\ 1 & \varphi_{P1} & \cdots & \varphi_{PL} \\ 0 & \varphi_{P1,1} & \cdots & \varphi_{PL,1} \\ \cdot & \cdot & \cdot & \cdot \\ 0 & \varphi_{P1,N} & \cdots & \varphi_{PL,N} \end{bmatrix}$$

    The procedure described above can be generalized by the application of intrinsically linear models. Such models are nonlinear, but they can be led to linear ones by simple transformations. Note among them the multiplicative function

$$\widetilde{F}(\underline{a}) = a_o \cdot \prod_{\ell=1}^{L} \varphi_\ell^{a_\ell} \qquad (13)$$

with the transformation

$$\ln \widetilde{F}(\underline{a}) = \ln a_o + \sum_{\ell=1}^{L} a_\ell \cdot \ln \varphi_\ell \qquad (14)$$

In the case of multiplicative function

$$\underline{F} = \left\{ \begin{array}{c} \ln F(\underline{x}_1) \\ F(\underline{x}_1),_1 \\ \cdots \\ F(\underline{x}_1),_N \\ \cdots \\ \vdots \\ \vdots \\ \ln F(\underline{x}_P) \\ F(\underline{x}_P),_1 \\ \cdots \\ F(\underline{x}_P),_N \end{array} \right\}, \qquad \underline{a} = \left\{ \begin{array}{c} \ln a_o \\ a_1 \\ \vdots \\ \vdots \\ a_L \end{array} \right\},$$

$$[\varphi] = \left[\begin{array}{cccc} 1 & \ln \varphi_{11} & \cdots & \ln \varphi_{1L} \\ 0 & \dfrac{F(\underline{x}_1)}{\varphi_{11}} \varphi_{11,1} & \cdots & \dfrac{F(\underline{x}_1)}{\varphi_{1L}} \varphi_{1L,1} \\ \cdot & & & \cdot \\ 0 & \dfrac{F(\underline{x}_1)}{\varphi_{11}} \varphi_{11,N} & \cdots & \dfrac{F(\underline{x}_1)}{\varphi_{1L}} \varphi_{1L,N} \\ \cdot & \cdot & & \cdot \\ \cdot & \cdot & & \cdot \\ \cdot & \cdot & & \cdot \\ 1 & \ln \varphi_{P1} & \cdots & \ln \varphi_{PL} \\ 0 & \dfrac{F(\underline{x}_P)}{\varphi_{P1}} \varphi_{P1,1} & \cdots & \dfrac{F(\underline{x}_P)}{\varphi_{P1}} \varphi_{PL,1} \\ 0 & \dfrac{F(\underline{x}_P)}{\varphi_{P1}} \varphi_{P1,N} & \cdots & \dfrac{F(\underline{x}_P)}{\varphi_{PL}} \cdot \varphi_{PL,N} \end{array} \right]$$

Test problem: two-bar truss (Svanberg 1987).

Consider the two-bar truss in Figure 1 under the external force F, where F_x = 24.8 kN, F_Y = 198.4 kN.

There are two design variables: x_1 (cm^2) is the area of the cross section of both bars and x_2 (m) is half of the distance between nodes 1 and 2. The vertical coordinate of nofe 3 is fixed: Y_3 = 1 m. The objective function is the weight of the structure and the constraint functions define stress in both bars, which must not be greater than 100 N/mm^2. For this simple problem all functions can be formulated analytically:

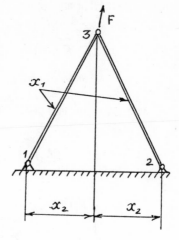

$$F_0(x_1, x_2) = c_1 x_1 \sqrt{(1 + x_2^2)}$$

$$F_1(x_1, x_2) = c_2 \sqrt{(1 + x_2^2)} \left(\frac{8}{x_1} + \frac{1}{x_1 x_2} \right) \leq 1$$

$$F_2(x_1, x_2) = c_2 \sqrt{(1 + x_2^2)} \left(\frac{8}{x_1} - \frac{1}{x_1 x_2} \right) \leq 1$$

where c_1 = 1.0, c_2 = 0.124 and A_1 = 0.2, A_2 = 0.1, B_1 = 4.0, B_2 = 1.6.
A feasible starting point has been chosen as follows: $x_1^{(o)}$ = 2.5, $x_2^{(o)}$ = 1.0. The simplest form of the multiplicative expression has been used for the approximate constraint function:

$$\widehat{F}_1(x, x_2) = a_0 x_1^{a_1} x_2^{a_2}$$

The objective function has been approximated by linear expression:

$$\widehat{F}_0(x_1, x_2) = a_0 + a_1 x_1 + a_2 x_2$$

The solution of this problem as given by x_1^* = 1.41, x_2^* = 0.38, $F_0(\underline{x}^*)$ = 1.51 has been obtained in 6 steps.

This test problem has been used by K. Svanberg (1987) to compare various optimization techniques. The minimum number of required iterations (k = 5) was obtained using the method of moving asymptotes (MMA).

References

Malkov, V.P.; Toropov,V.V.; Filatov,A.A. 1982: Simulation approach to optimization of deformable systems. In: Applied problems of strength and plasticity (in Russian), pp.62-69. USSR: Gorky University

Toropov, V.V. 1989: Simulation approach to structural optimization. Structural Optimization 1, 37-46

Svanberg, K. 1987: The method of moving asymptotes - a new method for structural optimization. Int. J. Numer. Meth. Eng. 24, 359-373

Modulef – A FEM Tool for Optimization

S. Kibsgaard
Institute of Mechanical Engineering
Aalborg University
Pontoppidanstraede 101, DK – 9220 Aalborg East, Denmark

Abstract

This paper deals with the development of a shape optimization system. The advantages of using publicly available sub-programs are discussed, and the problems that are associated with interfacing them, and how to solve these, are described. In particular we emphasize the treatment of the analysis part of such systems, and it is described how it is possible to obtain an analysis module that is easily integrated in a larger system concept by making improvements and extensions to the finite element program Modulef. The extensions made to Modulef (i) facilitate boundary shape variation, modelling the boundary by parametric curves defined by the position of master nodes, (ii) allows for evaluation of design sensitivities by means of numerical finite difference techniques or by semi-analytical differentiation, and (iii) facilitate the interfacing with other programs by using the Modulef concept of data structures for data transfer, storage, etc. Two examples, illustrating the system performance, will be presented.

Introduction

An evolution towards more and more automated design processes has evolved in modern times. In particular, most of the critical engineering components are now being developed in an interaction process between the designer and the computer, and commercially available computer programs for decision support, design aid, strength calculations etc. are now widely used in the engineering community.

A fairly new generation of such programs are now being developed under the common label "Integration", i.e., by integrating earlier programs, each performing complex tasks like the above, towards larger software packages that are able to support a steadily increasing part of the design process. Eschenauer [1], Braibant & Fleury [2], Esping [3], Santos & Choi [4] etc., have dealt with the development of this type of system.

In 1989 Kibsgaard et.al. [5] formulated a general concept of an integrated system for design, analysis and optimization, based on experience from the prototype shape optimization system CAOS developed by Rasmussen [6],[7] which combines computer aided design, finite element analysis and mathematical programming techniques into an integrated package. This strategy has so far lead to the development of new optimizer strategies, Rasmussen [8], studies of semi analytical design sensitivities, Kibsgaard [9], and the development of a new theory on reducing the errors in the semi-analytical method, Olhoff [10]. Furthermore the project has been extended by the integration of topology optimization as described by Bendsøe [11], Olhoff et.al. [12]; and Thomsen [13] has considered optimization of fibre layered composite discs, using the methods of Pedersen [14], to be integrated in the overall system.

One of the central elements in the concept is the analysis system used to calculate the structural response. The program must be capable of coping with all kinds of geometries and loadings, should be "easy" to interface with some sort of boundary parametrization technique, and it must be possible to implement code for sensitivity analysis. These are conflicting demands, on the one hand you wish to have the insight and confidence in the code that comes from writing it yourself, and makes changes easier. However on the other hand, it will take years of programming to get the desired high standard, generality and complexity that most publicly available finite element programs possess. The dichotomy can be solved using the Modulef program (see Bernadou et. al. [15]). Modulef is a non-commercial, but publicly available finite element program, fully portable, written in standard FORTRAN 77, and consisting of some 3000 subroutines. It has well defined, but a bit complicated, programming standards, data structures and interfaces, and the user has full access to the source code. Modulef has approximately 40 elements for static, linear elastic analysis, and well developed facilities for mesh generation, visualization, storage, solution etc. For these reasons the Modulef program is chosen for finite element discretization and analysis, and the programming standards of Modulef form the basis for corrections, adjustments and extensions in the overall concept, thus making the Modulef library fully integrateable in the optimization system.

This optimization system is so far only capable of handling 2D-structures, thus some of the considerations in the ensuing refer only to the treatment of structures that can be discretized in 2D, i.e. plane or axi-symmetric structures.

Modulef Standards

Two of the key elements in the Modulef programming standards are the use of strictly defined data structures, and the dynamic management of memory.

Dynamic management of memory implies that all data in the program are stored in one single array, called the super array, regardless of whether the data are characters, integers or real numbers of single or double precision. Each time data is to be stored, you simply allocate a segment of the super array using the management utilities of Modulef. These can also be used for retrieving, readdressing, renaming, or copying data, and for making previously used space available for other storing operations. Due to the utilities there is no need to keep track of addresses, lengths etc. of the data, and they can be used in different parts of the program, without having to pass them as arguments or commons. It is however preferable to pass them as arguments whenever possible, considering the time spent for utility operation.

The different data structures are described in INRIA [16]. They are all built according to a set pattern, each with individual features, depending on the purpose of the structure. They can be stored in the super array, on files with sequential access, on files with direct access, or partly in core, partly on file. The structures are used for storage or data-transfer for large datasets, describing input or output of a given program module. For example, geometric meshes are stored in a data structure of type NOPO, boundary conditions in type BDCL, stresses in type TAE, materials data in type MILI etc. The general data structure is built of a number of individual arrays (normally 6, but it varies from 4 to 23, depending on the complexity of the data stored). The first array describes the date and time of creation, username, type of the structure, name of the problem etc. The last array generally contains the data in question; and the arrays in between contain information about the last one, how it is stored, the number of elements in it, the type of variables etc. General management utilities performing basic tasks like inclusion in the super array, storage on file, copying data structures, updating data structures etc., are also included in Modulef, and there is also a number of specific tools for each individual type.

The above Modulef standards makes the concept very flexible, passing the super array as the only argument in the routine calls, is sufficient to make all data shareable between the individual modules of the program, and the generality of the data structures makes it possible to either store all datasets in central memory or on file, Thus giving a very flexible program-structure depending on the capacity of your computer and the desired computing speed.

Each data structure is self-consistent and self-explanatory in principle, thus making interfacing with other programs pretty simple. The self consistency however demands a lot of memory, because a lot of information is irrelevant at times, and a lot of identical information has to be stored in several data structures, that might be in core simultaneously. The many necessary utility routine calls also makes the program work somewhat slower, and the routines use some memory as well, to keep track of the length, name and address of stored data and data structures.

Design model

Many commercial finite element programs have been endowed with some sort of optimization facility, which normally only define design variables as the point coordinates and cross sectional dimensions that anyhow should have been entered in order to perform the structural analysis. This kind of system rarely gives the user the possibility to define shapes of lines, curves and interfaces as design boundaries. To integrate the latter possibilities in the present system we have defined a so-called design model. The design model of a structure is a representation of the geometry of the structure. The design model is to be used as the basis for the analysis model in calculating the structural response and the design sensitivities, and in updating the design, and it must therefore include sufficient information to form the model for structural analysis, as well as the model for optimization, including the definition of the design variables. We illustrate this with the example in fig. 1, the classical fillet problem, which among others has been treated by Bendsøe & Kikuchi [17], Soares et.al. [18] and Rasmussen [19].

To analyze the initial structure it suffices to use the points marked with crosses as basis for the analysis model, but if one of the contour lines is not straight it becomes necessary to have the shape of the non straight boundary curve as input for the analysis as well. This could be as a function expression or as coordinates of the finite element nodes on the boundary.

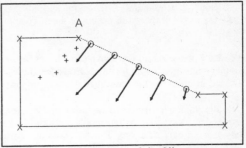

Fig. 1 *The initial geometry of the fillet.*

Likewise, if the contour lines are to remain straight in an optimization process, it suffices to use the position of some of the corner nodes, marked with crosses, as design variables. But if we want to perform the classical stress minimization, by changing the shape of the dotted contour line, it is necessary to use its shape as design variable for sensitivity analysis and optimization, either by using the coefficients of a function expression directly as design variables, or by using some geometrical entities like point coordinates, curvatures etc. as design-variables to determine the shape of the curve.

Based on successful experiences with the prototype system CAOS, Rasmussen [6],[7], the boundary parametrization strategy has been chosen as the so-called master node technique. The positions of the master nodes govern the design boundary shape by curve fitting, using functions of some specific type. Up to now polynomials, piecewise straight lines, cubic splines and B-splines are implemented. The master nodes, which control the shape of the dotted boundary of fig. 1 are encircled.

By including the corner points and the master nodes in the design model, we achieve full control of the shape of the structure in question, and it is possible to use the coordinates of either type of points as design variables.

Furthermore it must be possible to control the move direction and interval. This is done by defining vectors, that can be chosen as the move directions and intervals of one or more of the points in the design model. These modifier vectors can be defined using some of the existing points, or we can define and use some new points to be included in the design model. These are called modifier points, and they are shown in fig. 1 as the points marked with pluses. In the present example all the master nodes, and none of the corner points, are used as design variables. The modifier vectors are all defined with the corner point A as the base end, and different modifier points as the point end. The modifier vectors in question are marked symbolically with arrows at their corresponding master node.

The total design model thus consists of three parts. The first one defining the corner points, master nodes and modifier points together with a description of the curve types, and how they should fit, the second part describes the modifier vectors, using the points and nodes of the first part, and the third part describes the connexions between the modifiers and the corner points or master nodes. In this way it is possible to use a mixture of modifiers and points/nodes as one design variable, and it is much less expensive to link them together in this way in the problem formulation, than to impose extra constraints on the optimization problem

Mesh generation

The structures to be optimized by a general optimization system can take on various geometrical forms, and the mesh generation routines must therefore be able to generate well adapted element meshes for many kinds of geometrically complex domains. Thus, it must be possible to remesh the domain, as it changes with a change in the design variables, in order to perform a new structural analysis, and it must be possible to remesh the domain as a function of a given perturbation of one of the design variables, in order to perform the semi-analytical design sensitivity analysis, which requires an unchanged mesh topology.

It is also evident, that in order to be able to utilize all the analysis capabilities achieved by integrating Modulef, the mesh generation must be able to generate all kinds of meshes, e.g., segmented, triangular, quadrilateral, tetrahedral, pentahedral and hexahedral meshes, all with or without midside and interior nodes. Moreover the mesh generator must be able to re–number, smoothen, relocate, and to refine, etc., a given mesh.

Modulef possesses all the above capabilities, as described in George [20], thus making it unnecessary to implement further mesh facilities. However, it is essential to integrate the existing facilities with the design model, in order to take advantage of their properties.

The design model has been included in Modulef using one of the data structure facilities, termed an associated array. The design model is stored in such an array, and it is then possible to associate it to any type of data structure. The data structure in question is updated to contain all storage information about the associated array, and the management utilities will treat them as a connected whole. Associating the design model to a geometric net in a data structure of type NOPO thus makes the information available for all other parts of the program, as long as the data structure in question is in main memory. The only limitation is that the super–array must be in the calling arguments list of the subroutines. The different mesh routines are then able to read the data defining the design model, and to use these to generate the desired mesh, to remesh a domain, to perturb the mesh, etc.

All in all, the integrated design model / 2D–mesh generator is able to generate the mesh types indicated in fig. 2., The first column being the initial mesh generated in a domain by either mapping or free–mesh, the second column is a remesh of a perturbed domain, the third column a perturbed mesh of a perturbed domain and the fourth column is a smoothed perturbed mesh of a perturbed domain. The perturbation being grossly exaggerated to that of a finite difference sensitivity analysis (overall or semi–analytical).

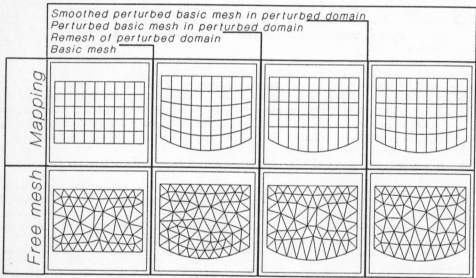

Fig. 2. *Mesh generation of a perturbed domain*

The mesh possibilities illustrated in fig. 2 gives the user various choices to perform finite difference based sensitivity analysis, either overall or semi–analytical. All in all there are five possible ways to remesh the perturbed domain – the free remeshing of a perturbed domain is the only method to be excluded, because it changes the topology of the mesh.

The remeshing of a domain, necessitated by changes in the design variables from one iteration to another, can also be done in most of the illustrated ways –only the unsmoothed perturbed mesh method is to be excluded, for mapping as well as for free mesh, due to the strong distortion of the elements along the boundary, that accrues from major shape changes.

The mesh facilities can of course be mixed according to the users
example, and they constitute a very powerful discretization tool i̶

Sensitivity Analysis

The sensitivity analysis has been programmed in a very general w̶
structures for storage, so this part of the programme is valid
contained in Modulef. Semi–analytical sensitivity analysis is imp̶
types, including beam, disc, plate, shell, solid, and axisymme̶
configurations. The semi–analytical method is implemented f̶
volume and compliance sensitivities are also implemented. W̶
element sensitivities, and further studies for other finite eleme̶

Interfacing

Modulef has postprocessing facilities for most
types of results, and for many graphical output
devices, as described in George et. al. [21]. There
is however a paucity of routines for stress visualization, and interface programs to Patran has had
to be implemented, so as to use this excellent
program for visualization of Modulef results. Due
to the documented data structures for storage, and
the management utilities, it has been a moderate
task to program the interface.

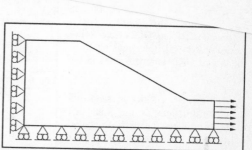

Fig. 3 *Fillet, support and loading*

Examples

The overall optimization system is still under
development, but the modular construction of
Modulef and the implemented extensions, makes
it very easy to build a small command file of
some 20–30 lines, that calls the required modules
for a given optimization task. The mathematical
programing problem is formulated according to the
bound formulation of Olhoff [22], and a simplex
routine is used as optimizer.

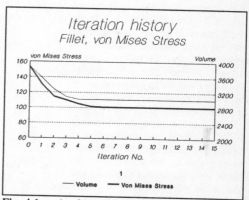

Fig. 4 *Iteration history for the fillet problem*

Fillet

The first example is the well known fillet problem described in the section on the design model.
The fillet is loaded with a unidirectional line
loading with a magnitude of 100 N/mm, and
supported as illustrated in fig. 3. The objective is
to minimize the maximum value of the "von Mises
reference stress", by changing the shape of the
sloped boundary segment. The master node move
directions and intervals are as illustrated by the
arrows of fig. 1.

The iteration history is shown in fig. 4, and the
final geometry and mesh is shown in fig. 5.

The fillet has been discretized using a free mesh
method, and the finite elements are 3 node triangular constant stress elements. The sensitivity
analysis has been performed by the semi–analytical method, based on an unsmoothed perturbed
basic mesh in the perturbed domain.

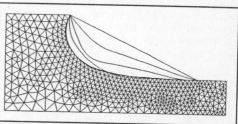

Fig. 5 *Final geometry and mesh for the fillet*

...rminated in 6-7 ...ove limit of 12%, ...re in good agreement ...he literature.

...plate joint
...cond example is also a fillet ...em, inspired by the Danish naval ...andard Flex 300" programme.

The Standard Flex is a multipurpose naval ship with an overall length of 54 meters and a displacement of 300 ton. The ships are entirely built in GRP–sandwich (GRP .eq. Glass fibre Reinforced Polyester), thus making them the largest GRP–laminate ships in the world.

One of the weak spots in sandwich plate constructions is the joint that connects orthogonal panels. A detailed figure of the traditional way to manufacture these T–joints is shown in figure 6.

The objective of the optimization process is to minimize the maximum tensile stress in the skin and reinforcing laminates, with a constraint on the core stresses, that they are not to exceed 1 N/mm². The design boundaries are the outer an inner contour of the reinforcing laminate, with 5 equally distributed master nodes on each contour, and the corner points in each end as design points. Each modifier vector is attached to two pairwise design points, so the thickness of the skin is forced to remain constant. The load and support conditions are illustrated on fig. 7, with the optimized geometry.

All the materials are computed as being isotropic. The skin with a modulus of elasticity, E = 15000 N/mm², and Poissons ratio ν = 0.26. Core: E = 120 N/mm², ν = 0,38. Polyester filling: E = 3500 N/mm², ν = 0.4.

The GRP–skins are meshed with a mapping method, using 4 node quadrilateral CST–elements. The core, the fillet and the polyester stopping are meshed with a free mesh algorithm, using 3 node triangular CST–elements. The initial mesh of the critical area is shown in the "blow up" in fig. 7, together with the final geometry of the structure.

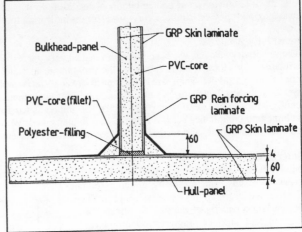

Fig. 6 *Typical Hull–Bulkhead joint (T–joint)*

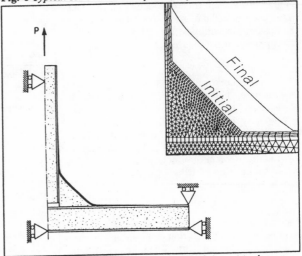

Fig. 7 *Final geometry, with a "blow up" of the critical area.*

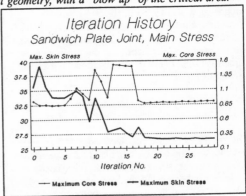

Fig. 8 *Iteration history for the sandwich plate joint*

The optimizer is a simplex routine with a move limit of 3%, this lead to the somewhat unstable converging iteration history of fig. 8. The skin stress is reduced by 25% in the optimized geometry shown in fig. 7. The same results were obtained with a move limit of 0.5% in app. 90 iterations, but avoiding the unstable convergence.

The optimized shape is somewhat surprising, one would expect much smaller curvatures of the reinforcing laminate in both ends. The reason for the large curvatures is probably due to a poor calculation of the stress field in the vicinity of the kinks, partly because of somewhat distorted elements, and partly because CST-elements are not to well suited for modelling slender structures like the laminate skin in question. A finer discretization and better elements is expected to give a smoother geometry of the reinforcing laminate.

Conclusions

The present paper has given a brief description of Modulef as a finite element tool for optimization, and some of its vital standards for programming. It ought to be clear, that to achieve the same analysis and mesh generation facilities by programming from scratch, would have taken significantly longer than the app. 2.5 man year so far involved in this project.

The mesh generation facilities are of vital importance to any finite element package, but even more important in an optimization system. The "Sandwich Plate Joint" example gives some insight in the kind of complex domains the mesh facilities are able to handle. The inclusion of free mesh algorithms in general optimization systems has made discretization much easier, and it is the authors belief that the smoothed perturbed free mesh of perturbed domains can satisfy any sensitivity convergence requirements quite as well as a mapping remesh of a perturbed domain, the latter being, at present, favored by many. As far as the author is concerned it is however unnecessary to remesh the domain interior for sensitivity analysis, in the belief, that if the discretization is so fine as to calculate the structural response satisfactorily, the perturbed basic mesh in perturbed domain strategy is sufficiently accurate for design sensitivity approximation.

Acknowledgement – The present work has received support from the Danish Technical Research Council (Programme of Research on Computer Aided Design)

References

[1]   H. Eschenauer, Numerical and Experimental Investigations on Structural Optimization of Engineering Designs, Research Laboratory for Applied Structural Optimization at the Institute of Mechanics and Control Engineering, University of Siegen, FRG. (Bonn+Fries, Siegen, 1986)

[2]   V. Braibant, C. Fleury, Shape Optimal Design using B–splines, (Computer Methods in Applied Mechanics and Engineering, Vol. 44, 1984, pp 247–267)

[3]   B. J. D. Esping, The Oasis Structural Optimization System (Computers and Structures, Vol. 23, 1986, pp 365–377)

[4]   J. L. T. Santos, K. K. Choi, Integrated Computational Considerations for Large Scale Structural Design Sensitivity Analysis and Optimization (Proceedings of a GAMM–seminar "Discretization Methods and Structural Optimization – Procedures and Applications" October 5–7, 1988, Siegen, FRG. Eds. H. A. Eschenauer and G. Thierauf, pp. 299–307. Lecture Notes in Engineering 42, Springer Verlag, New York, Heidelberg, Berlin, 1989)

[5]   S. Kibsgaard, N. Olhoff, J. Rasmussen, Concept of an Optimization System, (Proceedings of OPTI89 – First Int. Conf. on Computer Aided Optimum Design of Structures: Applications, Southampton, UK, June 1989 (Eds. C. A. Brebbia, S. Hernandez). Computational Mechanics Publications, Springer–Verlag, 1989)

[6]   *J. Rasmussen*, The Structural Optimization System CAOS (Structural Optimization, Vol. 2, No. 2, pp 109–116, 1990)

[7]   *J. Rasmussen*, Development of the Interactive Structural Shape Optimization System CAOS (in Danish), (Institute of Mechanical Engineering, Aalborg University, Special Report No. 1a, August 1989)

[8]   *J. Rasmussen*, Structural Optimization by Accumulated Function Approximation, (Institute of Mechanical Engineering, Aalborg University, Report No. 20, June 1990. Submitted to Int. Journal for Numerical Methods in Engineering)

[9]   *S. Kibsgaard*, Sensitivity Analysis – The Basis for Optimization, (submitted for WCCM–II, Sec. World Congress on Computational Mechanics, Aug. 27–31, 1990, Stuttgart, FRG)

[10] *N. Olhoff*, On Elimination of Inaccuracy in Semi– analytical Sensitivity Analysis, (submitted for Int. Conf. on Engineering Optimization in Design Processes. Sept. 3–4, 1990, Karlsruhe, FRG)

[11] *M. P. Bendsøe*, Optimal Shape Design as a Material Distribution Problem (Structural Optimization Vol. 1, pp 193–202, 1989)

[12] *N. Olhoff, M. P. Bendsøe, J. Rasmussen*, On CAD–integrated Structural Topology and Design Optimization, (submitted for WCCM–II, Sec. World Congress on Computational Mechanics, Aug. 27–31, 1990, Stuttgart, FRG)

[13] *J. Thomsen*, Optimization of Composite Discs, (Institute of Mechanical Engineering, Aalborg University, Report No. 21, June 1990. Submitted to Structural Optimization)

[14] *P. Pedersen*, On Optimal Orientation of Orthotropic Materials, (Structural Optimization Vol. 1, pp 101–106, 1989)

[15] *M. Bernadou, P. L. George, A. Hassim, P. Joly, P. Laug, A. Perronet, E. Saltel, D. Steer, G. Vanderborck, M. Vidrascu*, Modulef, A Modular Library of Finite Elements, (INRIA – Institut National de Recherce en Informatique et en Automatique, France, 1986)

[16] *INRIA*, Description des Structures de Donnes Modulef (in french), (Modulef report no. 2, Institut National de Recherce en Informatique et en Automatique, France, April 1987.)

[17] *M. P. Bendsøe, N. Kikuchi*, Generating optimal Topologies in Structural Design using a Homogenization Method, (Comp. Meth. in App. Mech. and Eng., Vol. 171, pp 197–224, 1988)

[18] *C. A. Mota Soares, H. C. Rodrigues, K. K. Choi*, Shape Optimal Structural Design using Boundary Elements and Minimum Compliance Techniques, (Journal of Mechanisms, Transmissions and Automation in Design, Vol. 106, pp 518–523, 1984)

[19] *J. Rasmussen*, Collection of Examples – CAOS Optimization System, (Institute of Mechanical Engineering, Aalborg University, Special Report No. 1c, 2nd. ed. June 1990)

[20] *P. L. George*, Modulef: Construction et Modification de Maillages (in french), (INRIA – Institut National de Recherce en Informatique et en Automatique, Rapportes Techniques No. 104, France, Feb. 1990)

[21] *P. L. George, A. Golgolab, B. Muller, E. Saltel*, Manuel d'Utilisation du Logiciel Graphique Fortran 3D (in french), (Modulef report no. 47, INRIA – Institut National de Recherce en Informatique et en Automatique, France, Mai 1989)

[22] *N. Olhoff*, Multicriterion Structural Optimization via Bound Formulation and Mathematical Programming, (Structural Optimization, Vol. 1, No. 1, pp 11–18 1989)

COMPUTER AIDED MULTICRITERION OPTIMIZATION SYSTEM IN USE

Andrzej Osyczka

Technical University of Cracow ul.Warszawska 24
31-155 Kraków POLAND

Abstract: In the paper software package called Computer Aided Multicriterion Optimization System (CAMOS) is briefly described. The system enables the designer to solve single and multicriterion optimization problems for nonlinear programming models with continuous, integer, discrete and mixed design variables. The system is designed to facilitate the interactive processes considering both input/output information arrangement and multicriterion decision making problems. Several engineering design examples are presented to show the advantages of the use of the system. These examples deal with design and optimization of helical springs, hydrostatic journal bearings, helical gearsets multiple clutch brakes. Multicriterion optimization models for all the examples are briefly described and the computer aided design sessions with CAMOS are presented. The structure of CAMOS and the subroutine which describes the design problem are arranged in the way that the analysis, optimization and decision making phases are one entity. The analysis and decision making phases can be supported by a graphical illustration of the problem formulation and the solution obtained. Both the system and problem oriented computer aided optimum design modules are coded in FORTRAN and prepared for an IBM PC/XT/AT.

1. Introduction

To enable the designer to participate actively in strategy of seeking the best design, most of optimum design methods are recently developed towards interactive on line use [1], [2], [3]. For multicriterion design optimization, these interactive processes refer to problem analysis, optimization routines and decision making environments. Graphics facilities should also be available so that the required data can be displayed for the designer.

In this paper the Computer Aided Multicriterion Optimization System (CAMOS) is briefly described [4]. The aim of this system is to combine analysis, optimization and decision modules into one entity. The possibilities of the system are shown by means of several optimum design examples.

2. Description of CAMOS

The software package CAMOS is prepared to solve a multicriterion optimization problem of the form.

$$\min_{x \in R^n} \ \{ \ f(x) \in R^k \ | \ g(x) \geq 0 \ , \ h(x) = 0 \ \} \tag{1}$$

where: x — vector of n design variables
$f(x)$ — vector of k objective functions
$g(x)$ — vector of m inequality constraints
$h(x)$ — vector of p equality constraints

To solve nonlinear programming problems with continuous variables the following single criterion optimization methods are used:

RS — Random Search method
DS — Direct Search method of Hooke and Jeeves
SM — Simplex Method of Nelder and Mead
VM — Variable Matrix method of Davidon, Fletcher and Powell

FT – Flexible Tolerance method

The random search method is used to
(i)   generate a good starting point for remaining iterative method,
(ii)   solve models with discrete, integer and mixed design variables,
(iii) generate a set of Pareto–optimal solutions for both continuous and discrete models.

    Each of the iterative method, i.e. DS, SM, VM and FT method can be used separately or together with the RS method in the way that the results obtained using RS method are the starting solutions for each iterative method.

    To solve multicriterion optimization problems the following methods are available in CAMOS
1. Min–max method
2. Global criterion method
3. Weighting min–max method
4. Weighting objective method
5. Normed weighting method
6. Method for generating a set of Pareto–optimal solutions (used only with RS method).

The solution of the multicriterion optimization problem is understood here as findig a Pareto optimal solution which verbally can be defined as the point x* for which no criterion can be improved without worsening at least one other criterion [5].
    The structure of CAMOS is such that the user prepares a problem dependent subroutine which has a computer aided design form. This subroutine is linked to the system to make an execution version of the program. The way the program is then executed is presented in Fig. 1.

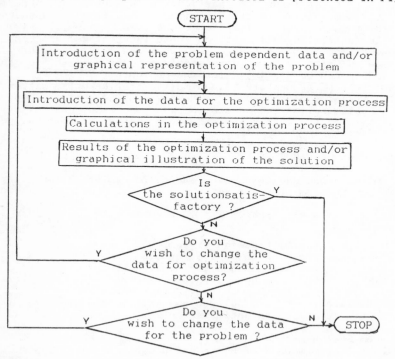

Fig. 1: General concept of CAMOS

CAMOS is designed especially for interactive use on mikrocomputers and it is coded in FORTRAN. HELP utilityy offers a special category of information on all data to be introduced.

3. Applications of CAMOS

3.1. Optimal design of helical compression springs

Optimization model

Vector of design variables is $x = [x_1, x_2, x_3, x_4]^T$ where:

x_1 — wire diameter

x_2 — mean coil diameter

x_3 — free length

x_4 — number of active coils

The forth design variable may be considered as the integer variable.

Vector of objective functions is $f(x) = [f_1(x), f_2(x), f_3(x)]^T$ where:

$f_1(x)$ — weight of the spring

$f_2(x)$ — outer diameter of the spring

$f_3(x)$ — length of the spring

Optimization results

An example of the final stage of a computer aided session with CAMOS for the above model is shown in Fig. 2. These results, made in the form of a hard copy of the screen present two solutions obtained for different weighting coefficients using weighting min-max method.

3.2. Optimal design of multiple clutch brakes

Optimization model

Vector of design variable is $x = [x_1, x_2, x_3, x_4]^T$ where:

x_1 — inner radius

x_2 — outer radius

x_3 — thickness of discs

x_4 — number of friction surfaces

The forth design variable is considered as the integer variable.

Vector of objective functions is $f(x) = [f_1(x), f_2(x), f_3(x), f_4(x)]^T$

where: $f_1(x)$ — weight of the brake

$f_2(x)$ — stoping time

$f_3(x)$ — number of friction surfaces

$f_4(x)$ — outer diameter

Optimization results

Similarly as for the spring design two final solutions obtained while working with CAMOS are presented in Fig. 3.

WEIGHTING MIN-MAX METHOD

WEIGHTING COEFFICIENTS
.2000 .6000 .2000
STARTING POINT CHOSEN BY THE SYSTEM FOR SEEKING THE OPTIMUM
.857970E+01   .732759E+02   .218848E+03
VALUE OF 1 OBJECTIVE FUNCTION =   .1183609E+01
VALUE OF 2 OBJECTIVE FUNCTION =   .7884743E+02
VALUE OF 3 OBJECTIVE FUNCTION =   .2001623E+03
VECTOR OF DECISION VARIABLES
.8266836E+01   .7058059E+02   .2001623E+03   .1100000E+02
VALUES OF INEQUALITY CONSTRAINTS
.2983774E+02   .1254959E-03   .2115257E+02   .5231376E+02
.5266836E+01   .1537800E+01   .9600000E+02   .2757981E+01
.9580371E+02   .8266836E+01   .7058059E+02   .2001623E+03
.9000000E+01   .3351909E+00   .3664809E+01

GRAPHICAL ILLUSTRATION OF THE SOLUTION
SCALE 10mm = ⌴   SCALE 1.0 ⫶ 1.0

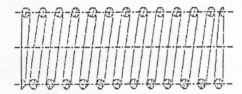

WEIGHTING COEFFICIENTS
.6000 .2000 .2000
STARTING POINT CHOSEN BY THE SYSTEM FOR SEEKING THE OPTIMUM
.857460E+01   .787090E+02   .229045E+03
VALUE OF 1 OBJECTIVE FUNCTION =   .1141959E+01
VALUE OF 2 OBJECTIVE FUNCTION =   .8530085E+02
VALUE OF 3 OBJECTIVE FUNCTION =   .1890451E+03
VECTOR OF DECISION VARIABLES
.8460248E+01   .7684060E+02   .1890451E+03   .9000000E+01
VALUES OF INEQUALITY CONSTRAINTS
.4095489E+02   .3366487E+01   .1469915E+02   .5838035E+02
.5460248E+01   .2082547E+01   .9600000E+02   .2723815E-02
.5863880E+03   .8460248E+01   .7684060E+02   .1890451E+03
.7000000E+01   .3999695E+01   .3051264E-03

GRAPHICAL ILLUSTRATION OF THE SOLUTION
SCALE 10mm = ⌴   SCALE 1.0 ⫶ 1.0

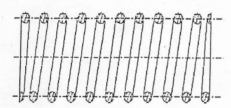

Fig. 2: Optimization session with CAMOS for spring design

```
----------------------------------------------------------
                    WEIGHTING MIN-MAX METHOD
----------------------------------------------------------
       WEIGHTING COEFFICIENTS
            .1000   .1000   .1000   .7000
VALUE OF  1 OBJECTIVE FUNCTION =     .7561983E+01
VALUE OF  2 OBJECTIVE FUNCTION =     .1526963E+01
VALUE OF  3 OBJECTIVE FUNCTION =     .3400000E+02
VALUE OF  4 OBJECTIVE FUNCTION =     .1079552E+00
VECTOR OF DECISION VARIABLES
      .6019776E-01    .1079552E+00    .1097939E-02    .3400000E+02
VALUES OF INEQUALITY CONSTRAINTS
      .5197760E-02    .2044830E-02    .2775741E-01    .9793932E-04
      .2440721E+00    .0000000E+00    .3400000E+02    .1127952E+06
      .2844598E+03    .2910910E+05    .7739929E+01    .1847299E+02
```

GRAPHICAL ILLUSTRATION OF THE SOLUTION

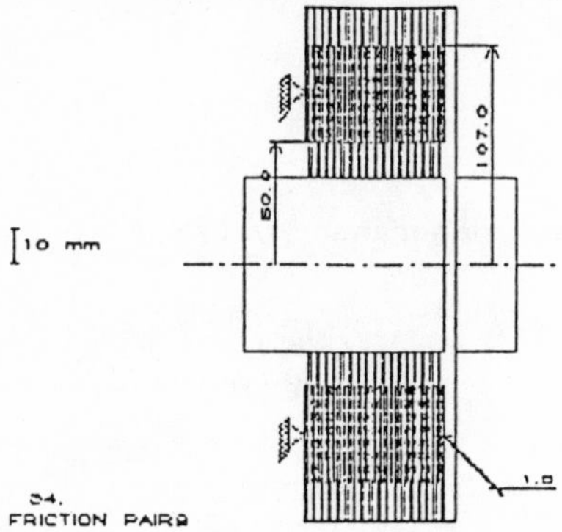

```
       WEIGHTING COEFFICIENTS
            .1000   .1000   .7000   .1000
VALUE OF  1 OBJECTIVE FUNCTION =     .1100462E+02
VALUE OF  2 OBJECTIVE FUNCTION =     .1796035E+01
VALUE OF  3 OBJECTIVE FUNCTION =     .3100000E+02
VALUE OF  4 OBJECTIVE FUNCTION =     .1085872E+00
VECTOR OF DECISION VARIABLES
      .6077404E-01    .1085872E+00    .1733085E-02    .3100000E+02
VALUES OF INEQUALITY CONSTRAINTS
      .5774042E-02    .1412836E-02    .2781312E-01    .7330849E-03
      .2285413E+00    .3000000E+01    .3100000E+02    .1135182E+06
      .2760856E+02    .3201980E+05    .7724399E+01    .1820391E+02
```

GRAPHICAL ILLUSTRATION OF THE SOLUTION

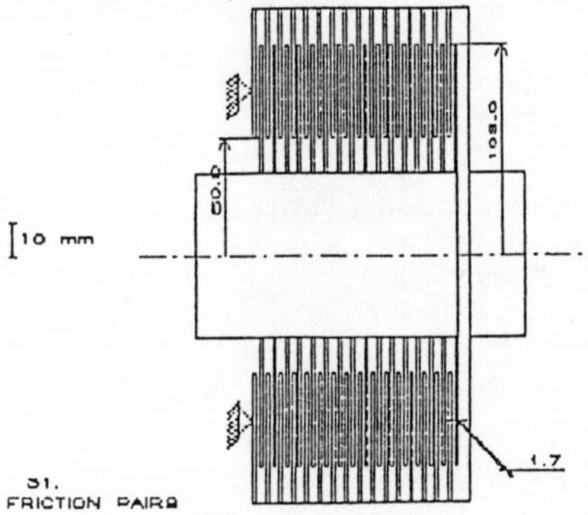

Fig. 3: Optimization session with CAMOS for multiple clutch brake design

3.3. Optimal design of helical gearsets

Optimization model

Vector of design variables is $x = [x_1, x_2, x_3, x_4, x_5]^T$ where:

x_1 — width of the gear rim

x_2 — diameter of the input shaft

x_3 — diameter of the output shaft

x_4 — number of teeth of the pinion weel

x_5 — modules of the gear weels

Vector of objective functions is $f(x) = [f_1(x), f_2(x), f_3(x)]^T$ where:

$f_1(x)$ — volume of the gear set

$f_2(x)$ — distance between the axes

$f_3(x)$ — width of the gear rim

Optimization results

For this problem an optimal solution generated by CAMOS is presented in Fig. 4.

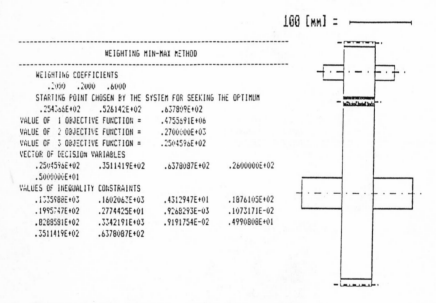

```
100 [MM] = ├─────────┤
```

```
----------------------------------------------------------------
                    WEIGHTING MIN-MAX METHOD
----------------------------------------------------------------
   WEIGHTING COEFFICIENTS
     .2000   .2000   .6000
   STARTING POINT CHOSEN BY THE SYSTEM FOR SEEKING THE OPTIMUM
     .2543o6E+02    .526142E+02    .637809E+02
VALUE OF 1 OBJECTIVE FUNCTION =    .4755691E+06
VALUE OF 2 OBJECTIVE FUNCTION =    .2700000E+03
VALUE OF 3 OBJECTIVE FUNCTION =    .2504596E+02
VECTOR OF DECISION VARIABLES
     .2504596E+02    .3511419E+02    .6378087E+02    .2600000E+02
   .5000000E+01
VALUES OF INEQUALITY CONSTRAINTS
     .1335988E+03    .1602063E+03    .4312947E+01    .1876105E+02
     .1995747E+02    .2774425E+01    .9268293E-03    .1073171E-02
     .8288591E+02    .3342191E+03    .9191754E-02    .4990808E+01
     .3511419E+02    .6378087E+02
```

Fig. 4: Solution of a gearset optimization problem

3.4. Optimal design of hydrostatic bearings

Optimization model for journal bearings

Vector of design variables is $x = [x_1, \ldots, x_9]^T$ where:

x_1 — orifice diameter

x_2 — supply pressure of the fluid

x_3 — angle of the circumferential land length

x_4 — angle of the circumferential pocket length

x_5 – depth of the pocket

x_6 – axial pocket land length

x_7 – radial clearance

x_8 – length of the bearing

x_9 – diameter of the bearing

Vector of objective functions is $f(x) = [f_1(x), f_2(x)]^T$ where:

$f_1(x)$ – total power loss in the bearing

$f_2(x)$ – size of the bearing

Optimization results

For this problem an axample of the optimal solution obtained on the plotter as the output of CAMOS is presented in Fig. 5.

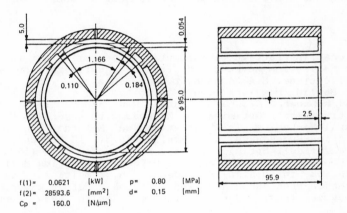

f(1)= 0.0621 [kW]   p= 0.80 [MPa]
f(2)= 28593.6 [mm²] d= 0.15 [mm]
Cp = 160.0 [N/μm]

Fig. 5: Computer graphic output for an optimal solution – journal bearing

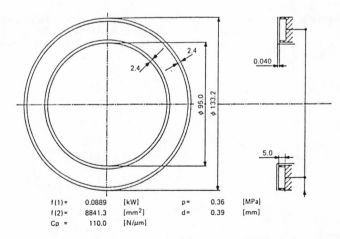

f(1)= 0.0889 [kW]   p= 0.36 [MPa]
f(2)= 8841.3 [mm²] d= 0.39 [mm]
Cp = 110.0 [N/μm]

Fig. 6: Computer graphic output for an optimal solution – thrust bearing

Similar model is created for hydrostatic thrust bearings for which an axample of final solution obtained on plotter is presented in Fig. 6. More detailed description of these problems is given in [6].

4. Final Remarks

The software package CAMOS is a user — friendly program which makes it useful for designer of various backgrounds. While running the program the designer is asked to provide only the basic information. Following the results, he obtains all the necessary information which enables him to make the right decision in design problems with conflicting criteria.

Each of the optimization problem introduced to CAMOS is prepared in a general, computer aided design form which as a module is attached to the system. The computer library of such modules is created and under development. This makes the design of any of the presented problems very easy and the whole design process is just a short session with a computer.

References

[1] I. S. Arora: Interactive Design Optimization of Structural Systems. In: Discretization Methods and Structural Optimizations-Procedures and Applications. H.A.Eschenauer and G.Thierauf (Eds), Springer-Verlag, Berlin, New York (1989) 10–16

[2] H. A. Eschenauer, E. Schäfer and H. Bernau: Application of Interactive Vector Optimization Methods with Regard to Problems in Structural Mechanics. In: Discretization Methods and Structural Optimization-Procedures and Applications. H.A.Eschenauer and G.Thierauf (Eds), Springer-Verlag, Berlin, New York (1989) 110–117

[3] H. A. Eschenauer, A. Osyczka and E. Schäfer: Interactive Multicriteria Optimization in Design Processes. In: Multicriteria Design Optimization-Procedures and Applications. H.Eschenauer, J.Koski and A.Osyczka (Eds), Springer-Verlag, Berlin, New York (1990) 71–114

[4] A. Osyczka: Computer Aided Multicriterion Optimization System (CAMOS). In: Discretization Methods and Structural Optimization-Procedures and Applications. H.A.Eschenauer and G.Thierauf (Eds), Springer-Verlag, Berlin, New York (1989) 263–270

[5] A. Osyczka: Multicriterion Optimization in Engineering with FORTRAN Programs, Ellis Horwood, Chichester, 1984

[6] J. Montusiewwicz and A. Osyczka: A Decomposition Strategy for Multicriteria Optimization with Application to Machine Tool Design. Engineering Costs and Production Economics. To apear.

DYNAMIC STRUCTURAL OPTIMISATION BY MEANS
OF THE U.A. MODEL UPDATING TECHNIQUE

ir. S.Lammens *, dr. ir. W.Heylen, prof. dr. ir. P.Sas **

K.U.Leuven, Mechanische Konstruktie en Produktie
Celestijnenlaan 300 B, B-3030 Heverlee, Belgium

ABSTRACT

In recent years several methods have been developed for updating finite element models with regard to their dynamic characteristics, by means of experimentally obtained vibration information. Many techniques use modal characteristics (resonance frequencies, damping factors, modal displacements, ...) for tuning physical parameters of the model (elasticity moduli, mass density, plate thicknesses, cross sectional characteristics, ...) in order to obtain a finite element model which shows the same dynamic features as the real structure. One of those techniques is the U.A. Model Updating Technique.

Since this technique not only uses physically interpretable model parameters to be modified, but also allows an easy selection and/or weighting of those parameters and of the dynamic characteristics to be matched, it is also applicable to dynamic structural optimization. If used as such, the desired vibration parameters serve as a goal for the structural optimization procedure. The algorithm will determine modifications of the structure, within the imposed boundary conditions, that will match this desired dynamic behaviour.

An example will illustrate this process.

1. INTRODUCTION

Today, modern mechanical constructions must fulfill a broad variety of goals on one hand, but on the other hand several constraints are imposed on the practical design concept. Typical technical goals can be weight minimisation, stress concentration minimisation, life time maximisation, etc. In order to describe the complex relationship between design parameters and goals, analytical models, especially finite element models, are used. Due to incorrect modeling, geometrical oversimplification and uncertainties on the finite element input data, perfect correlation between reality ("known" by measurements) and the analytical model is seldom found. The use of model updating procedures allows the designer to tune the analytical model in order to yield minimum differences between the experimental and analytical characteristics. The result is that more confidence can be given to further use of the analytical model, e.g. for structural optimisation calculations.

A procedure for updating dynamic finite element models (UA-model updating - User Acceptance) has recently been developed at K.U. Leuven. This UA-procedure formulates the problem as a minimisation problem, being approached with a non linear programming algorithm (ref. 1).

While updating tries to tune the dynamic characteristics of a finite element model to the experimentally obtained modal parameters, dynamic structural optimisation aimes at tuning the dynamic characteristics of a structure to values imposed by practical considerations. It is obvious that basically

---

* NFWO research assistant
** NFWO research associate

both problems lead to the similar formulations. The UA-model updating techniques can easily be adapted for solving structural optimisation problems.

After a short overview of the UA-modelupdating technique and its adaption to structural optimisation, this paper presents an elaborate example of model updating and structural optimisation.

2. UA-UPDATING AND STRUCTURAL OPTIMISATION

The UA-updating procedure uses as updating parameters the material and element properties of the finite element model. Such parameters are Young's modulus, mass density, cross sections, moments of inertia, etc. The user has to define the set of parameters (p_k, k=1,r) together with weighing factors w_k expressing the confidence of the user in the actual value of the parameters p_k ($w_k\approx1$: p_k is assumed to be quasi correct; $w_k\approx0$: p_k is assumed to be totally wrong). The procedure uses 3 criteria to change these parameters in order to obtain a better correlation between experimental results and finite element calculations:

a) Analytical eigenfrequencies should converge to the experimental frequencies. The frequency shifts due to changes of the updating parameters are predicted using the sensitivities of the eigenfrequencies with respect to selected parameter changes. The sensitivity coefficients are combined in a set of s linear equations for obtaining frequency correlation. This first order Taylor expansion can be written explicitly:

$$\sum_{k=1}^{r} (\frac{\partial f_{i,a}}{\partial p_k}) \Delta p_k = f_{i,e} - f_{i,a} \qquad i=1,s \qquad (1)$$

with:

r: number of selected updating parameters
s: number of frequency correlation requirements
p_k: kth selected updating parameter
Δp_k: unknown parameter change for updating parameter k
$f_{i,a}$: ith analytical eigenfrequency
$f_{i,e}$: ith experimental resonance frequency

b) The experimental modeshapes should be orthogonal to each other when wheighed with the analytical mass- and stiffnessmatrix. Due to the expected measurement and curve fitting errors on the modal vectors, perfect orthogonality is not required, but reduction of important off diagonal orthogonality matrix elements to a sufficiently low level is attempted. Acceptable levels of off diagonal orthogonality matrix elements are of the order of 10% when compared with corresponding main diagonal elements. These requirements are combined into two sets of linear equations:

$$\left| \sum_{k=1}^{r} (\frac{\partial mo_{ij}}{\partial p_k}) \Delta p_k + mo_{ij} \right| \leq \alpha_M \qquad i\neq j \qquad (2)$$

$$\left| \sum_{k=1}^{r} (\frac{\partial ko_{ij}}{\partial p_k}) \Delta p_k + ko_{ij} \right| \leq \alpha_K \qquad i\neq j \qquad (3)$$

with:

mo_{ij}: $\{\psi_{i,exp}\}^t [M] \{\psi_{j,exp}\}$

ko_{ij}: $\{\psi_{i,exp}\}^t [K] \{\psi_{j,exp}\}$

$\{\psi_{i,exp}\}$: normalized and expanded experimental mode vector i. These mode vectors are scaled to unity modal mass for computation of mo_{ij} and to unity modal stiffness for computation of ko_{ij}.

α_M: maximum value for off diagonal elements of the mass orthogonality matrix ($\alpha_M = 0.1$ typically)

α_K: maximum value for off diagonal elements of the mass orthogonality matrix ($\alpha_K = 0.1$ typically)

c) The iteration parameters must not exceed lower and upper bounds in order to restrict the updated models to the space of "user defined" acceptable models. Mathematically, this can be expressed by a set of linear constraints:

$$l_k \leq p_k + \Delta p_k \leq u_k \qquad k=1,r \qquad\qquad l_k: \text{lower bound on variable } p_k \qquad (4)$$
$$u_k: \text{upper bound on variable } p_k$$

Although the set of equations (1), (2) and (3) can be solved with a least square approximation, normally a non linear programming approach is used. The objective function that is used is:

$$W_f \sum_{n=1}^{s} W_n \left(\sum_{k=1}^{r} a_{nk}^f x_k - 1 + f_{n,a}/f_{n,e} \right)^2 + W_{Mo} \sum_{q=1}^{t} W_q \left(\sum_{k=1}^{r} a_{ijk}^M x_k - 1 + mo_{ij} \right)^2 +$$

$$+ W_{Ko} \sum_{p=1}^{t} W_p \left(\sum_{k=1}^{r} a_{ijk}^K x_k - 1 + ko_{ij} \right)^2 + W_p \sum_{k=1}^{r} x_k^2 \qquad (5)$$

with:

$x_k = w_k \Delta p_k/p_k$, weighted relative parameter change

$$a_{nk}^f = \frac{\partial f_{n,a}}{\partial p_k} \frac{p_k}{w_k f_{n,e}} \qquad\qquad a_{ijk}^M = \frac{\partial mo_{ij}}{\partial p_k} \frac{p_k}{w_k} \qquad\qquad a_{ijk}^K = \frac{\partial ko_{ij}}{\partial p_k} \frac{p_k}{w_k}$$

The last term in (5) is a penalisation of parameter changes. Constraints (4) still remain valid. This minimisation problem with constraints can be solved with standard algorithms. The weighing factors (W_f, W_{Ko}, W_{Mo}, W_n, W_m, W_p) are defined and optimised in a control strategy that ensures the stability of the global iteration proces. More details can be found in (ref. 1).

The equations and algorithms for structural optimisation are exactly the same. The parameter selection with weighting factors is identical. This selection expresses which material or element properties may change while tuning the selected eigenfrequencies $f_{i,a}$ of the finite element model to the frequencies $f_{i,e}$ which are defined by the user. In structural optimisation however equations (2) and (3) are not used. There are no orthogonality constraints. Only equations (1) and (4) will be used.

The only difference with the updating procedure is that during the iterations of the optimisation the eigenfrequencies $f_{j,a}$ of the finite element model that are not selected to be tuned to a fixed frequency, are checked if they do not violate following user specified constraints:

$$f_{j,a} \notin [f_{ln}, f_{un}] \qquad\qquad (6)$$

f_{ln} and f_{un} are the lower and upper bound of regions where no resonance frequencies of the structure are allowed. If $f_{j,a}$ violates the constraints (6) then $f_{j,a}$ becomes part of the set of selected eigenfrequencies $f_{i,a}$ and gets as targetvalue $f_{j,e}$:

$$f_{j,e} = f_{ln} - 0.75 \text{ Hz} \qquad \text{or} \qquad f_{j,e} = f_{un} + 0.75 \text{ Hz}$$

The objective function for the non-linear programming approach becomes:

$$W_f \sum_{n=1}^{s} W_n \left(\sum_{k=1}^{r} \overset{f}{a}_{nk} x_k - 1 + f_{n,a}/f_{n,e} \right)^2 + W_p \sum_{k=1}^{r} x_k^2 \qquad (7)$$

with:

$x_k = w_k \, \Delta p_k / p_k$, weighted relative parameter change

$$\overset{f}{a}_{nk} = \frac{\partial f_{n,a}}{\partial p_k} \frac{p_k}{w_k \, f_{n,e}}$$

constrained by equations (4) and (6).

3. ELABORATE EXAMPLE

The subject of this example was an aluminium tennisracket. The goal of the optimisation calculations was to tune the first 2 resonance frequencies to 120 Hz and 190 Hz and to avoid resonance frequencies in the range between 340 Hz and 370 Hz.

First a simplified finite element model of the racket was developed and the dynamic characteristics were calculated. An experimental modal analysis test was performed. Next the finite element model was updated for the frequency range of interest using the UA-updating procedure. With the updated model the structural optimisation calculation was performed. Finally the results of this optimisation were verified on the real structure.

3.1. Finite Element Model

Figure 1 gives a simplified view of the geometry of the structure. A 219 dof beam model was developed. Only the connection between handgrip and frame was modeled with shell elements. The honeycomb structure that fills the beams of the frame and the handgrip (figure 1, cross section B-B) was not taken in account in the definition of the element properties (cross sectional area and 2nd moments of area). This error had to be corrected by the updating procedure.

3.2. Experimental results

An experimental modal analysis test was performed. The racket was free-free suspended and was excited with a hammer in point 12 (figure 2) in the z-direction. In each point the 3 translational dof's were measured. For the estimation of the modal parameters a multiple degree of freedom technique was used. The resonance frequencies and damping values were determined with a least squares time domain parameter estimation technique. The corresponding modeshapes were identified by means of a least squaresfrequency domain parameter estimation technique.

3.3. Modelupdating

Table 1 gives the analytical and experimental frequencies and the frequency differences; the correlation between the corresponding modeshapes is quantified by the modal assurance criterion (MAC-value). The 4th analytical modeshape, a compression of the racket, was not identified in the measurement test. Probably this modeshape was not excited by the input impulse in the z-direction (figure 2). The analytical frequencies are systematically too high. This indicates a global underestimation of mass or a global overestimation of stiffness. As already mentioned, the honeycomb structure (figure 1, cross section B-B) was neglected in the finite element model. It is expected that this neglect causes the mass and the stiffness of the racket to be underestimated by the finite element model. The MAC-values are acceptable.

The first step of the updating procedure is the selection of the updating parameters. The racket was divided in 5 subdomains (figure 3). These subdomains are parts of the racket that have homogeneous material and element properties. Some of this properties were selected as updating parameters. Some smaller updating cases that were performed, showed that the most effective set of updating parameters was the following set:

subdomain 1: ρ, Ax, Iy, Iz	ρ: mass density
subdomain 2: ρ, Ax, Ix, Iy, Iz	Ax: cross sectional area
subdomain 3: ρ, Ax, Ix, Iy, Iz	Ix: 2nd moment of area in torsion
subdomain 4: ρ	Iy: 2nd moment of area about the y-axis (bending)
subdomain 5: ρ	Iz: 2nd moment of area about the z-axis (bending)

Normally the whole structure should have a uniform mass density. In this case however, changes of the mass density were necessary to correct for the neglect of the mass of the honeycumb structure. As a consequence no uniform mass density was required. All updating parameters got the same weighting factor, because there was no reason to assume that one parameter was more accurate than an other one. 4 iterations were considered.

Because of noise on the measurement data it is useless to require perfect matching of the experimental and the analytical model. Frequency differences up to 3% are considered acceptable. Table 2 shows the updated frequencies, the frequency differences and the MAC-values after each iteration. The results after 2nd iteration are the best. The first 3 frequency differences are less then 3% and all MAC-values are acceptable. After the 2nd iteration the process begins to diverge.

Figure 4 shows the variations of the updating parameters after the 2nd iteration in percent of their original value. The values of ρ and Ax have increased, especially in subdomains 2 and 3. The values of Ix, Iy and Iz have decreased. Since the original frequencies were too high, it is normal that the stiffness parameters have decreased, but considering the presence of the honeycumb structure an increase of the stiffness parameters with respect to the original model was expected. Apparantly neglecting the honeycumb structure was not the only important error in the simplified, original finite element model; the estimates for bending stiffness Iy and Iz were too high.

The updated model was used as a base for the structural optimisation calculation.

3.4. Structural Optimisation

As already mentioned above the goals of the optimisation procedure were to tune the first 2 frequencies to 120 Hz and 190 Hz and to avoid resonance frequencies in the range between 340 Hz and 370 Hz.

To allow a verification of the result by measurements on the existing structure, one had to choose optimisation parameters that were easy to apply. Therefore only the mass density of a few parts of the structure (figure 5) were chosen as optimisation parameters, because it is easy to add mass to that elements on the existing structure.

10 iterations were asked. Table 3 shows the variations of the resonance frequencies during the updating process. After the 9th iteration the requirements are perfectly fulfilled. however it makes no use to ask for such an accurate result. The finite element model is not a perfect model of the real structure. Frequency shifts up to 5% can be expected.

Notice that after the 2nd iteration the 3th frequency is in the range between 340 Hz and 370 Hz. The UA-procedure automatically corrects this violation of the constraints.

Figure 6 shows how the optimisation parameters change during the optimisation procedure. After the conversion of these mass densities to masses of the elements following results are found:

    subdomain 8: add 37.1 g
    subdomain 9: add 0 g
    subdomain 10: add 0.9 g
    subdomain 11: add 2.5 g

3.5. Verification of the Results

This results were applied on the real structure. A new experimental modal analysis test was performed. This test yielded following results for the first 4 resonance frequencies: 126 Hz, 194 Hz, 373 Hz and 477 Hz.

The frequency differences between calculated and measured frequencies are less than 5%. Considering the simplified finite element model and the low number of optimisation parameters that were used, these results are very good.

4. CONCLUSION

The first part of the paper discusses the adoption of the UA-model updating strategy to structural optimisation calculations. The second part proves,on the basis of a practical example, the validity of the UA strategy for structural optimisation calculations.

[1] T. Janter, "Construction oriented updating of dynamic finite element models using experimental modal data.", PhD dissertation 89D1, K.U. Leuven, Feb. 1989.

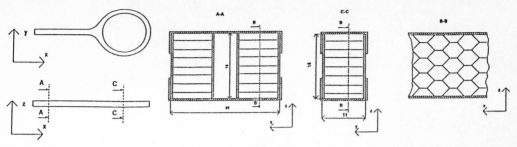

Figure 1: simplified view of the racket

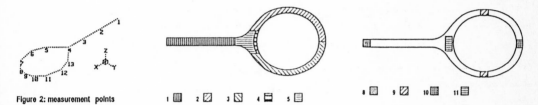

Figure 2: measurement points

Figure 3: subdomains for updating

Figure 5: subdomains for optimisation

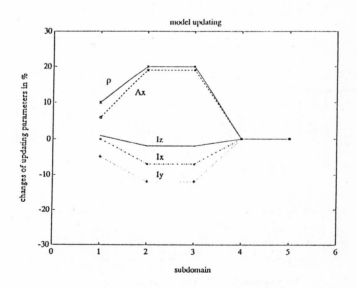

Figure 4: variations of the updating parameters

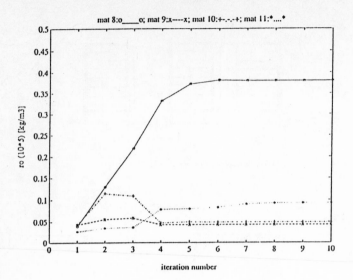

mat 8:o____o; mat 9:x---x; mat 10:+-.-.+; mat 11:*....*

Figure 6: variations of the optimisation parameters

exper. freq. (Hz)	F.E.M freq. (Hz)	freq. diff. (%)	MAC- value (%)	description of the modeshape
139.67	162.66	16.5	91	bending around the y-axis
217.90	249.83	14.7	72	bending around the z-axis
401.98	486.78	21.1	78	2nd bending around the y-axis
-	580.20	-	-	compression in the x-direction
492.73	642.02	30.3	86	torsion

Table 1: experimental frequencies, F.E. frequencies, differences between experimental and F.E. frequencies,MAC-values and description of the modeshapes.

Iteration 1			Iteration 2			Iteration 3			Iteration 4		
f (Hz)	Δf (%)	MAC (%)	f (Hz)	Δf (%)	MAC (%)	f (Hz)	Δf (%)	MAC (%)	f (Hz)	Δf (%)	MAC (%)
140.79	0.80	92	136.09	2.56	92	132.71	4.98	92	136.53	2.24	79
225.29	3.85	73	220.71	1.29	73	217.27	0.29	74	223.29	2.47	57
426.57	6.12	80	413.60	2.89	80	404.88	0.72	80	415.90	3.46	66
540.57	9.71	74	518.66	5.26	86	480.39	2.50	74	579.53	0.18	78

Table 2: eigenfrequencies (f), differences between the eigenfrequencies and the experimental resonance frequencies (f) and MAC-values after each iteration.

Original	1	2	3	4	5	6	7	8	9	10	Target
136.09	122.94	119.60	121.67	120.28	120.12	120.01	120.00	120.00	120.00	120.00	120.00
220.71	203.91	196.33	192.98	190.61	190.18	190.04	190.01	190.01	190.00	190.00	190.00
413.60	379.79	369.61	374.28	371.01	370.69	370.69	370.73	370.74	370.75	370.75	-
518.66	453.00	452.19	480.63	478.54	478.59	478.59	478.68	478.71	478.71	478.71	-

Table 3: eigenfrequencies after each iteration of the optimisation process.

OPTIMIZATION WITH STABILITY CONSTRAINTS – SPECIAL PROBLEMS

OPTIMAL I-SECTION OF AN ELASTIC ARCH
UNDER STABILITY CONSTRAINTS

Bogdan Bochenek and Michał Życzkowski

Institute of Mechanics and Machine Design,
Technical University of Cracow,
Warszawska 24, 31-155 Kraków, POLAND.

1. INTRODUCTORY REMARKS.

Although problems of optimal design of arches against buckling have been investigated for almost forty years and numerous solutions to the problem exist (c.f. Gajewski and Życzkowski, 1988) there are still subjects open for discussion. One of them, that is very important from a practical point of view, is optimization of thin-walled arches. It can be proved that in many cases even prismatic thin-walled (I- or box-section) arches are much more efficient than solid ones with optimal mass distribution. On the other hand, for thin-walled structures the stability analysis is more complicated and, in particular, the consideration of local stability is required. Probably the only papers dealing with such problems are due to Bochenek (1988) and Mikulski (1988). The first one considers the problem of parametrical and variational optimal design of box-section arches with respect to overall and local stability, whereas the second paper presents some solutions concerning optimization of arches with variable I-sections. Mikulski does not consider local stability constraints.

2. OUT-OF-PLANE BUCKLING OF AN I-SECTION ARCH.

We consider a plane elastic circular I-section arch which is loaded by a uniformly distributed radial pressure p^* of Eulerian behaviour aftter buckling (direction constant in space). The arch axis is assumed to be inextensible whereby the prebuckling state is momentless, so that only the axial force $N_{zo}^* = -p^* R^*$ differs from zero before buckling. With a view to describe the critical state of the arch, the extended version of the equations previously used for solid cross-section (e.g. Bochenek and Gajewski, 1989) is derived. The extension is due to introduction of the moment of restrained torsion

$$M_\omega^* = D_\omega^* (\gamma''' - \alpha'' / R^*) \tag{1}$$

and the bimoment

$$B^* = -D_\omega^* (\gamma'' - \alpha' / R^*) \tag{2}$$

along with the definition of the variable twist as

$$\tau^* = \gamma' - \alpha / R^* . \tag{3}$$

Here α and γ are the angles of rotation, R^* is the radius of the arch, and D_ω^* is the restrained torsion rigidity. Moreover, the effect of the prebuckling compression implies the following modified formula for torsional rigidity (Vlasov, 1959) :

$$C^* = C_s^* - (B_x^* + B_y^*) p^* R^* / (E A^*) \tag{4}$$

In (4) C_s^* is the free torsion rigidity, B_x^* and B_y^* are the flexural rigidities, and A^* is the cross-sectional area. The cross-sectional area A^* for the given volume V^* and the length of the arch l^* equals

$$A^* = V^* / l^* . \tag{5}$$

However, for the reasons of numerical convenience we treat (5) as an inequality constraint

$$A^* \leq V^* / l^* = A_o^* \tag{6}$$

though it is expected that the optimal A^* is equal to A_o^* (constraint is active).

The out-of-plane buckling is the only form of overall loss of stability which needs to be considered for the I-section arch, because the cross-sectional torsional rigidity is small. After some rearrangements, the set of eight first order differential equations governing the critical state of the arch (assuming fixed load direction in the course of buckling) takes the form

$$v' = -\alpha \qquad\qquad \alpha' = M_x/B_x - \varepsilon\,\gamma$$

$$\gamma' = \tau + \varepsilon\,\alpha \qquad\qquad M_x' = K - \varepsilon^2\alpha\,p - \varepsilon\,M_z \qquad (7)$$

$$M_z' = \varepsilon\,M_x \qquad\qquad K' = 0$$

$$\tau' = -2\,\varepsilon^2 R^2 B/D_\omega \qquad\qquad B' = M_z - C\,\tau$$

where

$$C = C_s/\psi - (B_x + B_y)\,p/(12\,\Phi\,R^2). \qquad (8)$$

All quantities, namely the displacement v, the angles of rotation α and γ, the twist τ, and the increments of the internal forces, i.e., bending moment M_x, total twisting moment M_z, shear force K, and bimoment B are dimensionless and defined with reference to the coordinate axes, that are normal, binormal and tangent to the undeformed axis of the arch. The definitions of these variables

$$v = v^*/l^* \quad,\quad \tau = \tau^*/l^* \quad,\quad K = K^*\,l^{*2}/B_o^* \quad,\quad M_x = M_x^*\,l^*/B_o^* \quad, \qquad (9)$$

$$M_z = M_z^*\,l^*/B_o^* \quad,\quad B = B^*/B_o^*$$

dimensionless global critical loading

$$p_g = p^*\,R^{*3}/B_o^* \qquad (10)$$

and the dimensionless geometrical characteristics

$$B_x = B_x^*/B_o^* \;,\; B_y = B_y^*/B_o^* \;,\; C = C^*/B_o^* \;,\; C_s = C_s^*/B_o^* \;, \qquad (11)$$

$$D = D^*l^{*2}/B_o^* \;,\; \Phi = A^*/A_o^* \;,\; \psi = B_o^*/C_o^* \;,\; R = R^*/\sqrt{A_o^*} \;,\; \varepsilon = l^*/R^*$$

are now introduced with dimensional quantities marked with asterisks. The quantity B_o^* of the dimension of bending rigidity is defined by $B_o^* = E\,A_o^{*2}/12$.

We consider an arch with clamped ends, and the boundary conditions for the state equations (7) are taken to be

$$\alpha(0) = \gamma(0) = v(0) = \tau(0) = \alpha(1/2) = M_z(1/2) = K(1/2) = \tau(1/2) = 0 \qquad (12)$$

These conditions are set up for s=0 and s=1/2 owing to the symmetry of the structure in the prebuckling state and with a view to identify the symmetric form of out-of-plane buckling, which is known to be the critical global buckling mode for an arch of the type considered. The independent variable s is defined as s^*/l^*, and is measured along arch axis.

After introducing the dimensionless cross-sectional dimensions

$$b = b^*/\sqrt{A_o^*} \; , \; t_1 = t_1^*/\sqrt{A_o^*} \; , \; h = h^*/\sqrt{A_o^*} \; , \; t_2 = t_2^*/\sqrt{A_o^*} \qquad (13)$$

where b, t_1, h, t_2 are flange width, flange thickness, web depth, web thickness respectively, the geometrical characteristics for the considered I-section may be presented in the form

$$B_x = 2t_1 b^3 \; , \; B_y = 6bh^2 t_1 + h^3 t_2 \; , \; C = 2bt_1^3 + ht_2^3 \; , \; D_\omega = t_1 b^3 h^2 \qquad (14)$$

3. LOCAL INSTABILITY.

For an arch with I-section, loss of stability of the web or flange may occur in addition to overall buckling. Approximate values of the critical loads associated with these local instabilities may be calculated using the following simple models. With a view to describe loss of stability of the web, the "column model" (Fig.1) is introduced. The critical load value for the case of load application in centre of the cross-section is given by (Timoshenko,1936)

$$p_v^* = \overline{\chi}_2 E \, t_2^{*3}/(12h^{*2}) \qquad\qquad \overline{\chi}_2 = 18.7 \qquad (15)$$

or for dimensionless quantities

$$P_v = \chi_2 t_2^3/h^2 \qquad\qquad \chi_2 = \overline{\chi}_2 R^3 \; . \qquad (16)$$

With a view to obtain more apropriate formula for web buckling load the "annular plate model" could be used but at the cost of much more complex stability analysis.

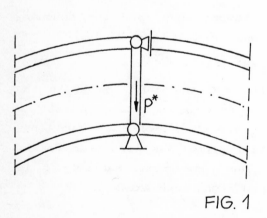
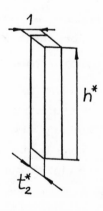

FIG. 1

For the flange instability we adopt the "shell model" indicated in Fig.2, and the critical load may then be calculated using an expression proposed by Hayashi (1971):

$$P_f^* R^* / A^* = \bar{\chi}_1 E \, t_1^{*2} / (2b^*)^2 \qquad \bar{\chi}_1 = 4.2 \qquad (17)$$

or $\qquad P_f = \chi_1 (2bt_1 + ht_2) \, t_1^2 / b^2 \qquad \chi_1 = 3\bar{\chi}_1 R^2$. $\qquad (18)$

Relative to the original expression, we have replaced half of the flange (supported-free shell) by the four times longer supported-supported shell. This assumption is based on the fact that there is a similar deflection distribution for a cantilever and a clamped-clamped beam (loaded by uniform load) if the clamped-clamped beam length is four times the length of the cantilever.

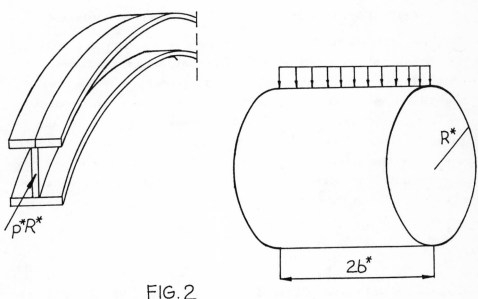

FIG. 2

4. THE OPTIMAL DESIGN PROBLEM.

The optimization problem is to determine the values of the arch cross-section dimensions so as to maximize the smallest critical load from among the three instability loads considered

$$\max \min (\ p_g \ , \ p_f \ , \ p_v \ ) \tag{19}$$

subject to

$$2bt_1 + ht_2 \leq 1 \tag{20}$$

The total volume of the arch and its length and radius, are considered to be given. Adopting the so-called bound formulation of a max-min problem, see e.g., Olhoff (1988), the problem stated above can be transformed into following

$$\min \lambda \tag{21}$$

subject to

$$1/p_g - \lambda \leq 0 \ , \qquad 1/p_f - \lambda \leq 0 \tag{22}$$
$$1/p_v - \lambda \leq 0 \ , \qquad 2bt_1 + ht_2 \leq 1$$

where a parameter λ, which plays the role of an additional design variable, is introduced.

5. ALTERNATIVE FORMULATION OF THE OPTIMIZATION PROBLEM.

In the case where the cross-section dimensions are taken to be design variables and where optimization is performed with respect to only one type of overall buckling mode, we can assume that the constraints (22) will be active throughout in the optimal solution. This will assure that there are no singularities (like, e.g., h tending to infinity and t_2 tending to zero) in the optimal solution. With this in mind, we have also considered the following modified (alternative) formulation of the current problem:

$$\max p_g \tag{23}$$

subject to

$$2bt_1 + ht_2 \leq 1 \ , \quad p_g - p_f = 0 \ , \quad p_g - p_v = 0 \tag{24}$$

In comparison with the original problem (19,20) or (21,22), the new one both contains inequality and equality constraints and therefore requires a slightly different numerical treatment.

6. THE NUMERICAL METHOD.

In order to solve the mathematical programming problems formulated above, a numerical method based on the so-called Method of Moving Asymptotes (MMA) is applied. This method applies to inequality constrained mathematical programming problems and is clearly described in the paper by Svanberg (1987), so details will be omitted for reasons of brevity. Here we only need to explain how the MMA method has been adopted for the problem with equality constraints. The possible method of dealing with equality constraints is to treat them as a set of, in general, nonlinear equations. Solving this set we can reduce the total number of design variables and then perform minimization with respect to remaining variables and subject to only inequality constraints.

7. RESULTS.

Numerical results have been obtained for both optimal design formulations, and as expected for the problem (21,22) all constraints become active at the optimum, so that solutions for both problem formulations become identical. Figure 3 shows the optimal solution for arch steepness parameters $\varepsilon = \pi$. The distance between the ends of the arch was chosen to be 1m and the total volume $10^{-5} m^3$. The transverse dimensions are in the figure multiplied by five.

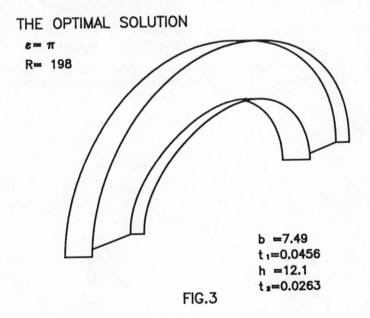

THE OPTIMAL SOLUTION

$\varepsilon = \pi$

$R = 198$

b $= 7.49$
$t_1 = 0.0456$
h $= 12.1$
$t_2 = 0.0263$

FIG.3

To check the validity of our assumption, we computed a posteriori the critical loads associated with in-plane buckling of the optimized arches, and found that they are more than ten times higher than the considered out-of-plane ones. This confirms that out-of-plane buckling is the only type of global instability that needs to be considered for analysed I-section arches. Thereby necessity of multimodality which has been proved for solid rectangular cross-section (Bochenek and Gajewski, 1989) is here unlikely irrespective of the arch steepness parameter.

References

Bochenek,B., (1988): On optimal thin-walled arches against buckling, GAMM Wissenschaftliche Jahrestagung 1988, Wien.
Bochenek,B. ; Gajewski,A., (1989): Multimodal optimization of arches under stability constraints with two independent design functions, Int.J.Solids Struct., 1, 67-74.
Gajewski,A.; Zyczkowski,M.,(1988): Optimal structural design under stability constraints, Dordrecht/ Boston/ London , Kluwer Academic Publishers.
Hayashi, T. et al.,(1971): Handbook of structural stability, Sec.4 (Shells), Tokyo,Column Research Committee of Japan, Corona Publishing.
Mikulski,L. (1988): Statische Stabilitat der elastischen Bogen und deren optimale Gestaltung, GAMM Wissenschaftliche Jahrestagung 1988, Wien.
Olhoff,N.,(1989): Multicriterion structural optimization via bound formulation and mathematical programming, Structural Optimization, 1, 11-17.
Svanberg,K.,(1987): Method of Moving Asymptotes - a new method for structural optimization, Int.J.Num.Meth.Eng., 24, 359-373.
Timoshenko,S.P.,(1936): Theory of elastic stability, , New York, Mc Graw Hill Book Co.
Vlasov,V.Z.,(1959): Thin-walled elastic beams, Moskva, Fizmatgiz, (in Russian).

APPLICATION OF OPTIMIZATION PROCEDURES TO SUBSTRUCTURE SYNTHESIS

H.H. MÜLLER-SLANY

Research Laboratory for Applied Structural Optimization
Institute of Mechanics and Control Engineering
University of Siegen, D-5900 Siegen, Germany

Abstract: Substructure Techniques are applied to the modelling of elasto-mechanical systems. The aim is the generation of simple mathematical models for the dynamic analysis of the complex system. This paper shows how complex systems can be modelled by means of simple lumped-mass-substructures. For the single substructures simple yet highly accurate adapted models are to be generated in an optimization process. In order to solve this vector optimization problem, a hierarchical optimization procedure is developed. The model adaptation and substructure synthesis of a tree structure is presented as an example.

1. SYSTEM MODELLING BY MEANS OF SUBSTRUCTURE SYNTHESIS

The modelling of complex dynamic systems is carried out using methods of the substructure analysis and synthesis. This procedure is characterized by a separation of the complex system into substructures in order to generate substantially reduced mathematical substructure models from which the global equations of motion of the complex system can be reconstructed. The papers of GREIF/WU [1] and GREIF [2] provide a comprehensive review of the mathematical modelling concept of the "Component Mode Synthesis" in its various forms. The procedures differ in the consideration of coupling points between the single substructures at the model synthesis to the mathematical complex model.

Here, a concept is to be introduced which uses very simple but dynamically precise lumped-mass-models for the substructures on the basis of the physical modelling. After joining them at the coupling points, these physical substructure models form a simple, substantially condensed physical complex model to which the conventional analysis methods of system dynamics can be applied.

An important advantage of this generation procedure of a simple physical substitute system is the realistic modelling. In the engineering design process the following advantages are achieved for specific problems:

- Simple, easily presentable physical interpretation of the system dynamics (e.g. computer animation);
- Clear, easily realizable and flexible consideration of system variations:
    - Changing of parameters in partial structures,
    - change of the geometry of the system,
    - variation of links,
    - manageable system description and numerical simulation in mini-computer systems (e.g. task of the controller synthesis).

In order to reach this goal, the development of a physical modelling strategy using mathematical programming procedures was necessary. Hereby, the substantially condensed substructure models have to adopt those physical properties of the real subsystems which gain important physical influence in the substructure synthesis.

For that reason, this concept enables the system adaptation to experimentally achieved or desired dynamic properties. The expression *adapted model* is introduced for the modelling for a substructure. In order to realize the model adaptation, structural optimization procedures are used. The substructure synthesis is presented successfully for tree structures.

2. A SUBSTRUCTURE ELEMENT AS AN ADAPTED MODEL

The adapted model is a highly condensed substitute system for complex elasto-mechanical systems which, at the same time, has to fulfill special demands on accuracy [3]. It is realized as a simple physical substitute system which is build of beam elements and concentrated point masses. An adaptation process transfers chosen properties of the real system into the adapted model, the latter being a highly flexible tool for the detailed analysis in the succeeding numerical simulation.

The choice of the properties of the real system which are to be adapted depends on the simulation objective which the adapted model is generated for. Table 1 shows a compilation of the dynamic properties of an adapted model as a lumped-mass-system. The *mass-geometric properties* are of fundamental importance: Total mass, position of the centre of mass, elements of the inertia tensor. The correct reproduction of these parameters in the adapted model determines the correct behaviour of the rigid-body motion and the realistic values of the stress resultants or of the output in control systems.

criterion	Lumped-Mass-System with p points of mass	influence on dynamics
1. Mass-geometric properties		
total mass	$M = \sum_{i=1}^{p} m_i$	rigid-body motion, global forces and moments in the system
centre of mass	$I_{os} = [I_x, I_y, I_z]^T$	
inertia tensor	$\underline{\theta}_s = \begin{bmatrix} \theta_{xx} & \theta_{xy} & \theta_{xz} \\ \theta_{xy} & \theta_{yy} & \theta_{yz} \\ \theta_{xz} & \theta_{yz} & \theta_{zz} \end{bmatrix}$	
2. Elasto-static properties		
deformation behaviour	$I = K^{-1} F$ $I_j = \underline{K}^{-1} \mathrm{diag}[m_1,\dots,m_p]\omega_j^2 g_j$	conditions for correct eigenmodes
3. Modal properties		
Eigen-frequencies	ω_j , $j=1,\dots,k$	correct eigenmodes
Eigenmode	g_j , $j=1,\dots,k$	
linear momentum	$P_j = \sum_{i=1}^{p} m_i \omega_j g_j$, $j=1,\dots,k$	correct stress resultants in synthesis problems
angular momentum	$L_{oj} = \sum_{i=1}^{p} I_{oi} \cdot P_j$, $j=1,\dots,k$	
kinetic energy	$T_j = \omega_j^2 g_j^T \mathrm{diag}[m_1,\dots,m_p]g_j$ $j=1,\dots,k$	vibration energy of the system

Table 1:   Dynamic properties of the adapted model as a lumped-mass-system

The *modal* properties are the second important area of adaptation for the *adapted model*. It can be observed that a substructure synthesis from adapted models which fulfill the adaptation conditions of the eigenfrequencies and eigenmodes only does not achieve a considerably precise model of the complex system. The conformity of the vectors of *linear* and *angular momentum* of the substructure model and of the original substructure in the state of the investigated eigenvibrations is a demand which gains a highly important influence on the result.

The task of the model adaptation, i.e. the reconstruction of the mass and stiffness matrices with the given spectral data of the system, has been dealt with in the literature in various ways: by means of matrix-oriented procedures of *inverse eigenvalue problems* [4], [5], or by means of *modal perturbation methods* in the range of the redesign-problem [6]. Both procedures are, for the time being, unable to adapt the model to the linear and angular momentum properties of the original. In our case, a solution procedure basing upon the minimization of an error functional under application of the mathematical programming has been chosen.

3. GENERATING THE ADAPTED MODEL AS AN OPTIMIZATION PROBLEM: THE HIERARCHICAL OPTIMIZATION PROCEDURE

The adapted model for a substructure is designed as a lumped-mass-model the parameters of which are combined in the design vector $\mathbf{x} \in \mathbb{R}^n$: Geometric values, coordinates, cross-sectional areas, area moments of inertia and masses. The design variables x_i, $i = 1, ..., n$, have to be chosen in such a way that the lumped-mass-system adopts the same dynamic properties as the real system (Table 1). In an adaptation process, this is achieved by minimizing the error expressions of the scalar and vectorial dynamic properties. The basics of the adapted model are explained in Table 2.

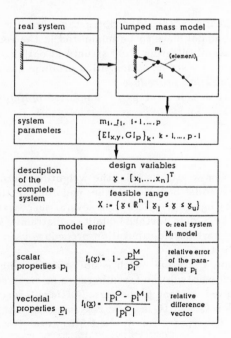

Table 2:   The adapted model as a lumped-mass-system

The model adaptation can be formulated as a vector optimization problem as in the case of structural optimization problems [7], [8]. The adaptation problem can largely be solved by the program structure of the 3-columns-concept of the structural optimization introduced by ESCHENAUER. Special aspects of the problem then occur in the problem-specific area of the so-called optimization model. Table 3 shows the vector optimization problem for the model adaptation.

$$
\min_{\underline{x} \in X} \{ \underline{f}(\underline{x}) \}
$$

$$
X := \{ \underline{x} \in \mathbb{R}^n \mid \underline{x}_L \leq \underline{x} \leq \underline{x}_U \}
$$

$$
\underline{f}(\underline{x}) = \begin{bmatrix} \left(1 - \frac{m^M}{m^O}\right)^2 \\[4pt] \left(\frac{|\underline{r}_{\ddot{o}s}^O - \underline{r}_{\ddot{o}s}^M|}{|\underline{r}_{\ddot{o}s}^O|}\right)^2 \\[4pt] \sum_{x,y,z} \left(1 - \frac{\theta_{ij}^M}{\theta_{ij}^O}\right)^2 \\[4pt] \sum_{i=1}^{k} \left(1 - \frac{\omega_i^M}{\omega_i^O}\right)^2 \\[4pt] \sum_{i=1}^{k} \left(\frac{|\underline{q}_i^O - \underline{q}_i^M|}{|\underline{q}_i^O|}\right)^2 \\[4pt] \sum_{i=1}^{k} \left(\frac{|\underline{p}_i^O - \underline{p}_i^M|}{|\underline{p}_i^O|}\right)^2 \\[4pt] \sum_{i=1}^{k} \left(\frac{|\underline{L}_i^O - \underline{L}_i^M|}{|\underline{L}_i^O|}\right)^2 \end{bmatrix}
$$

M: model
O: real system

$\|\underline{q}_i^O\| \cdot \|\underline{q}_i^M\| \cdot 1$

Table 3   Optimization problem for the adaptation process

A large number of single objective functions, each describing the error for a system property between the original and the model, have to be minimized in the vector optimization problem of the model adaptation. Differing from structural optimization problems, the formulation of this problem as a scalar optimization problem has to follow a new strategy: the *hierarchical optimization procedure*.

The hierarchical optimization procedure uses a strategy of simplified preference functions $p[\mathbf{f}(\mathbf{x})]$:

$$
p[\mathbf{f}(\mathbf{x})] = \sum_{i=1}^{m} w_i\, f_i(\mathbf{x}) \tag{1}
$$

by setting the weighting factors $w = 1$ or $w = 0$. By this, the weighting factors act as control coefficients. Until the adapted model is achieved, the modelling in the hierarchical optimization is structured by varying the numbers w in several successsive steps. Each step is especially characterized by adding further equality constraints which fix the already adapted properties. The hierarchical optimization procedure used for substructure modelling is shown in Table 4.

This procedure requires an optimization algorithm which is able to solve highly nonlinear optimization problems and which works in the feasible range in every main iteration in order to secure already adapted system properties. Sufficient results were achieved with the optimization algorithm **QPRLT** (Quadratic Programming with Reduced Line Search Technique). The Research Laboratory for Applied Structural Optimization at the University of Siegen developed this method as a hybrid algorithm from a generalized reduced gradient and a sequential quadratic procedure [9].

271

1st optimization step	
adaptation: mass-geometry	$w_{2,3} = 1$
constraints: mass, bounds	$w_{1,4,5,6,7} = 0$

Initial design

2nd optimization step	
adaptation: eigenfrequencies	$w_4 = 1$
constraints: mass-geometry, bounds	$w_{1,2,3,5,6,7} = 0$

3rd optimization step	
adaptation: linear and angel. moment.	$w_{6,7} = 1$
constraints: mass-geom., eigenfr., bounds	$w_{1,2,3,4,5} = 0$

final design · *adapted substructure model*

Table 4:   Hierarchical optimization procedure for substructure models

4. EXAMPLE: SUBSTRUCTURE SYNTHESIS FOR A TREE STRUCTURE

The tree structure in Fig. 1 consists of equal structural elements (Fig. 2) with a rectangular hollow-section. For the substructure elements acc. to Fig. 2 an adapted lumped-mass-model with 8 point-masses is to be determined which is suitable for the substructure synthesis according to Fig. 3.

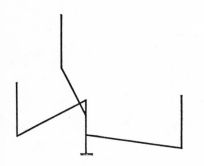

Fig. 1:   Tree structure

Fig. 3:   Synthesis structure

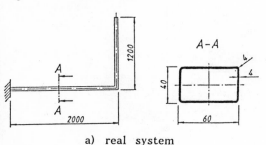

a) real system

Fig. 2: Substructure element

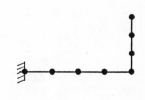

b) lumped-mass-model

The substructure model has 28 design variables:

$$\mathbf{x} = [x_1, \ldots, x_{28}]^T \tag{2}$$

$x_1 \cdots x_5$: coordinates,
$x_6 \cdots x_{13}$: mass points,
$x_{14} \cdots x_{28}$: area moments of inertia.

The adapted substructure model has to correspond with the original up to the 8th eigenfrequency $f_8 = 200,7$ Hz. For a successful substructure synthesis, the errors of the mass-geometric values, of the first 8 eigenvalues, and of the vectors of linear and angular momentum in the state of the first 8 eigenvibrations have to be minimized in the adaptation process. In order to solve the vector optimization problem, the hierarchical optimization procedure according to Table 5 is introduced.

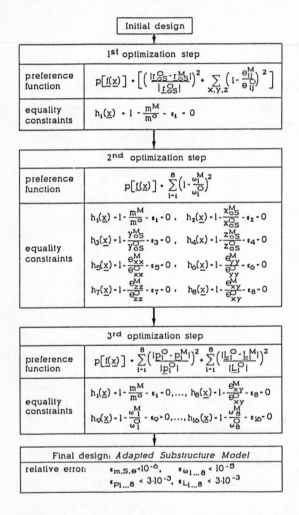

Table 5: Hierarchical optimization procedure for the adaptation of the substructure model

The substructure synthesis with the *final design* of the adapted substructure model provides a complex system with minimal dynamic errors compared to the original system. Frequency errors and mode errors are compiled in Table 6. Thereby, three modelling concepts are compared:

1) The conventional lumped-mass-systemwith equally distributed point-masses and uniform bending stiffness of the beams.
2) The synthesis structure made of substructure models which have been adapted up to the 8th eigenvibration (f_8 = 200.7 Hz) of the substructure.
3) Equal to model 2) with an adaptation up to the 6th eigenvibration, f_6 = 81.8 Hz.

Hereby, the mode error is the value of the difference vector

$$\Delta y_i = |y_i^O - y_i^M|, \quad \|y_i^O\| = \|y_i^M\| = 1 \tag{3}$$

with O: original, M: model, and i: number of the mode. In order to determine the vectors, 9 points of the complex structure marked in Fig. 4 are taken into consideration.

mode	f_i [Hz]	frequency error [%]			mode error [%]		
i		LM	AM 8	AM 6	LM	AM 8	AM 6
1	2,65	0,51	0,01	0,01	0,07	0,02	0,04
2	3,25	0,75	0,01	0,01	0,28	0,03	0,07
3	3,63	0,85	0,01	0,02	0,39	0,04	0,06
4	4,92	1,14	0,02	0,02	0,79	0,08	0,12
5	5,34	0,80	0,01	0,01	1,37	0,07	0,15
6	5,83	0,79	0,01	0,01	1,04	0,07	0,15
7	13,89	0,10	0,05	0,05	1,81	0,19	0,44
8	15,39	5,49	0,10	0,06	7,99	0,54	2,04
9	15,83	5,18	0,08	0,06	15,18	0,72	2,33
10	16,92	1,74	0,04	0,00	23,02	0,88	1,77
11	18,45	2,38	0,04	0,03	12,18	0,43	2,31
12	18,65	2,36	0,04	0,04	8,43	0,53	1,49
13	20,04	2,46	0,02	0,01	6,36	0,62	2,36
14	22,54	1,20	0,04	0,09	10,35	0,68	1,68
15	54,00	2,84	0,03	0,24	7,94	1,22	1,20
16	59,44	2,98	0,08	0,16	17,78	1,88	5,84
17	62,45	2,57	0,05	0,07	10,47	2,85	6,31
18	64,58	0,01	0,05	0,05	20,37	2,08	6,48
19	68,68	3,52	0,02	0,05	20,15	2,29	7,47
20	74,55	4,00	0,08	0,09	14,85	2,18	4,89
21	98,44	3,16	0,09	0,89	8,89	3,43	4,03
22	128,78	4,21	0,50	3,33	19,24	6,74	13,74
23	149,00	8,43	0,08	7,28	23,16	15,34	10,62
24	150,25	7,71	0,02	6,80	69,73	16,74	21,29
25	156,96	6,31	0,13	5,92	70,88	15,96	14,51
26	177,02	4,10	0,37	5,61	40,75	15,92	15,64
27	185,10	5,04	0,09	6,33	78,44	19,35	17,07
28	192,88	7,46	0,38	7,08	94,14	17,42	15,45
29	197,43	7,81	0,10	7,02	58,32	9,48	14,04
30	198,69	7,79	0,14	7,27	39,94	10,75	19,23

Table 6:   Frequency and mode errors of the synthesis structure
LM: Conventional lumped-mass-system,
AM 8:   Adaptation of the substructure model up to the 8th mode, f_8 = 200.7 Hz
AM 6:   Adaptation of the substructure model up to the 6th mode, f_6 = 81.8 Hz

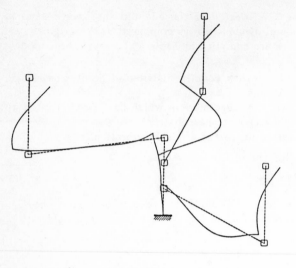

□ 9 selected points
for mode error
calculation
(see (3) and table 6)

Fig. 4: 20th eigenform of the complex structure

5. CONCLUSION

The application of optimization procedures generates efficient and highly accurate *adapted models*. The connected vector optimization problem is solved by means of a hierarchical optimization strategy. If the vectors of linear and angular momentum in the state of eigenvibrations are considered, these adapted models are very suitable for the substructure synthesis. Speaking in terms of physics, this adaptation means a correct reproduction of the inner forces at the coupling points of the synthesis structure in the state of eigenvibrations. In the case of tree structures, the frequency error Δf is, up to the adaptation frequency value, less than 1%.

REFERENCES

[1] GREIF, R.; WU, L.: Substructure Analysis of Vibrating Systems. The Shock and Vibration Digest. 1983(15), No. 1, p 17-24.

[2] GREIF, R.: Substructuring and Component Mode Synthesis. The Shock and Vibration Digest. 1986(18), No. 7, p 3-8.

[3] MÜLLER-SLANY, H.H.: Generierung angepaßter dynamischer Ersatzsysteme mit Optimierungsprozeduren. ZAMM 70(1990).

[4] BIEGLER-KÖNIG, F.: Inverse Eigenwertprobleme. Dissertation. Universität Bielefeld, 1980.

[5] GLADWELL, G.M.L.: Inverse Problems in Vibration. Martinus Nijhoff Publishers, Dordrecht, 1986.

[6] SANDSTRÖM, R.E./ANDERSON, W.J.: Modal Perturbation Methods for Marine Structures. SNAME Transactions, Vol. 90, 1982, p 41-54.

[7] ESCHENAUER, H.; POST, P.U.; BREMICKER, M.: Einsatz der Optimierungsprozedur SAPOP zur Auslegung von Bauteilkomponenten. BAUINGENIEUR 63 (1988) 515-526.

[8] ESCHENAUER/KOSKI/OSYCZKA: Multicriteria Design Optimization -Procedures and Application.. Springer Verlag, Berlin/Heidelberg/New York, 1990.

[9] BREMICKER, M.: Dekompositionsstrategie in Anwendung auf Probleme der Gestaltoptimierung. Dissertation, Universität Siegen. Fortschritt-Berichte VDI, Reihe 1, Nr. 173. VDI-Verlag, Düsseldorf, 1989.

MULTIMODAL OPTIMIZATION OF UNIFORMLY COMPRESSED CYLINDRICAL SHELLS

Antoni Gajewski

Institute of Physics,
Technical University of Cracow,
ul.Podchorążych 1, 30-084 Kraków, Poland

1. Introductory remarks

The survey by Krużelecki & Życzkowski (1985) contains nearly all papers devoted to the optimal structural design of shells up to 1984 (606 references). Recent topics can be found in the book by Rozvany (1989). In particular, the eighth chapter of the monograph by Gajewski & Życzkowski (1988) concernes the optimal structural design of shells under stability constraints. Only a few papers deal with variational optimization problems, among them one should mention papers by Andreev, Mossakovsky & Obodan (1972), Solodovnikov (1974), Medvedev (1980,1981), Ryabtsev (1983a,b) and Levy & Spillers (1989) in which unimodal formulations were considered.

A possibility of the multimodal optimization of a cylindrical shell was observed by Medvedev & Totsky (1984).

The present paper deals with a similar problem. However, a more accurate linear boundary value problem is used for the calculation of the consecutive eigenvalues up to 20 ones. Sensitivity analysis in conjunction with Pontryagin's maximum principle and an appropriate iterative numerical procedure allow to improve the shell wall thickness in successive iterations.

2. Stability of the cylindrical shell

Assuming the momentless precritical state of the uniformly radially compressed cylindrical shell its bifurcational stability can be determined by the linear boundary value problem in the form of the canonical set of ordinary differential equations:

$$Y'_i = A_{ij} Y_j, \qquad \mathbf{Y} = (u, v, w, \varphi, M, Q, Z, T)^T, \quad i=1..8, \quad j=1..8, \qquad (2.1)$$

where:

$$A_{12} = \nu n \varkappa, \qquad A_{13} = \frac{\nu \varkappa X_2}{X_1}, \qquad A_{15} = \frac{-\eta^2 \varkappa^3}{\phi X_1}, \qquad A_{18} = \frac{-\eta^2 \varkappa^2}{\phi X_1},$$

$$A_{21} = \frac{-n\varkappa}{X_3}, \qquad A_{24} = -\frac{n\eta^2 \phi^2}{4X_3}, \qquad A_{27} = \frac{2\eta^2 \varkappa^2}{(1-\nu)\phi X_3}, \qquad A_{34} = 1,$$

$$A_{43} = \frac{\nu \varkappa^2 (n^2 -1)}{X_1}, \qquad A_{45} = \frac{12\varkappa^4}{\phi^3 X_1}, \qquad A_{48} = \frac{\eta^2 \varkappa^3}{\phi X_1},$$

$$A_{51} = \frac{-(1-\nu)n^2 \phi^3 X_4}{6\varkappa X_3} + \frac{P}{\varkappa}, \qquad A_{54} = \frac{(1-\nu)n^2 \phi^3 X_4}{6\varkappa^2 X_3}, \qquad (2.2)$$

$$A_{56} = 1, \qquad A_{57} = \frac{n\eta^2 \phi^2}{4X_3}, \qquad A_{62} = \frac{-(1-\nu^2)n\phi}{\eta^2} + nP,$$

$$A_{63} = \frac{-(1-\nu^2)\phi}{\eta^2 X_1} \left[X_6 + \frac{n^2(n^2 -2)}{12}\eta^2 \phi^2 X_5 \right] + n^2 P,$$

$$A_{65} = \frac{\nu \varkappa^2 (n^2 -1)}{X_1}, \qquad A_{68} = -\frac{\nu \varkappa X_2}{X_1}, \qquad A_{72} = \frac{(1-\nu^2)n^2 \phi}{\eta^2} - n^2 P,$$

$$A_{73} = \frac{(1-\nu^2)n\phi}{\eta^2} - nP, \qquad A_{78} = \nu n \varkappa, \qquad A_{81} = \frac{-(1-\nu)n^2 \phi^3 X_4}{6X_3} + n^2 P,$$

$$A_{84} = \frac{(1-\nu)n^2 \phi^3 X_4}{6\varkappa X_3} - \frac{P}{\varkappa}, \qquad A_{87} = \frac{-n\varkappa}{X_3}.$$

For simplicity the following additional quantities have been introduced:

$$X_1 = 1 - \frac{\eta^2 \phi^2}{12}, \qquad X_2 = 1 - \frac{n^2 \eta^2 \phi^2}{12}, \qquad X_3 = 1 + \frac{\eta^2 \phi^2}{4},$$

$$\tag{2.3}$$

$$X_4 = 1 + \frac{\eta^2 \phi^2}{16}, \qquad X_5 = 1 - \frac{\eta^2 \phi^2}{12(1-\nu^2)}, \qquad X_6 = 1 - \frac{\eta^4 \phi^4}{144(1-\nu^2)}.$$

The state variables u,v and w are dimentionless displacements (in relation to the length l), φ is the slope, M,Q,Z and T are auxiliary variables related to generalized stresses, P is the dimensionless radial pressure.

The following parameters have been introduced as well:

$$\eta = \frac{h_o}{R}, \qquad \varkappa = \frac{l}{R}, \qquad P = \frac{(1-\nu^2)q}{\eta^3 E} \tag{2.4}$$

The buckling mode was assumed to have n waves in the circumferential direction. Therefore, originally partial differential equations were reduced to ordinary ones. The dimentionless thickness ϕ (the control function) is assumed to be a function of the dimentionless axial coordinate $x=\xi/l$, namely $\phi(x)=h(x)/h_o$, where $h_o=V_{min}/2\pi Rl$, so as to satisfy the constant volume condition:

$$\int_0^1 \tilde{\phi}(x) \, dx = 1, \tag{2.5}$$

for any admissible control function $\phi=\tilde{\phi}(x)$.

In order to distinguish the symmetric and antisymmetric buckling modes in the longitudinal direction x, the state equations (2.1) should be complemented by suitable boundary conditions. In the case of simply supported edges they are as follows:

$$v(0) = w(0) = M(0) = T(0) = u(1/2) = \varphi(1/2) = Q(1/2) = Z(1/2) = 0 \tag{2.6}$$

for symmetric buckling (even number m of half-waves in x-direction),

$$v(0) = w(0) = M(0) = T(0) = v(1/2) = w(1/2) = M(1/2) = T(1/2) = 0 \tag{2.7}$$

for antisymmetric buckling (odd number m of half-waves in x-direction).

The boundary value problem (2.1), (2.6) or (2.7) is self-adjoint and, for the shell of constant thickness, is quite equivalent to that obtained by Flügge (1967). For example, if the following parameters are assumed: $m=1, n=2, \varkappa=2, \nu=0.3$ and $\eta=0.05$, the state equations (2.1) and the boundary conditions (2.6) determine the critical loading $P=20.4968$ which is equal to the most accurate of all the results given by Flügge(1967).

However, the general state equations (2.1) may be simplyfied (mainly by reason of the inequality $\eta\phi \ll 1$) with the assumptions:

$$X_1 = X_2 = X_3 = X_4 = X_5 = X_6 = 1, \tag{2.8}$$

where the equation $X_2 = 1$ is a good approximation for a relatively small number of the circumferential waves n. The simplified boundary value problem leads to the critical value $P=20.4660$, so that the error is negligible.

3. Optimal structural design

The aim of this paper is to determine the optimal thickness of the shell $\phi = \phi^*(x)$ which maximizes the lowest critical load under constant volume condition (2.5) and some additional geometrical constraints. Sensitivity analysis in conjunction with Pontryagin's maximum principle is chosen as the solution method. Therefore, the Hamiltonian connected with self-adjoint boundary value problem (2.1), (2.6) or (2.7) and the constant volume condition (2.5) takes the form:

$$H = \frac{\varkappa^4}{\phi^3}\left[-B_1 - B_2\phi^2 + B_3\phi^3 + (B_4 - \Lambda)\phi^4 + B_5\phi^5 + B_6\phi^6\right] + \ldots, \tag{3.1}$$

where the terms without the control function ϕ have been ommitted, and where:

$$B_1 = 12M^2, \qquad B_2 = \left(\frac{\eta}{\varkappa}\right)^2\left[T^2 + \frac{2}{1-\nu}Z^2\right] + \frac{2}{\varkappa^3}MT,$$

$$B_3 = \frac{1}{\varkappa^4}\left\{2\nu\varkappa(nv+w)T - 2\nu\varkappa^2(n^2-1)wM + 2n\varkappa uZ + 2\varphi Q + \right.$$

$$\left. +P\left[\frac{2}{\varkappa}u\varphi - 2nvw - n^2(u^2+v^2+w^2)\right]\right\}, \tag{3.2}$$

$$B_4 = \frac{1-\nu^2}{\eta^2 \varkappa^4} (nv+w)^2, \qquad\qquad B_5 = \frac{n\eta^2}{2\varkappa^4} \varphi Z ,$$

$$B_6 = \frac{(1-\nu)n^2}{6\varkappa^6} \left[(\varphi-\varkappa u)^2 + \frac{1}{2}(1+\nu)(n^2-2)\varkappa^2 w^2 \right] .$$

The constant Lagrange multiplier Λ should be determined from the condition (2.5).

In the case of an N-modal optimization problem a new improved control function can be calculated from the formula:

$$\phi^{(1)} = \phi^{(0)} + \delta\phi, \qquad\qquad (3.3)$$

where $\phi^{(0)}$ denotes the normalized control function known from the previous step. The global increment $\delta\phi$ can be determined as a linear combination of the gradients g_i, connected with particular modes of buckling:

$$\delta\phi = \varepsilon(x) \left[\mu_1 g_1(x) + \mu_2 g_2(x) + \ldots + \mu_N g_N(x) - 1 \right], \qquad (3.4)$$

where:

$$g_i(x) = \frac{\varkappa^4}{\phi^4} (3B_1 + B_2\phi^2 + B_4\phi^4 + 2B_5\phi^5 + 3B_6\phi^6), \qquad (3.5)$$

and an arbitrary function $\varepsilon(x)$ (the gradient step) is assumed to be constant.

The small changes of particular eigenvalues are determined by the formulae:

$$\delta P_i = \int_0^1 g_i(x) \, \delta\phi(x) \, dx \qquad\qquad (3.6)$$

By equating one of the variations δP_i, in particular δP_1, to the remaining (N-1) values and making use of the normalization condition (2.5) one can determine the constant multipliers μ_i.

4. Numerical calculations, results and conclusions

Starting from a cylindrical shell of constant thickness (for $\varkappa=2$, $\nu=0.3$, $\eta=0.05$) the control function $\phi(x)$ was being improved according to the rule (3.3) for the unimodal approach, connected with the lowest eigenvalue $P(m,n) = P(1,4)$. For a certain value of the gradient step ε the consecutive ten iterations lead to the control functions presented in Fig.1. The evolution of the corresponding 19 values of the critical loads is demonstrated as well. It can be observed that in the 11-th iteration 7 eigenvalues are nearly equal to each other, i.e. $P(1,4)$, $P(1,6)$, $P(1,7)$, $P(1,8)$, $P(2,6)$, $P(2,7)$, $P(2,8)$.

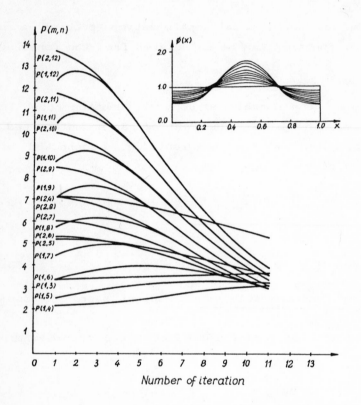

Fig.1. Evolution of eigenvalues in 11 iterations

From this point on 7-modal optimization would be necessary. However, the multimodal procedure is rather complicated for effective calculations because of a changeable number of the degree of modality. Therefore, further optimization process was continued as a unimodal one with respect to the lowest eigenvalue at each iteration step. Such a method of calculation is, of course, very laborious but it can be performed automatically. The consecutive solutions presented in Fig.2 suggest that the wall thickness of the shell should be bounded

by an upper limit at least. One of the possible solutions is
shown in Fig. 4. Of course, geometrical constraints can be used
in arbitrarily chosen intervals of the x-axis. In such a case
the optimization procedure leads to a ribbed shell.

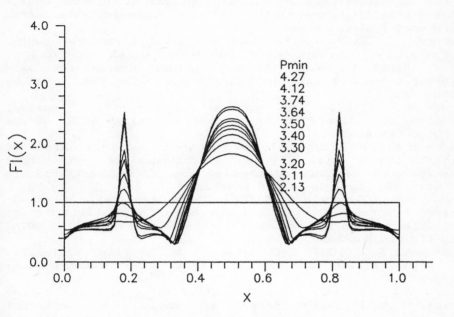

Fig.2. The wall thicknesses in consecutive iterations

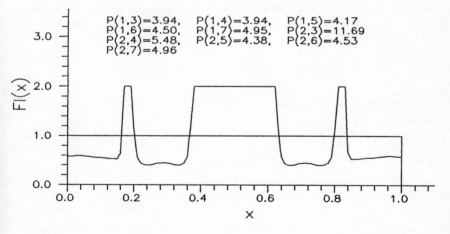

Fig 3. One of the possible solutions with geometrical
constraints

REFERENCES

Andreyev L.V., Mossakovsky V.I., Obodan N.I. (1972): On optimal thickness of a cylindrical shell loaded by external pressure, Prikl.Mat.Mekh., **36**,4, 719-725 (in Russian).

Flügge W. (1967), Stresses in shells, Springer, New York.

Gajewski A., Życzkowski M. (1988): Optimal structural design under stability contraints, Kluwer Academic Publishers, Dordrecht-Boston-London.

Krużelecki J., Życzkowski M.(1985): Optimal structural design of shells - a survey, Solid Mech.Archives, **10**, 101-170.

Levy R., Spillers W.R. (1989): Optimal design for axisymmetric cylindrical shell buckling, J.Eng.Mech., **115**,8, 1683-1690.

Medvedev N.G. (1980): Some spectral singularities of optimal problems of variable thickness shells, Dokl.AN SSSR, Ser.A, **9**,59-63 (in Russian).

Medvedev N.G.(1981): Optimal control technique in problems of stress state, stability and vibrations of orthotropic shells with variable thickness, Vopr.Optim.Proekt.Plastin i Obol., Saratov, 48-50 (in Russian).

Medvedev N.G., Totsky N.P. (1984): On multiplicity of eigenvalue spectrum in optimal problems of stability of variable thickness cylindrical shells, Prikl.Mekh. **20**,6, 113-116 (in Russian).

Rozvany G.I.N. (1989): Structural design via optimality criteria, Kluwer Academic Publishers, Dordrecht-Boston-London.

Ryabtsev V.A. (1983a): On optimization of a circular cylindrical shell with prescribed mass under external axisymmetric step-wise pressure, Prikl.Mekh. **19**,3,117-120 (in Russian).

Ryabtsev V.A. (1983b): Optimization of a cylindrical shell with prescribed mass under external pressure, Izv.AN SSSR, Mekh.Tverd.Tela. 6,124-129 (in Russian).

Solodovnikov V.N. (1974): Algorithm for calculation of variable thickness of a shell optimal with respect to stability, Dinam.Sploshn.Sredy **19/20**, Novosibirsk, 118-128 (in Russian).

The paper was partly supported by grant CPBP 02.01 of the Polish Academy of Sciences.

The 'Cut and Try' Method in the Design of the Bow

B. W. Kooi, Department of Biology, Free University
P.O. Box 7161, 1007 MC Amsterdam, The Netherlands
E-mail kooi@bio.vu.nl

1 Introduction

In history man used bows which differ much in shape as well as applied materials. *Simple* bows made out of one piece of wood, straight and tapering towards the ends have been used by primitives in Africa, South America and Melanesia. In the famous English longbow the different properties of the sapwood and heartwood were deliberately put to use. Eskimoes used wood together with cords plaited of animal sinews and lashed to the wooden core at various points. The Angular bow found in Egypt and Assyria are examples of *composite* bows. In these bows more than one material was used. In Asia the bow consisted of wood, sinew and horn. These bows reached their highest development in India, in Persia and in Turkey. In the 1960's composite bows of maple and glass fibres, or later carbon fibres, imbedded in strong synthetic resin were designed. Today almost all bows seen at target archery events are of this type of bow.

Bowyers (manufacturers of archery equipments) relied for the design of the bow heavy upon experience. The performance of the bow was improved by 'try and cut' method. In the 1930's bows and arrows became the object of study by scientists and engineers, Hickman, Klopsteg and Nagler, see [1] and [2]. There work influenced strongly the design and construction of the bow and arrow. Experiments were performed to determine the influence of different parameters. They also made mathematical models. As part of modelling simplifying assumptions were made. Hence only bows with specific features could be described.

In [3] and [4] we dealt with the mechanics of the different types of bow: *non-recurve*, *static-recurve* and *working-recurve* bows. The developed mathematical models are much more advanced, so that more detailed information was obtained giving a better understanding of the action of rather general types of bow.

In Section 2 of this paper the problem is formulated. All design parameters are charted accurately and quality coefficients are identified. The importance of the application of dimensional analysis is emphasized. In Section 3 the performance of different types of bow are compared. Roughly speaking the design parameters can be divided into two groups. One determines the mechanical performance of the bow. Within certain limits, these parameters appear to be less important as is often claimed. The other group of parameters concerns the strength of the materials and the way these materials are used in the construction of the bow. It turns out that the application of better materials and that more of this material is used to a larger extent, contribute most to the improvement of the bow.

2 Formulation of the problem

In essence the bow proper consists of two elastic *limbs*, often separated by a rigid middle part called *grip*. Because the bow is usually held vertical or nearly vertical, we can speak of the upper limb and of the lower limb. The back of a limb is the side facing away from the archer, the belly the opposite side. The bow is braced by fastening a string between both ends of the limbs. The distance between the grip on the belly side and the string in that situation is called the brace height or *fistmele*. After an arrow is set on the string, called *nocking*, the archer pulls the bow from braced situation in full draw. This action is called *drawing*. Then, after *aiming*, the arrow is loosed or released, called *loosing*.

We are concerned with bows of which the limbs move in a flat plane, and which are symmetric with respect to the line of aim. The bow is placed in a Cartesian coordinate system $(\underline{x},\underline{y})$, the line of symmetry coinciding with the $\underline{x}$-axis and the origin O coinciding with the midpoint of the bow, see Figure 2.1. We assume the limbs to be inextensible and that the Euler-Bernoulli beam theory holds. The total length of the bow is denoted by $2\underline{L}$. In our theory it will be represented by an elastic line of zero thickness, along which we have a length coordinate $\underline{s}$ measured from O, hence for the upperhalf we have $0 \le \underline{s} \le \underline{L}$.

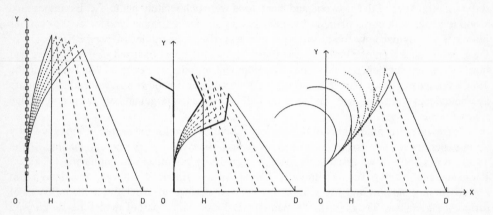

Figure 2.1 Three types of bow: the a) non-recurve bow, b) static-recurve bow and c) working-recurve bow.

This elastic line is endowed with bending stiffness $\underline{W}(\underline{s})$ and mass per unit of length $\underline{V}(\underline{s})$. The geometry of the unstrung bow is described by the local angle $\theta_0(\underline{s})$ between the elastic line and the $\underline{y}$-axis, the subscript 0 indicates the unstrung situation. $\underline{L}_0$ is the half length and $2\underline{m}_g$ the mass of the grip. The length of the unloaded string is denoted by $2\underline{l}_0$, its mass by $2\underline{m}_s$. We assume that the material of the string obeys Hooke's law, the strain stiffness is denoted by $\underline{U}_s$. Note that whether the length of the string or the brace height denoted by $|\underline{OH}|$ fixes the shape if the bow in braced situation.

The classification of the bow we use, is based on the geometrical shape and the elastic properties of the limbs. The bow of which the upper half is depicted in Figure 2.1.a is called a *non-recurve* bow. These bows have contact with the string only at their tips $(\underline{s} = \underline{L})$ with coordinates $(\underline{x}_t,\underline{y}_t)$. There may be concentrated masses $\underline{m}_t$ with moment of inertia $\underline{J}_t$ at each of the tips, representing for instance horns used to fasten the string.

In the case of the *static-recurve* bow, see Figure 2.1.b, the outermost parts of the limbs are stiff. These parts are called *ears*. Its mass and moment of inertia with respect to the centre of gravity of the ear $(\underline{x}_{cg},\underline{y}_{cg})$ are denoted by $\underline{m}_c$ and $\underline{J}_c$, respectively. The flexible part $\underline{L}_0 \le \underline{s} \le \underline{L}_2$ is called the *working part* of the limb. In the braced situation the string rests upon string-bridges, see Figure 2.1.b. These string-bridges are fitted to prevent the string from slipping past the limbs. The place of the bridge of the upper limb is referred to as $(\underline{x}_b,\underline{y}_b)$.

With a *working-recurve* bow the parts near the tips are elastic and bend during the final part of the draw, see Figure 2.1.c. When drawing such a bow the length of contact between string and limb decreases gradually until the point where the string leaves the limb, denoted by $\underline{s} = \underline{s}_w$, coincides with the tip $\underline{s} = \underline{L}$ and remains there during the final part of the draw. In Figure 2.1 the bow is pulled by the force $\underline{F}(\underline{b})$, where the middle of the string has the $\underline{x}$-coordinate $\underline{b}$. To each bow belongs a value $\underline{b} = |\underline{OD}|$ for which it is called fully drawn indicated by a subscript 1. The force $\underline{F}(|\underline{OD}|)$ is called the *weight* of the bow and the distance $|\underline{OD}|$ is its *draw*. By releasing the drawn string at time $\underline{t} = 0$ and holding the bow at its place, the arrow, represented by a point mass $2\underline{m}_a$ is propelled. The arrow leaves the string when the acceleration of the midpoint of the string becomes negative. This moment is denoted by $\underline{t}_l$ and the muzzle velocity of the arrow is referred to as $\underline{c}_l$.

A shorthand notation for a bow and arrow combination is introduced with

$$\underline{B}(\underline{L},\underline{L}_0,\underline{W}(\underline{s}),\underline{V}(\underline{s}),\theta_0(\underline{s}),\underline{m}_a,\underline{m}_t,\underline{J}_t,\underline{m}_c,\underline{J}_e,\underline{m}_g,\underline{x}_{cg},\underline{y}_{cg},\underline{x}_{b0},\underline{y}_{b0},\underline{x}_{t0},\underline{y}_{t0},\underline{L}_2,\underline{U}_s,\underline{m}_s,|\underline{OH}| \text{ or } \underline{l}_0;$$
$$|\underline{OD}|,\underline{F}(|\underline{OD}|),\underline{m}_b), \tag{2.1}$$

where $\underline{m}_b$ is the mass of one limb excluding the mass of the grip.

Note that the last two mentioned parameters are added to the list artificially. This implies that both functions $\underline{W}(\underline{s})$ and $\underline{V}(\underline{s})$ are constrained. We consider the values of these functions for $\underline{s} = \underline{L}_0$ to be already fixed by both constraints. The first constraint concerning the weight, is an implicit relationship between a number of parameters of which $\underline{W}(\underline{s})$ is one of them, and the weight $\underline{F}(|\underline{OD}|)$ of the bow. The second constraint is just

$$\underline{m}_b = \int_{\underline{L}_0}^{\underline{L}_2} \underline{V}(\underline{s}) \, d\underline{s} + \underline{m}_c . \tag{2.2}$$

and for a given mass of the bow the value $\underline{V}(\underline{L}_0)$ is derived easely. This shows that both functions are considered to be the product of a function $\underline{W}(\underline{s})/\underline{W}(\underline{L}_0)$ and $\underline{V}(\underline{s})/\underline{V}(\underline{L}_0)$ of the length coordinate $\underline{s}$ into IR and a parameter $\underline{W}(\underline{L}_0)$ and $\underline{V}(\underline{L}_0)$ with dimensions.

2.1 Dimensional analysis

In this paragraph the 24 parameters of Equation (2.1) are considered as elements of a dimensional space π, spanned by a fixed system of units E_1 = length in cm, E_2 = force in kgf, E_3 = mass in kg and E_4 = time in .03193 sec (see later on). According to the Pi-theorem one can write (2.1) in the form with the dimensionally independent parameters $|\underline{OD}|$, $\underline{F}(|\underline{OD}|)$ and $\underline{m}_b$, referred to as $\underline{A}_1$, $\underline{A}_2$ and $\underline{A}_3$ (dimensional base) and dimensionally dependent parameters $\underline{L}, \dots , |\underline{OH}|$ or $\underline{l}_0$, referred to as $\underline{B}_1, \dots , \underline{B}_{21}$. With $j = 1(1)\,21$ and $i = 1(1)\,3$ we have,

$$\underline{B} = B(B_1, \dots ,B_{21}) \prod_{i=1}^{3} A_i{}^{a_i} , \quad \underline{B}_j = B_j \prod_{i=1}^{3} A_i{}^{a_{ji}} , \quad A_i = A_i \prod_{k=1}^{4} E_k{}^{z_{ik}} , \tag{2.1.1}$$

where $a_{ji}, a_i \in IR$ and $B_j \in IR^+$. We have for example

$$\underline{L} = L \, |\underline{OD}| , \quad \underline{K} = K \, \underline{F}(|\underline{OD}|) , \quad \underline{m}_a = m_a \, \underline{m}_b , \tag{2.1.2}$$

and for the functions of the length coordinate $\underline{s}$

$$\underline{W}(\underline{s}) = W(s) \, |\underline{OD}|^2 \cdot \underline{F}(|\underline{OD}|) , \quad \underline{V}(\underline{s}) = V(s) \, \frac{\underline{m}_b}{|\underline{OD}|} . \tag{2.1.3}$$

Observe that these functions of $\underline{s}$ are also transformed to functions of the dimensionless length coordinate s. Also the angle $\theta(\underline{s})$ between the elastic line and the $\underline{y}$-axis will be transformed to $\theta(s)$, where we should have used a new symbol. With respect to dimensional analysis this yields no added difficulties. Finally we have

$$|\underline{OD}| = |OD| \ cm, \quad \underline{F}(|\underline{OD}|) = F(|OD|) \ kgf, \quad \underline{m}_b = m_b \ kg \ . \tag{2.1.4}$$

So, quantities with dimension are labelled by means of a underscore '$\_$' and quantities without the underscore are the associated dimensionless quantities.

The unit of time is already fixed by the choice of the other 3 units: cm, kgf and kg. For the time $\underline{t}_l$, the moment the arrow leaves the string, we have

$$\underline{t}_l = t_l \ (L, \ldots, |OH| \ or \ l_0) \cdot \sqrt{\frac{m_b \cdot |\underline{OD}|}{\underline{F}(|\underline{OD}|)}} = t_l \cdot \sqrt{\frac{m_b \cdot |\underline{OD}|}{F(|\underline{OD}|)}} \cdot \sqrt{\frac{kg \cdot cm}{kgf}} \ . \tag{2.1.5}$$

This means that the unit of time equals .03193 sec.

2.2 Quality coefficients

The purpose for which the bow is used has to be considered in the definition of a cost-function which could be optimized in order to obtain the 'best' bow. However, it appears to be very difficult to define such a unique cost-function. Therefore we introduce a number of quality coefficients which can be used to judge the performance of a bow and arrow combination. The static quality coefficient q is given by

$$q = A = \frac{A}{|\underline{OD}| \cdot \underline{F}(|\underline{OD}|)} \ , \tag{2.2.1}$$

where A is the dimensionless energy stored in the elastic parts of the bow, the working parts of the limb and the string, by deforming the bow from the braced position into the fully drawn position. The dynamic quality coefficients are the efficiency η and the muzzle velocity referred to as v. They are defined by

$$\eta = \frac{m_a \cdot c_l^2}{A}, \quad v = \sqrt{\frac{q \cdot \eta}{m_a}} \ . \tag{2.2.2}$$

Observe that by definition these quality coefficients are dimensionless. This means that the sensitivities of these coefficients with respect to the elements of the dimensional base $|\underline{OD}|$, $\underline{F}(|\underline{OD}|)$ and $\underline{m}_b$ can be obtained directly, without solving the governing equations of motion which constitute the mathematical model again. The selection of the dimensional base is not unique. The motivation to take draw, weight, and mass of one limb is the following. The maximum draw and weight depend on the stature of the archer. His 'span' determines the maximum draw and his strength the maximum weight, so both have physical limitations. In practice the minimum mass of one limb has technical limitations which will be the subject of the next section. In our approach the elements of the dimensionless basis are selected based on the limitations which make the optimization problem well posed. Hence, the choice of the dimensional base is coupled to the formulation of the quality coefficients. The advantage of this technique is that with the comparison of different bows, taking the quantities $|OD|$, $F(|OD|)$ and m_b equal to 1, yields interpretable results for the quality coefficients.

2.3 The construction of the bow

The limbs of the bow are considered as a beam of variable cross-section $\underline{D}(\underline{s})$ made out of one material with density ϱ and Young's modulus $\underline{E}$. With the Euler-Bernoulli hypothesis the normal stress $\sigma(\underline{s},\underline{t},\underline{r})$ depends linear on $\underline{r}$, the distance from the neutral axis which passes through the centroid of the cross-section. We assume that the maximum bending moment for each $\underline{s}$ as a function of time $\underline{t}$ occurs in the fully draw situation. Then we have, when $\underline{A}_b$ denotes the elastic energy in the limbs of the bow in fully drawn situation,

$$\frac{A_b}{m_b} = 2 \int_{L_0}^{L} \int_{\underline{D}(\underline{s})} 1/2 \frac{\sigma_1^2(\underline{s},\underline{r})}{E} \, d\underline{D}(\underline{s}) \, d\underline{s} \Big/ \int_{L_0}^{L} \varrho \, C(\underline{s}) \, d\underline{s} \, , \qquad (2.3.1)$$

where $C(\underline{s})$ is the area of the cross-section. The stress σ_1 is the resulting normal stress due to the bending moment in the fully drawn bow, indicated by the subscript 1.

We define now two useful quality coefficients

$$\delta_{bv} = 1/2 \frac{\sigma_w^2}{\varrho E} \, , \qquad a_D = \frac{A_b}{2 m_b \cdot \delta_{bv}} \, , \qquad (2.3.2)$$

where σ_w is the working stress of the material, equal to the yield point or the ultimate strength divided by factors of safety. The quantity δ_{bv} is the amount of energy per unit of mass which could be stored in the material. In Table 2.3.1 an indication of this quantity of some materials used in making bows is given.

material	σ_w kgf/cm^2 x 10^2	E kgf/cm^2 x 10^5	ϱ kg/cm^3 x 10^{-6}	δ_{bv} kgf cm/kg
steel	70.0	21.0	7800	1300
sinew	7.0	.09	1100	25000
horn	9.0	.22	1200	15000
yew	12.0	1.0	600	11000
maple	10.8	1.2	700	7000
glassfibre	78.5	3.9	1830	43000

Table 2.3.1 Mechanical properties and the energy per unit of mass δ_{bv} for some materials used in making bows, see also [5].

The dimensionless coefficient a_D is generally smaller than 1 for two reasons. Firstly, the tensile and compression stresses in the outermost fibres of the limb may be less than the working stress in the fully drawn bow. With the design of the limbs one has to assure stability of the limb, without tendency to twist or distort laterally when the bow is drawn. In practice this is accomplished by the requirement that the width of a limb may not become smaller than the thickness. Not all material near to the tips is used then to the full extent.

The stress in the fibres near the neutral axis is smaller than the working stress and this reduces the coefficient a_D too. Suppose the bow is in a homogeneous stress-state, then

$$a_D = \int_{L_0}^{L} \frac{I(\underline{s})}{\underline{c}^2(\underline{s})} \, d\underline{s} \Big/ \int_{L_0}^{L} C(\underline{s}) \, d\underline{s} \, , \qquad (2.3.3)$$

where $\underline{e}$ is the distance between the outermost fibres and the neutral axis. If the material has the same strength in tension and compression, it will be logical to choose shapes of cross-section in which the centroid is at the middle of the thickness of the limb, equal to 2 $\underline{e}(\underline{s})$. $\underline{I}(\underline{s})$ is the moment of inertia of the cross-section with respect to the neutral axis. The quantity $2\underline{I}/\underline{e}$ is called the section modulus. In handbooks the magnitude of the moment of inertia and the section modulus are tabulated for various profile sections in commercial use. We stick at the usage of the defined coefficient a_D which is dimensionless and follows in a straightforward manner from our statement of the problem.

We consider the quantity α_D defined by $\alpha_D(\underline{s}) = (\underline{I}(\underline{s}))/(\underline{C}(\underline{s})\,\underline{e}(\underline{s})^2)$, $L_0 \le \underline{s} \le L$, for various shapes of cross-section of limbs. For a bow with similar cross-sections at different values of $\underline{s}$ we have $a_D = \alpha_D$. The English longbow possessed a D-shape cross-section, the belly side approximately formed by a semicircle and the back side being a rectangular. When the radius of the semicircle equals the half of the thickness of the limb α_D equals .255, so smaller than .333 for a rectangular and more than .25 for a elliptical shape. Steel bows, for instance the Seefab bow invented in the 1930's, were on the principle of a flattened tube. We assume that the inner diameter equals k times the outer diameter for any line through the centre of the ellipses. For k = .9 the magnitude if a_D becomes .4525, so larger than the other mentioned values, as to be expected, because relatively more material is placed near the outermost fibres.

There is still another technique to increase the value of a_D. In ancient Asiatic bows, horn and sinew, together with wood, were used on the belly and back side, respectively. Horn is a superb material for compressive strength and sinew laid in glue has a high tensile strength, see Table 2.3.1. The Young's modulus of both materials is rather small, but the permissible strain is very high. The space between the two materials near the outermost fibres is filled up with light wood, which has to withstand the shearing stresses. In modern bows horn and sinew are replaced by synthetic plastics reinforced with fibreglass or carbon. So, in composite bows not only better materials are used, but they are also used in a more profitable manner. For composite bows we define equivalent quantities for the Young's modulus and density for a simple bow which has the same mechanical action as the limb of the composite bow. If these magnitudes are substituted in the product $a_D \cdot \underline{\delta}_{bv}$ it can be substantially larger than the magnitude of the product for simple wooden bows. The importance of this product follows from the equation for the stored energy $\underline{A}_b$ per mass of the limb $\underline{m}_b$ and, when we neglect the elastic energy stored in the string, the equation for the muzzle velocity

$$\underline{c}_l = \sqrt{2 \,\frac{q}{A_b} \cdot \frac{\eta}{m_a} \cdot a_D \cdot \underline{\delta}_{bv}} \ . \tag{2.3.4}$$

Hence, the muzzle velocity of an arrow depends on: the static quality coefficient q divided by A_b, the amount of elastic energy stored in the fully drawn limbs, the efficiency divided by the dimensionless mass of the arrow and finally on the product $a_D \cdot \underline{\delta}_{bv}$. The first term shows that the amount of energy in the elastic parts of the bow in braced situation should be as small as possible. For q equals the energy stored in the fully drawn bow minus this energy in the braced bow. A large q implies, however, heavy limbs and this in turn implies a small m_a, because this is the arrow mass divided by the mass of one limb.

The efficiency depends largely on the masses of the arrow and string. This influence can be assessed from a simplified model of the string moving in straight lines between the points of attachment and the arrow. The resulting distribution of the kinetic energy along the string indicates that 1/3 of the mass of the string should be concentrated at the middle where the arrow meets the string. Hence,

$$\max \eta \approx \frac{m_a}{m_a + 1/3 \cdot m_s},$$

(2.3.5)

is an approximation of the maximum attainable efficiency. The quotient η/m_a increases with decreasing m_a and this holds probably also to a certain extent for a real bow. However, there are limits, for the arrow has to be strong enough to withstand the acceleration force. Further, every archer knows that it is not allowed to loose a fully drawn bow without an arrow. In that case the efficiency equals 0 and the bow or string can even break. For small arrow masses our assumption, the maximum bending moment to be equal to the value in the fully drawn situation, is obviously violated. There seems to be an optimum for the mass of the arrow.

In order to reduce the mass of the string the applied materials should be strong. Man-made fibres such as Dacron and Kevlar are used. The maximum force determines with the strength of the material the minimum mass of the string. This maximum force will certainly not occur in the fully drawn situation.

3 Results and conclusions

In this section we start with a sensitivity study for a straight-end bow. In our mathematical model derived in [4], the action of a bow and arrow combination is fixed by one point in a 24 dimensional parameter space. First of all we deal with the 21 dimensionless parameters. Three of them are functions, viz. $W, V, \theta : [L_0, L] \rightarrow IR$. These functions are written in a simple form:

$$W_n(s) = W_n(L_0) \cdot (\frac{L-s}{L-L_0})^{\beta_n}, L_0 \leq s \leq \varepsilon_n, \ 1/3 \ W_n(L_0), \varepsilon_n \leq s \leq L,$$

and

$$V_n(s) = V_n(L_0) \cdot (\frac{L-s}{L-L_0})^{\beta_n}, L_0 \leq s \leq \varepsilon_n, \ 1/3 \ V_n(L_0), \varepsilon_n \leq s \leq L.$$

with $\varepsilon_1 = L$, $n = 1$, $\varepsilon_n = L - (L-L_0) \cdot 3^{-1/\beta_n}$, $n = 2,3$ and $\beta_1 = 0$, $\beta_2 = 1/2$, $\beta_3 = 1$. (3.1)

The shape of the unstrung bow is given by

$$\theta_0(s) = \theta_0(L_0) + \kappa_0 \cdot (L-s)/(L-L_0), \qquad L_0 \leq s \leq L.$$

(3.2)

Under this description, these functions is fixed by only three parameters β, $\theta_0(L_0)$ and κ_0. The two string-parameters, the mass m_s and the stiffness U_s, are for a particular material fixed by the number of strands. An increase of this number, ns makes the string stiffer but also heavier. We start with a straight-end bow described by Klopsteg in [1]. This bow is referred to as the KL-bow (Figure 2.1.a). In shorthand notation introduced in (2.1), it is represented by

$$KL(1.286,.1429,n=1,\theta_0(L_0)=0,\kappa_0=0,.0769,0,0,0,0,0,0,1.286,0,1.286,0,1.286,$$
$$1.286,131,.0209,|OH| = .214)$$

(3.3)

The sensitivity coefficients, i.e. the partial derivatives of the quality coefficients with respect to the design parameters, are presented in Table 3.1. These sensitivity coefficients were calculated with the classical approach with finite-difference approximations.

	L	L_0	β	$\theta_0(L_0)$	κ_0	m_a	m_t	J_t	ns	IOHI
$W(L_0)$	4.24	−3.4	.45	2.6	1.81	.0	.0	.0	.0	−.33
$V(L_0)$	−1.4	1.4	.70	.0	.0	.0	.0	.0	.0	.0
q	.07	−.05	.0	−.15	−.18	.0	.0	.0	.0	−.39
η	−.11	.15	−.06	.46	.3	3.4	−1.1	.0	.0	.0
v	.0	.07	.07	.25	.0	−6.7	−1.6	.0	.0	−.83

Table 3.1 Sensitivity coefficients for the straight-end KL-bow.

We conclude that the mass of the arrow is the most important parameter for the efficiency and for the muzzle velocity. Further tip-masses should be avoided because they reduce the efficiency.

Representations of different types of bows used in the past and in our time form clusters in the parameter space. In the preceding sensitivity analysis one cluster, that for a straight-end flatbow, was analysed. In what follows we consider other types of bow: another non-recurve bow, the Angular bow, to be called the AN bow (Figure 3.1.a), two Asian types of static-recurve bow, to be called the PE (Figure 2.1.b) and TU bow (Figure 3.1.b) and two working-recurve bows, one with an extreme recurve, to be called the ER bow (Figure 2.1.c) and a modern working-recurve bow to be called the WR bow (Figure 3.1.c).

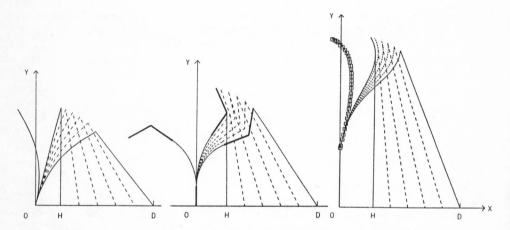

Figure 3.1 The static deformation shapes of a) the AN-bow, b) the TU-bow and c) the WR-bow.

All the quality coefficients for these types of bow are shown in Table 3.2. The results indicate that the muzzle velocity is about the same for all types. So, within certain limits, these dimensionless parameters appear to be less important than is often claimed. The efficiency of strongly recurved bows is rather bad.

In Table 3.2 the values of the quality coefficients for a modern working-recurve bow are also given. The shape of the bow for a number of draw-lengths is shown in Figure 3.1.c. Figures 3.2.a and 3.2.b give the shape of the limb and the string, before ($0 \le t \le t_l$) and after ($t_l \le t$) arrow exit. Observe the large vibrations of the string after the arrow leaves the string, which imply that the brace height has to be large.

Bow	q	η	ν	m_a	m_s	$W(L_0)$	$V(L_0)$	A_b	$q/A_b \cdot \eta/m_a$
KL-bow	.407	.765	2.01	.0769	.0209	1.409	1.575	0.5155	7.85
AN-bow	.395	.716	1.92	.0769	.0209	0.2385	2.300	0.5493	6.70
PE-bow	.432	.668	1.94	.0769	.0209	0.2304	1.867	0.5879	6.38
TU-bow	.491	.619	1.99	.0769	.0209	0.1259	1.867	1.0817	3.65
ER-bow	.810	.417	2.08	.0769	.0209	0.3015	2.120	1.4150	3.10
WR-bow	.434	.729	2.23	.0629	.0222	2.5800	1.95	0.6930	7.25

Table 3.2 Dimensionless quality coefficients for a number of bows.

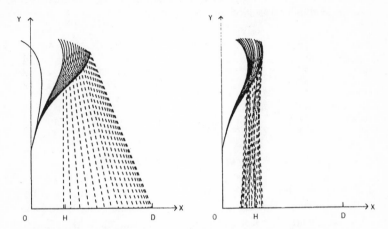

Figure 3.2 Dynamic deformation shapes of the WR-bow for a) $0 \le t \le t_l$ and b) $t_l \le t$.

These results show that the modern working-recurve bow is a good compromise between the non-recurve bow and the static-recurve bow. The recurve yields a good static quality coefficient and the light tips of the limbs give a reasonable efficiency. Note that the mass of the arrow of the modern working-recurve bow is smaller than the other values mentioned This accounts for a smaller efficiency, but also larger muzzle velocity.

How can these dimensionless quantities be used in the design of a bow?. In practice the manufacturer wants to design a bow with a specified draw $|OD|$ and weight $F(|OD|)$ using available materials with given Young's modulus $\underline{E}$ and density $\underline{\rho}$. The use of Equation (2.1.3) gives a value for $\underline{I}(\underline{s})$. The thickness of the limb associated with the distance between the outermost fibres and the neutral axis $\underline{e}$ is then fixed by

$$\underline{e}(\underline{s}) = \frac{I(\underline{s})}{M_1(\underline{s})} \cdot \sigma_w = \frac{W(\underline{s})}{M_1(\underline{s})} \cdot \frac{\sigma_w}{E} \cdot |OD| , \qquad (3.4)$$

These calculations can be done after the solution of the static equations yielding $M_1(\underline{s})$. After the selection of the shape of a cross-section of the limbs $\underline{D}(\underline{s})$, the width is fixed. To ensure stability this width should not be taken smaller than the thickness of the limb. The area of the cross-section $\underline{C}(\underline{s})$ can then be calculated. $\underline{C}(\underline{s})$ and the density $\underline{\rho}$ jointly determine the mass of the limbs $\underline{m}_b$. This completes the design of the limbs.

Hence, for a homogeneous stressed bow two parameters given in (2.1), the mass distribution V(s) and the total mass of the limb $\underline{m}_b$, depend on other parameters of (2.1) and

Young's modulus $\underline{E}$, density $\underline{\rho}$, working stress $\underline{\sigma}_w$ of the material and finally the shape of a cross-section $\underline{D}(\underline{s})$. More parameters of (2.1) are dependent in practice. The parameters concerning the ears, $\underline{m}_e$, $\underline{J}_e$, $\underline{x}_{cg}$, $\underline{y}_{cg}$, $\underline{x}_{b0}$, $\underline{y}_{b0}$, $\underline{x}_{t0}$, $\underline{y}_{t0}$, $\underline{L}_2$, are strongly related and there is a relationship between the string parameters $\underline{U}_s$, $\underline{m}_s$ and the strength of the material used for the string together with the maximum force in the string. This force is not known from the static calculations, so an initial guess has to be made, which must be checked after the dynamic calculations.

We considered a number of different bows used in the past and in our present. In Table 3.3 we give values for the parameters with dimension, weight, draw and mass of one limb, for a number of bows described in the literature. Estimations of a_D are also given. These approximations have to be rough for lack of detailed information.

Reference	Type	$F(|OD|)$	$|OD|$	$2m_b$	$2L$	δ_{bv}	A_b	a_D
[1]	flatbow	15.5	71.12	0.325	182.9	9000	.52	.20
[4]	longbow	46.5	74.6	0.794	189.4	9000	.52	.25
[6]	steelbow	17.2	71.12	0.709	168.9	1300	.52	.69
[7]	Tartar	46.0	73.66	1.47	188.0	20000	.59	.07
[8]	Turkish	69.0	71.12	0.35	114.0	20000	1.1	.77
[4]	modern	12.6	71.12	0.29	170.3	30000	.70	.07

Table 3.3 Parameters with dimension for a number of bows and an estimation of a_D.

These results indicate that the short Turkish bow is made of a combination of good materials which are used to the full extent. This explains the good performance of these bows in flight shooting, and not the mechanical performance of these bows; see Table 3.2. The modern materials are the best, but the calculated value of a_D for the modern bow suggests that they are used only partly. The calculated efficiency of the modern working-recurve bow correlates well with values given in the literature. The value of 72.9% is, however, rather low. The maximum for the efficiency according to Equation (2.3.5) from the mass of the string determined by the parameters given in Table 3.2 is about 90%. Hence, it seems to be possible to improve this quality coefficient with the help of the mathematical model presented here.

References

[1] C.N. Hickman, F. Nagler, P.E. Klopsteg, Archery: the technical side, National Field Archery Association, 1947.
[2] P.E. Klopsteg, Bows and arrows: a chapter in the evolution of archery in America, Smithsonian Institute, Washington, 1963.
[3] B.W. Kooi, On the Mechanics of the Bow and Arrow, Thesis, Rijksuniversiteit Groningen, 1983.
[4] B.W. Kooi, The Design of the Bow, to be published.
[5] J.E. Gordon, Structures or Why Things Don't Fall Down, Pelican Books, 1978.
[6] R.P. Elmer, Target Archery, Hutchinson's Library of sports and pastimes, London, 1952.
[7] S.T. Pope, Bows and Arrows, University of California Press, Berkeley, 1974.
[8] Sir Payne-Gallwey, The Crossbow, The Holland Press, London, 1976.

COMPUTER SIMULATED SELF-OPTIMIZATION OF BO

H. Huber-Betzer, C.Mattheck
Nuclear Research Center Karlsruhe, D-7500
Institute for Material and Solid State Rese

Results

Fig

Introduction

Biological load carriers are structures which are well adapted to their main loading conditions.

In the architecture of spongy bone, for instance, the orientation of the network of trabecula is supposed to be able to take the load best and with a minimum of weight.

The trabecula themselves are arranged in alignment with the trajectories of the principal stresses [1,2]. Changes in the loading situation result in a disturbance of an optimum condition. But biological structures are able to react and to restructure. They build new load-bearing elements, where they are needed or they eliminate stress-shielded regions. After these modeling processes the new system will be adapted to an optimum for the new situation and be used to its full capacity again.

But not only the inner trabecular architecture is able to adapt to a new loading situation. Biomechanical studies on the stress distribution in healthy bones showed that the outer shape and geometry of the long bones are designed in order to prevent overloaded or stress-shielded parts, too. So the morphology will change and the structure will alter if the former equilibrium is disturbed. Homogeneously distributed stresses along the surface areas are aspired. The effects will be shown with the help of a finite-element analysis.

Method

Using a new method (**CAO - Computer-Aided-Optimization**) for shape optimization [3,4] the adaption of different bony structures to changes in the loading situation is shown.

The principle is to simulate growth at surface areas of high stress levels and to allow the structure to shrink and to remove material in stress shielded regions compared to a given reference stress. In the following analysis the Von-Mises-stress distribution is used as a criterion for the loading situation. The action will be shown examplary by braces of a framework and the shape of long bones healing in malposition.

The aim is to reach a design which endures the given loads best with homogeneously distributed stresses at the surfaces regions e.g. without any notch stresses.

The following examples are plane strain FE-structures built of 4 node bilinear elements. An isotropic material behaviour is assumed. The Young's modulus and the Poisson's ratio are uniform for each model.

.1 shows on the left the initial FE-structure of a brace in a framework, the loading case and the boundary conditions. The model consists of two parallel horizontal beams and a slanted beam connecting these two beams. The pressure acts in a vertical direction for which the shape of the connecting brace is not an optimum design. The results are bending moments and inhomogeneously distributed stresses at the surface which cause the connecting brace to straighten up, shown by different stages of the contour of the structure. The surface of the structures is allowed to swell during the procedure in regions of high stresses and to shrink in regions of low stresses compared to the reference stress: σ_o. High stress levels are found in the transition of the horizontal to the slanted beam for the starting structure. It could be shown that similar to the remodeling behaviour of bone, stress peaks will be reduced first by adding material and lateron unneccessary material will be removed. The slanted brace is rotating and the terminal shape is a vertical brace positioned in direction of the applied pressure. The diagram and the von-Mises stress distribution show that the stresses along the contour are now homogenized (fig.1).

The second example are vertical braces connecting horizontal beams. The vertical load acting along the center line of the vertical brace would use the structure to its full capacity. So that situation would be an optimum for the vertical brace. The value of the homogeneously distributed stress for the above mention loading case is used as the reference stress for the loading case given in fig.2. The force is now acting at a certain distance from the center line of the stem and causes bending and the disturbance of the former optimum situation. Overloaded and stress shielded regions are produced. So the brace is drifting to reach an optimum for this loading situation and to obtain homogeneously distributed stresses on the surface. The results of the development are also represented for different stages. At last the stresses along both contours of the brace are homogenized.

Another example is a broken femur healed in malposition. The x-ray (fig.3d) shows the present-day situation. The simulation has been started with a structure of an assumed situation a short time after the fracture (fig.3a). The given load was a bending moment. In a first step only growing of the surface layer was allowed in order to reduce the maximum stresses. Fig.3b shows the adapted structure where the stress peaks have already disappeared. But parts of the bone are still stress-shielded. In a second step removal of material was allowed, too. The bone also grew into a state of homogeneously distributed stresses (fig.3c), into an optimized shape in the sense of a compromise design.

The same effect is demonstrated for the model of a broken fibula. Starting again from an assumed former situation the finite element structure was loaded by a single force on the top of the model in direction of the foot-point. The place where the force is applied is a second fracture site where two parts of the bone are in contact. The loading case causes bending moments in the model and stress peaks so that growth in overloaded region appears. The bone at the place where the force acts grows excessively and the calculated shape fits the real contour well. Increasing the contact area leads to a reduction of the high stress level at this upper point. The entire calculated shape fits the real structure well with respect to the chosen simplifications of the model.

Discussion

The aim of the study was to show the development of bony structures under the influence of pathological changes with the help of a new shape optimization method. The ability of biological load carriers to adapt their morphology to the current loading situation could be simulated. Braces of frameworks in spongy bone for instance are able to rotate into directions and drift to places where they are needed. The possibility to generate frameworks which perform their task best, with minimum weight should be also of great interest to engineering structures.

Considering the outer shape of long bones it could be observed that bone attaches material first where it is necessary to prevent failure and after stabilization the low-weight design is aspired and material removed. The comparison of the x-rays and the calculations makes clear that the remodeling of the bones in the examples hasn't yet finished. Low-loaded regions near the fracture side are supposed to decrease in future until a homogeneous stress state is reached.

The criterion of a homogenous von-Mises stress distribution seems to be completely sufficient for the prediction of bone growth.

The exact shape of the bones couldn't be found in all details because the finite element models are simplified two-dimensional structures which don't consider the respective depth nor the variation of the Young's modulus. But the tendency of bone adaption, growth and reduction of material in relation to the stress distribution could be described satisfactory.

Acknowledgements

The authors are grateful to the students B.Büstgens, A.Majorek and M.Mittwollen who helped performing the finite element studies.

References

[1] J.Wolff,
    'Über die Bedeutung der Architektur der spongiösen Substanz für die
    Frage vom Knochenwachsthum',
    Centralblatt für die Wissenschaften, No.54, 849-851, Dez.1869

[2] J.Wolff,
'Das Gesetz der Transformation der inneren Architektur der Knochen bei
pathologischen Veränderungen der äußeren Knochenform',
Sitzungsbericht der königlich preussischen Akademie der Wissenschaften
zu Berlin, Berlin 1884

[3] C.Mattheck,
'Engineering components grow like trees',
Materialwissenschaft und Werkstofftechnik 21 (1990) 143-168

[4] C.Mattheck, S.Burkhardt,
'A new method of structural shape optimization based on biological growth',
Int.J.Fatigue 12 (1990) 185-190

Fig.1 Adaptive rotation of trabecular bone

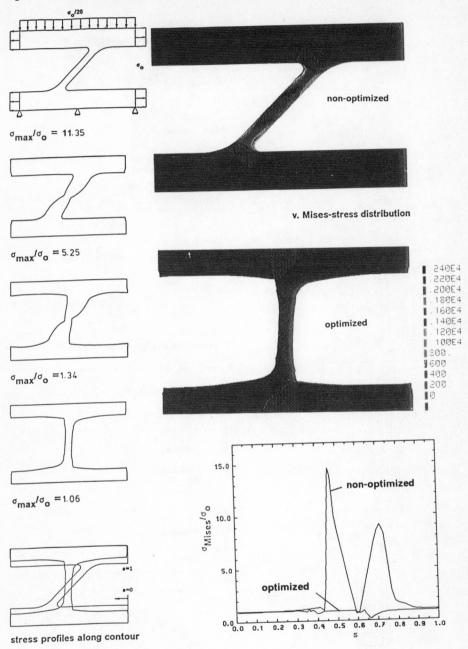

$\sigma_{max}/\sigma_o = 11.35$

$\sigma_{max}/\sigma_o = 5.25$

$\sigma_{max}/\sigma_o = 1.34$

$\sigma_{max}/\sigma_o = 1.06$

stress profiles along contour

non-optimized

v. Mises-stress distribution

optimized

Fig.2 Adaptive drifting of trabecular bone

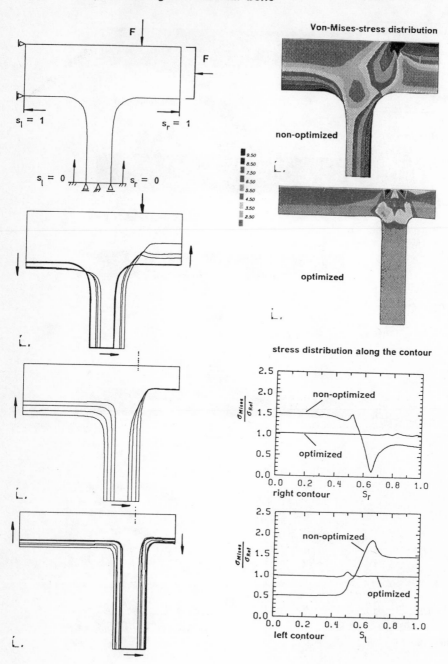

TRUSS OPTIMIZATION USING KNOWLEDGE BASE

Adam BORKOWSKI, Stanisław JÓŹWIAK, Małgorzata DANICKA

Institute of Fundamental Technological Research

Warsaw, Poland

1. Introduction

It is commonly accepted, that advances in computer technology should be exploited in a way that stimulates creativity of a designer. The CAD systems have already released him from the tedious drawing and the FEM-based programs overtook time-consuming strength calculations. One should not overlook, however, the circumstance that both improvements concern the second stage of the design process: the detailed planning and verification of an already chosen concept.

It is well known, that possible refinement of a given structural layout (e.g. an optimization of cross-sections or reinforcement), though generally welcome, give benefit in cost or weight of the structure of the order 5 - 20%. On the contrary, selection of the structural scheme particularly suitable for a given application may lead to a double or tripple gain.

It is not astonishing, therefore, that research efforts shift nowadays from classical design support tools, like structural optimization via mathematical programming, towards expert systems that help the designer in taking conceptual decisions. One of the earlier attempts in that direction has been taken by the second author in his paper [1], where structural optimization in classical sense was augmented by certain heuristics concerning the initial values of the design variables and the treatment of constraints. That effort has been further continued and the present paper presents an integrated system including a generator of initial designs. The latter uses experience stored in an appropriate database.

2. Domain Knowledge in Computer Programs

All computer programs have a certain amount of domain knowledge incorporated into them in the form of algorithms, rules, formulae and program constants. Depending on the form in which the knowledge is represented, two cathegories of computer programs can be considered.

In conventional programs domain knowledge is applied at the early stage of the program development, usually when the solution algorithm is formulated. It cannot be separated from the program itself. The improvement of the knowledge is difficult and requires modification of the source code or of the algorithm. The algorithm precisely defines what actions must be performed to obtain the solution. Each execution is carried out in exactly the same way, unless the designer modifies the program or the data.

Another cathegory of programs are, so called, knowledge-based programs, in which domain knowledge is separated from the processing mechanism. It is represented explicitly in the form of knowledge base. The domain expert concentrates on formulating relations and facts concerning the considered problem, while another specialist – knowledge engineer – is involved in developing the inference mechanism.

Both programming styles are applied in structural design. Strength calculations are successfully performed using conventional programs, as all stages of the solution process are defined well and all situations can be considered in the algorithm. In some stages of the design process knowledge based approach can be more suitable. One of such stages is generating the initial design. The computer support during this stage of the design process is weak and the stage itself is often omitted in the discussion. It is usually assumed, that the initial design is "somehow" generated, although it is generally acknowledged, that the most significant decisions concerning the structure are undertaken in this stage (i.e. type of structure, material, topology, supports, etc.). During the remaining stages adopted values are only modified and improved.

Two kinds of knowledge are involved in creating the initial design. The first one represents information enclosed in books, codes of practice, design manuals, etc. These recommendations often give only lower or upper bounds on the values of the design parameters and, being very useful during verification of the existing designs, are quite inefficient when a new structure must be dimensioned. In the real-life design, selection of the parameters is usually based on some heuristics, which constitute the second kind of knowledge involved in the process. The heuristics are the results of previously acquired experience.

3. Exploiting Experience

A person becomes a good designer in two stages: first, one has to accumulate a domain specific knowledge, then, one has to acquire sufficient experience. The first stage is usually accomplished during the professional education, whereas the second one requires practice in the design office. Few people manage to become outstanding designers. They are able to find innovative solutions, that do not follow the trends dictated by the previous experience, or even contradict them. This kind of the creative process seems to be, fortunately, beyond the scope of automatization [2].

On the contrary, it is quite possible at present to stimulate the full exploitation of the experience currently available in the specific domain of the design activity. Restricting the latter to Structural Engineering, we observe the following:

a) Both control data (loads, heights, spans, etc.) and the design parameters (coordinates, thicknesses, etc.) are discrete in their nature. This circumstance, coming from practical requirement of modularity and being cumbersome in the classical approach to structural optimization, becomes favourable when the AI approach is adopted.

b) A large but finite set of already accepted solutions (design experience) can be separated into smaller subsets by applying a suitable taxonomy of the domain. Organized into a relational database, this experience becomes available to each designer. This process is particularly beneficial in large design offices, where employees can take into consideration solutions worked out by leading experts.

c) A reasonable estimate of the solution not available at present in the database can be obtained in two ways: either by relaxing constraints of the query or by interpolating between adjacent projects.

d) The system is learning in the sense of improving quality of its proposals with increasing number of entries in the database. This is due to the circumstance that a more dense distribution of the base points in the design space improves the estimate. The proposed approach has been implemented differently than the case-based algorithm reported by Arciszewski and Ziarko [3], but both systems exhibit similar learning capability.

4. Implementation

A prototype computer program has been implemented on MS/DOS machine. The present version of the program is restricted to planar trusses, but few changes will allow us to expand it for structures in bending. The program integrates the following specialized units:
- the *input module* for direct graphical description of the structure, loading and design variables;
- the *analysis module*, that calculates structural response to the given load (linear elastic behaviour is assumed);
- the *optimization module* for minimum weight design with either cross-sectional areas or nodal coordinates as design variables;
- the *knowledge-based module* for the access to the database and, upon request, for the automatic generation of feasible design.

Procedural part of the system has been implemented in Pascal, whereas the expert part is written in Prolog.

4.1. Optimization Module

In the developed program, execution of the optimization unit is optional and the initiative is left to the user, who can proceed from interactive design to optimization in any stage of the process. The decision variables can be introduced when the initial design is formulated or during the modification.

The structure mass has been selected as objective function. The cross sectional areas and joint coordinates can be the decision variables. The constraints have been adopted according to the design code. They concern stresses in elements, maximum joint displacement, minimum slenderness of the compression members. Additional constraints concern the rate of the change of the decision variables and result from the method of the solution used in the program. The nonlinear programming problem is solved using the simplex method. The optimization can be performed in two modes. In the first, the optimization is carried until the final solution is found. In the other mode the user indicates the number of iterations, after which partial results are displayed. These concern the values of the decision variables and the value of the mass of the structure. Decision variables are presented displaying the modified shape of the structure and using colors to indicate suggested change in the cross sectional areas of the elements. Beside the values of decision variables, the change of the mass is also presented in the form of a bar chart. After checking the partial results, the user can:

− continue the optimization,
− reject the results and generally modify the structure,
−   accept the partial results as final values of the decision variables.

The main idea is to allow the user to control the optimization. An experienced designer can stop the process, if it proceeds in the wrong direction (unaccepted changes of the shape) or if program seems "locked" (nonsignificant mass changes).

4.2. The Knowledge-Based Module

The knowledge−based module consists of two parts:
− the database,
− the rule base.

The database contains all previously accepted solutions (stored automatically), as it seems the simplest and most natural way of retaining the knowledge and experience applied by the user during the design process. The following attributes serve as key−words in the database: function, shape, type of bracing, span, height in midspan, height at the support, and loading. Three of them, namely, the function, the span and the load must be specified by the user. Other attributes are optional. In the case when more than one solution with given attributes can be found, the list of selected files is displayed. Appropriate design is then visualized and modified to fit actual demands.

The initial design can be also generated using rule based module. The requirements and constraints, coming from design codes and recommendations can be relatively easily represented in the form of IF...THEN... rules and coded in the form of rule base. Some rules are presented in Fig.1.

(i)   IF
        function : roof   AND   span > 15m
        THEN
        shape : trapeze
(ii)   IF
        function : roof   AND   loading > 300kN
        THEN
        shape : trapeze
(iii)   IF
        function : roof   AND   initial shape : trapeze   AND
                                                 roof slope > 10%
        THEN
        new shape : triangle   AND   height at the support = 0
(iv)   IF
        function : column   AND   initial shape : constant   AND
                                                 loading > 300kN
        THEN
        new shape : variable

Figure 1. *Sample rules coded in the rule base.*

Similarly as when using the database module, the user indicates the attributes of the structure. The inference engine selects the most rational values of the missing attributes, basing upon the given rules and checks the consistency of the attributes stated by the user. In the case of inconsistency the parameters are modified and an adequate communicate is displayed.

After the set of attributes is complete, the expert module performs sizing of the structure, which goes on as follows. First, an attempt is made to find an identical structure in the database. If it is not possible, then the search is repeated iteratively with decreasing demand for similarity at each subsequent iteration. The type of bracing and the height are early discarded from the constraints, whereas the function and the span of the truss are preserved. The cross-sectional areas of the elements applied in the selected design are then adopted as initial values in the considered structure. After the selection has been completed, the project is displayed graphically and the user can proceed with possible modifications. Typical roof trusses generated by the module are presented in Fig.2.

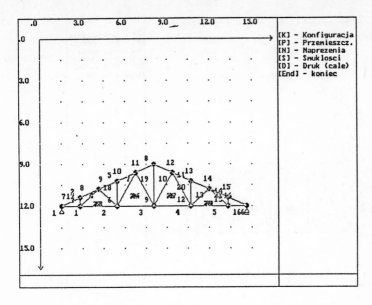

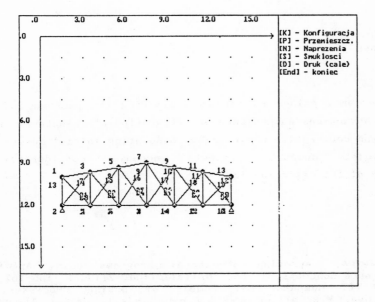

Figure 2. *Typical roof trusses generated by the expert module.*

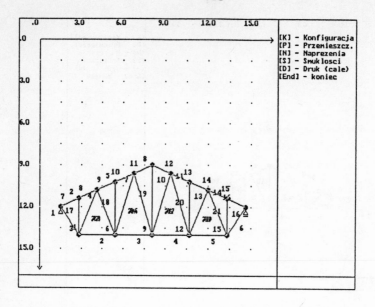

Figure 2 — continued.

5. Conclusions

The experience gained so far in exploiting the package indicates
clearly the advantages gained through integration of computer graphics,
procedural and declarative programming. Generating initial design based
upon previously approved projects increases considerably the
productivity of the design office.

References

[1] S. F. Jóźwiak, Improving Structural Programs Using Artificial
     Intelligence Concepts, J. Eng. Optimization, 2(1987), 155–162.
[2] G. E. Nevill, Computational Models of Design Processes, in:
     S.L.Newsome, W.R.Spiller and S.Finger (eds.), Design Theory '88
     (Springer, New York, 1989), 82–116.
[3] T. Arciszewski and W. Ziarko, Structural Optimization: a Case-Based
     Approach, ASCE J. of Computing in Civil Engineering, to appear.

OPTIMAL REDESIGN PROCESS SIMULATED BY VIRTUAL DISTORTIONS

Jan Holnicki-Szulc

Institute of Fundamental Technological Research, Polish Academy of
Sciences, Warsaw, Poland

Jacek T.Gierliński

WS Atkins Engineering Sciences, Woodcoate Grove, Ashley Road,
Epsom, KT18 5BW, England

1.   The Concept of VDM Simulation of Redesign

Following the concept of simulation of structural redesign by
a fictitious field of virtual distortions (cf.Ref.1) let us define
the *initial*, *modified* and *distorted* structural configurations
(Fig.1). The initial configuration corresponds to the structure
with the cross-sectional areas of elements A_i and to the states of
strains ε_i^L and internal forces $R_i^L = A_i \sigma_i^L$ (σ_i^L denotes stresses) due
to the defined external load. Analogously, the modified
configuration corresponds to the structure with the modified
cross-sectional areas of elements A_i^M and to the states of strains
ε_i^M and internal forces $R_i^M = A_i^M \sigma_i^M$ due to the same external load.

On the other hand, the distorted configuration corresponds to
the unchanged initial material distribution A_i, to the states of
strains ε_i^D and internal forces $R_i^D = A_i \sigma_i^D$ due to the unchanged
external load superposed with some additional load through virtual
distortion state ε_i^o.

The concept of simulation of redesign through virtual
distortions requires the equivalence of deformations and internal
forces for both configurations: modified and distorted:

$$\varepsilon_i^M = \varepsilon_i^D \equiv \varepsilon$$
$$R_i^M = R_i^D \equiv R$$

(1)

Therefore, for each element the simulation condition takes the following form:

$$A^M E \varepsilon = A E (\varepsilon - \varepsilon^o)$$ (2)

where E_i denotes Young modulus and $\sigma_i = E_i(\varepsilon_i - \varepsilon_i^o)$ describes the constitutive relation for elastic body with distortions.

Let us decompose the final (modified, distorted) state of deformations:

$$\varepsilon_i = \varepsilon_i^R + \varepsilon_i^L = \sum_j D_{ij} \varepsilon_j^o + \varepsilon_j^L$$ (3)

where ε_i^R denotes the *initial* deformations caused in the initial configuration of the structure by distortions ε_i^o and D_{ij} denotes the influence matrix describing initial deformations ε_i^R caused in the element i by the unit distortion $\varepsilon_j^o = 1$ imposed to the element j.

Substituting Eq.2 to Eq.1 the simulation conditions take the following form:

$$B_{ij} \varepsilon_j^o + \zeta_i \varepsilon_i^L = 0$$ (4)

where:

$$B_{ij} = \zeta_i D_{ij} + \delta_{ij}$$

$$\zeta_i = (A_i^M - A_i) / A_i$$

δ_{ij} – the Kronecker symbol.

Having determined the influence matrix D_{ij}, all modifications of structural deformations due to structural redesign (determined by the vector ζ_i) are described by the initial state of deformations: $\varepsilon_i^R = \sum_j D_{ij} \varepsilon_j^o$, where ε_j^o has to be calculated from the simulation equation (3).

The computational efficiency of the presented method arises from the fact that the technique of local corrections to the state of stresses and deformations is considerably cheaper then renewed global analysis (with stiffness matrix reformation). The effect of saving of the computer time is particularly high if redesign modificates only a small part of structural elements.

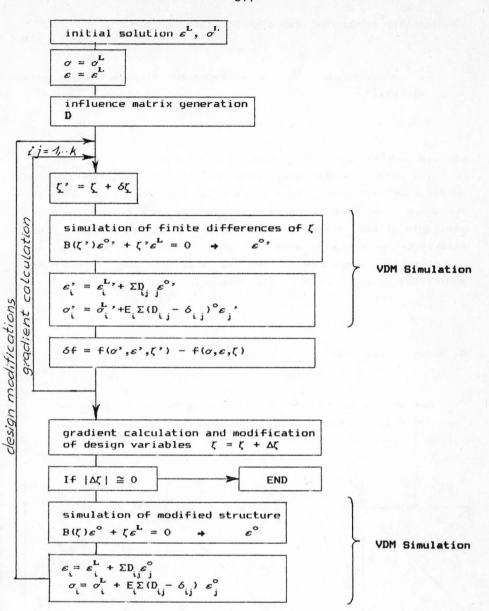

Table 1 *The algorithm of optimal redesign simulated by VDM*

2. Algorithm of Optimal Redesign Simulated by VDM

Let us consider l objective functionals (and side constraints):

$$f^i = f^i (\sigma(\zeta_j), \varepsilon(\zeta_j), \zeta_j) \qquad \begin{array}{l} i=1,2,\ldots l \\ j=1,2,\ldots k \end{array} \qquad (5)$$

where k denotes the number of design variables.

The main computational cost of the optimization process corresponds to the gradient calculations for the objective functions and side constraints. In the presented approach (cf.Table 1) the gradient calculation is done by the finite difference method simulated by virtual distortions. The same *VDM simulation* procedure is used to simulate subsequent design modifications and gradient calculations.

3. Simple Truss Example

Let us discuss the simple truss example shown in the Fig.1. The numerical process of structural redesign to minimize the volume of material:

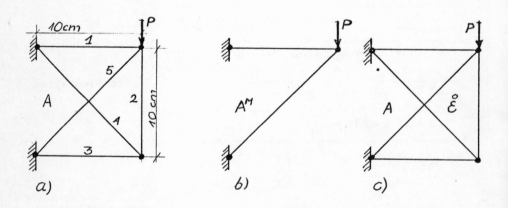

Fig.1 *Example of truss structure: (a) initial configuration; (b) modified (final) configuration; (c) distorted configuration*

i	$A_i \sigma_i$ [100N]	ε_i	A_i [cm]	ε_i^o	$A_i \sigma_i^R$ [100N]
Initial configuration:					
1	5.578	0.056	1.0	0.0	0.0
2	-4.422	-0.044	1.0	0.0	0.0
3	-4.422	-0.044	1.0	0.0	0.0
4	6.254	0.063	1.0	0.0	0.0
5	-7.888	-0.079	1.0	0.0	0.0
Final configuration:					
1	10.000	0.052	1.923	-0.047	4.422
2	0.000	-0.039	0.000	-0.039	4.422
3	0.000	-0.039	0.000	-0.039	4.422
4	0.000	0.039	0.000	0.039	-6.254
5	-14.142	-0.052	2.720	0.092	-6.254

Table 2 *Results of simulation process for the truss structure*

$$\min V \tag{6}$$

$$V = \sum_i 1_i A_i^M = \sum_i 1_i A_i (1 + \zeta_i)$$

subject to side constraints:

$$| \sigma_i | \leq \bar{\sigma} \tag{7}$$

was computed assuming $A_1 = .. = A_5 = 1cm^2$, P=1000N, $E_1 = .. = E_5 = 100MPa$, $\bar{\sigma}$ = 5.2MPa. The results are shown in Table 2.

4. Computational Efficiency of Sensitivity Analysis

Let us consider an objective function

$$f = f (u, \zeta) \tag{8}$$

and its derivative

$$df/d\zeta = \partial f/\partial\zeta + z^T du/d\zeta \tag{9}$$

where z is a vector with components $z_i = \partial f/\partial u_i$.

The first term in Eq.(9) is usually zero or easy to obtain, so we discuss only the computation of the second term. Differentiating the equilibrium equations for a finite element model:

$$K u = p \tag{10}$$

with respect to ζ we obtain:

$$K \, du/d\zeta = dp/d\zeta - dK/d\zeta \, u \tag{11}$$

or, premultiplying Eq.11 by $z^T K^{-1}$ we obtain:

$$z^T \, du/d\zeta = z^T K^{-1} (dp/d\zeta - dK/d\zeta \, u). \tag{12}$$

Calculation of $z^T du/d\zeta$ may be performed in two different ways (cf.Ref.2);

a) the direct sensitivity approach consists of solving Eq.11 for $du/d\zeta$ and then calculating the scalar product with z,

b) the adjoint sensitivity approach determines an adjoint vector $\underline{\lambda}$ as the solution of the system

$$K \, \underline{\lambda} = z \tag{13}$$

and then calculates $df/d\zeta$ from Eq.9 written in the form:

$$df/d\zeta = \partial f/\partial \zeta + \lambda^T (dp/d\zeta - dK/d\zeta \, u). \tag{14}$$

The cost of gradient calculation can be now estimated (neglecting the cost of K' calculation). Using subscripte m to denote the number of multiplications and α to denote the number of additions the cost of direct sensitivity analysis is equal:

a)
$$k(n_m^2 + n_\alpha^2) \quad \text{for calculation of } K'u$$
$$k(n_m^2 + n_\alpha^2) \quad \text{for calculation of } du/d\zeta = -K^{-1}(K'u)$$
$$kl(n_m + n_\alpha) \quad \text{for calculation of } df/d\zeta = z \, du/d\zeta$$

where k is the number of design variables, l is the number of objective functions and n is the number of degrees of freedom. Similarly, the cost of adjoint sensitivity analysis is equal:

b)
$$k(n_m^2 + n_\alpha^2) \quad \text{for calculation of } K'u$$
$$l(n_m^2 + n_\alpha^2) \quad \text{for calculation of } \underline{\lambda} = K^{-1}z$$
$$kl(n_m + n_\alpha) \quad \text{for calculation of } df/d\zeta = -\underline{\lambda}^T (K'u).$$

On the other hand, the numerical cost of gradient calculation by the VDM simulation can be estimated as follows (neglecting the cost of initial calculation of the influence matrix D_{ij}):

$k \ k_m + k^3 (k_m + k_a)$ — the crude approximation for the solution of the simulation problem $B(\zeta')\underline{\varepsilon}^{\circ\prime} = -\zeta'\underline{\varepsilon}^L$ where $\zeta' = \zeta + \delta\zeta$

c) $k(k_m^2 + k_a^2)$ — for calculation of $u' = D\varepsilon^{\circ\prime}$

$k(n_m + n_a)$ — for calculation of $du/d\zeta = (u' - u)/\delta\zeta$

$kl(n_m + n_a)$ — for calculation of $df/d\zeta = \partial f/\partial u \ du/d\zeta$.

Now, we can estimate when the cost of VDM simulation is cheaper then another sensitivity approach. This estimation determines the coefficient α such, that for $k < \alpha n$ the VDM approach is more economical. If $k<l$ then the direct sensitivity approach is more economical than the adjoint one. Then, the comparison of costs of computations (eg. taking only number of multiplications into account) a) and c) leads to the following inequality:

$$n^2\alpha^3 + n\alpha^2 + \alpha + 1 - 2n < 0. \tag{15}$$

We can check that for $n=100$ the solution of (15) is $\alpha<0.27$ and for $n=1000$ the solution is $\alpha<0.12$. Analogously, if $l<k$ then the adjoint sensitivity approach is more economical than the direct one. Then, the comparison of costs of computations b) and c) leads to the inequality

$$n^2\alpha^4 + n\alpha^3 + \alpha^2 + (1-n)\alpha - 1 < 0. \tag{16}$$

We can check that for $n=100$ the solution of (16) is $\alpha<0.25$ and for $n=1000$ the solution is $\alpha<0.10$.

Summing up, it can be said that for redesign process with only 10 - 25% of modificable elements the VDM simulation is more economical then the other sensitivity analysis methods. It is worth noting that the response of a structure for redesign is automatically calculated during the VDM sensitivity analysis and no extra computation is needed.

5. Conclusions

The VDM simulation approach is particularly effective technique if a small number of design variables is modificated in a repetitive iteration process. For example, a shape control of structural boundary can be treated as a problem with a small area of modificable boundary layer.

In the case of the structure with all modificable elements the non-gradient approach with virtual distortions generated in all structural elements in an iterative way can be applied. For example, the idea of iterative optimal remodeling with shifting material from overloaded to underloaded spots (Ref.3,4) can be used.

The VDM simulation technique allows to combine various modification problems within the same approach. For instance, the problem of optimal plastic design involves simultaneous simulation of modifications of both material distribution and constitutive properties (permanent plastic deformations) cf.Ref.5.

References

1. J.Holnicki-Szulc, J.T.Gierliński, Structural Modifications Simulated by Virtual Distortions, Int.J.Num.Meth.Eng.,vol.28, pp.645-666, 1989

2. R.Haftka, Elemants of Structural Optimisation, M.Nijhoff, Boston, 1985

3. N.Olhoff, J.E.Taylor, On Optimal Structural Remodeling, J.Optim.Th.Appl., vol.27,No.4,pp.571-582,1979

4. J.Holnicki-Szulc, Optimal Structural Remodeling-Simulation by Virtual Distortions, Comm.Appl.Num.Meth.,vol.5,pp.289-298,1989

5. J.Holnicki-Szulc, J.T.Gierliński, Optimisation of Skeletal Structures with Material Nonlinearities, Proceedings of 1st International Conference on Computer Aided Optimum Design of Structures, OPTI/89, Southampton, June 1989, edit.C.A.Brebia /S.Hernandes, Comp.Mech.Publ/Springer-Verlag

ON OPTIMAL STRUCTURES IN ENGINEERING

P. Knödel
Versuchsanstalt für Stahl, Holz und Stei
University of Karlsruhe, Kaiserstr. 12,
D-7500 Karlsruhe 1, FRG

1.   Introduction

1.1 Optimal structures in engineering

In many applications of structural engineering, the designer wishes to
achieve an 'optimal' structure. 'Optimal' is then spoken about in
terms like 'minimum weight with maximum performance' or 'minimum costs
with maximum performance'. The first and most strict of those demands
defines light-weight structures, like they are being used in the air-
and spacecraft engineering, where things have to be airbourne and
money does (almost) not matter. The second of those demands defines
light-weight structures of structural engineering (tanks and silos,
trussed roofs, profiled sheeting), where things should not move after
site-erection and money matters a lot.

Both of those have created light-weight structures, which are so in-
credible thin-walled (shells with wall-thickness vs. diameter far less
than 1/500 /10, 14/) and slender, that they are likely to buckle elas-
tically and thus besides strength-design creating a new class of problems in engineering mechanics /9, 11/.

1.2 Optimal structures in nature

Of course nature has developed light-weight structures, some of them
have become proverbial. An egg shell has a ratio wall-thickness vs.
diameter of about 100, and a part of a spiders leg (tibia), which is
almost a perfect circular cylindrical tube, has a ratio wall-thickness
to diameter of about 1/20 to 1/60 /4/.

One can see from those examples, that thin-walled structures of nature
are almost one order of magnitude thicker than thin-walled structures

neering. Thus nature avoides problems of elastic instability,
h under load control might cause sudden and catastrophic collapse.
nd if it has to be an extreme light structure, it will be not made up
of a single, very thin-walled shell: A bundle of tubes will make up a
relatively thick-walled non-isotropic shell, so that neither the global dimensions of the shell, nor the measures of the single tubes are
'thin-walled', at a total weight almost as low as of the isotropic
shell.

Up to now, we are not able to build such structures at due expenses.
We do build sandwich structures, which serve well by transferring
shearing forces through a comparable thick foam core whith comparable
poor mechanical properties, where excellent bending stiffness is provided by high-strength deck liners. But to build a structural element
like a bone of a birds wing, sturdy by its outer dimensions, but extreme light by its inner filigree-trusswork, without any of those thin
'needles' being in danger of buckling - that is far beyond our possibilities in manufacturing.

This paper discusses briefly some aspects of optimal structures in
engineering from a civil engineer's point of view.

2.   Criteria

In this section some criteria are outlined, with respect to which a
structure could be optimized.

2.1 Loading

When designing a structural component, one has to have a certain im-
pression of the kind of loading which will act onto this part. If the
complete load pattern is known, the structure can be designed to be
optimal for this load pattern, e.g. constant stress will be governing
each part of the component.

In many cases however, it is unexpectedly difficult to determine the
actual load pattern:

- The loading of the structure may be of stochastic character, e.g.
  due to wind or granular goods inside a silo (see /8, 14/). If the
  stochastic properties are known very well, it is at least possible
  to predict a certain propability for the crucial combination of

gusts, that will cause maximum bending moment in the shaft of a guyed mast /15/.

- Load is induced by adjacent structural parts. The actual loading pattern is then determined by the design and realization of the joints. Remainders of friction, where a hinge is designed, uneven foundations, where continuous support is designed, as well as geometric imperfections and misalignment /12/ can change the assumed loading pattern completely.

2.2 Durability

2.2.1 Fatigue

If the loading of a structural part is varying in time, fatigue problems become important. In order to prevent early fatigue cracking, it is essential to avoid geometrical notches. This means e.g. for welded steel structures, to design smooth transitions at different wall thicknesses, and to use flush-ground butt welds only, where for static loading untreated fillet welds would do /13/.

2.2.2 Corrosion

The medium, which is acting onto the structure (as primary action) or which is simply surrounding the structure but affecting it (as secondary action), can require special considerations. In such cases, where it is not possible to chose a resistant material, the structure has to be coated, which again calls for a special type of design: smooth surfaces, all gaps between adjacent parts sealed or tight-welded, all inaccessible areas closed airtight - where the choice of a proper coating system is another problem.

2.3 Serviceability

The deflections of a structure must be limited, even if (at first) strength is not affected (water ditches on roofs, second order effects with columns). Eigenfrequencies of building components have to be kept within certain ranges, in order to avoid people feeling uneasy.

2.4 Maintenance

Maintenance can be very costly during service life of a structure, e.g. painting works for a fine-skeletal steel structure like the Eiffel-tower. This requires small surfaces to be repainted, all parts to be easily accessible, possibilities to change parts during service (e.g. guying cables).

2.5 Quality Control

Very often it is neglected, that even the necessities of quality control affect the designing process of a structure:

- If a designer decides to use prestressed concrete for a bridge, he must be aware, that prestressing is a fine and efficient technology, if the cables are properly grouted after tensioning. If this is not ensured by quality control measures, the high-strength wires, which are very sensitive to stress corrosion, will be likely to fail within a few years.

- It is well known, that shot peening induces compressive residual stresses in the surface of a weld and thus inhibits the development of cracks with fatigue loading. Therefore this might be used in a controlled manufacturing process of e.g. machine parts, but up to now it is impossible to use this for welded steel construction, because there is no way to check the state of treatment of every single weld (comp. /16/).

3.   Structural Safety

3.1 Design-Load and Purpose

One purpose, which the designer defines as (main-)task of the structure, is followed in designing the structure, which means to model loading pattern and structural response only uni-dimensionally /17/. However, most structures have to serve several purposes (of declining importance), and the designer is likely to disregard some, which later on might prove to be important for the performance of the structure.

In 'reality' various influences, which are seldom known from the very beginning of the design process, cause the requirements for the structural behaviour to be multi-dimensional. Only a wholistic approach to the structural task with a thorough consideration of all 'ifs' or 'buts' can prevent from 'forgetting' one important aspect while optimizing with regard to another.

3.2 Shape

If one optimizes a structural member for compressive forces and/or bending in random direction, one might end up with a relatively thin-walled cylindrical shell. This shape, optimal (almost only, see /3/) by design, exhibits a high capability to bear compressive stresses in compression and bending.

Unfortunately, we are not able to manufacture such optimal designed shapes under usual conditions of steel building. The sheets get pre-buckles and residual stresses by milling, misalignments by assembling, and more prebuckles and residual stresses by joint-welding. Again unfortunately, a cylindrical shell is extremely sensitive to these imperfections (even if these are within the requirements of the respective codes /12/), so that the overall bearing capacity can be reduced as low as 15 % compared to the designed 'optimum'.

3.3 Material

In 'conventional' design, engineering materials are modelled mostly as homogenious, isotropic continuum. This means, that makroscopic theories are used for the constitutive laws, which are sufficient for application with most metal structures. Even fiber-reinforced materials are treated this way: if the fibres are non-orientated, isotropic properties are assumed, and if the fibres are orientated e.g. in different layers, each layer is treated as an orthotropic, but again homogenious continuum (/1/, comp. /2/).

Real materials are non-homogenious, e.g. a weld with micro-pores (or any other admissible defect) or a steel that has been subjected to ageing, which means a decomposition of the elements and consecutively different properties of the grains themselves and the grain boundaries.

4. Conclusions

Focussing to the shape-optimization of an engineering structure, one is likely to concentrate on one single loading pattern, which is held for the main loading case. Giving shape to the material along the flow of forces in a way, which leads to uniform stress in the entire structure, and reducing the total mass in a way, that this uniform stress is equal to the admissible stress of the material and the actual slenderness, results in an 'optimal' structure.

However, if this structure is really optimal (which means best possible, literally), any variation will make the structural performance worse, whereby variations might be slight changes in the loading conditions as well as slight differences between the manufactured shape and the designed shape. Indeed, the structure has been optimized successfully, but it has become a very sensible, single-purpose structure - not just of the kind we try to achieve in engineering usually.

Therefore, non-optimal structures in terms of shape-optimization are multi-purpose structures, which are robust against inevitable tolerances in manufacturing and changes in loading. They are likely to be optimal in a global sense of economics, where all costs of design, manufacture, quality control, operation and maintenance are understood, including certain levels of risk assessment.

5. Acknowledgements

For valuable discussions on this matter the author is indebted to Mr. G. Steidl, M. Amann and M. Prinz. Literature on structural research in biology has been supplied by Prof. H. Buggisch. Finally, the author wishes to acknowledge the support of the German Science Foundation for his studies on the stability of shells.

6. References

/1/ Barth, F.G.: Microfiber Reinforcement of an Arthropod Cuticle - Laminated Composite Material in Biology. Z. Zellforschung 144 (1973), pp 409-433.

/2/ Barth, F.G.: Skelett bei Gliederfüßern - Leben im Panzerkleid. bild der wissenschaft 9/1978, pp 32-44.

/3/   Blachut, J.: Combined Axial and Pressure Buckling of Shells having Optimal Positive Gaussian Curvature. Computers & Structures Vol. 26, No. 3, pp 513-519, 1987.

/4/   Blickhan, R.: Dehnungen im Aussenskelett von Spinnen. Diss. Frankfurt 1983.

/5/   Cox, H.L.: Comment on /17/. Int. J. Mech. Sci. 15 (1973), pp 855-857.

/6/   Deutscher Stahlbau-Verband: Stahlbau Handbuch - Für Studium und Praxis. 2. Auflage, Band 1, Stahlbau-Verlags-GmbH, Köln 1982.

/7/   Eibl, J. (ed): Proc., 2nd Int. Conf. "Silos - Forschung und Praxis, Tagung '88". SFB 219, University of Karlsruhe 10.-11. Oct. 1988.

/8/   Hampe, E.: Möglichkeiten und Grenzen der stochastischen Ermittlung des Silodrucks. pp 479-490 in /7/.

/9/   Knödel, P., Maierhöfer, D.: Zur Stabilität von Zylindern unter Axiallast und Randmomenten. Stahlbau 58 (1989), H. 3, S. 81-86.

/10/   Knödel, P., Schulz, U.: Buckling of Silo Bins loaded by Granular Solids. pp 287-302 in /7/.

/11/   Knödel, P., Thiel, A.: Zur Stabilität von Zylinderschalen mit konischen Radienübergängen unter Axiallast. Stahlbau 59 (1990), in print.

/12/   Knödel, P., Wolfmüller, F.: Geometric Deviations and Structural Behaviour of Tanks and Silos. Colloquium "Requirements to the Tank-Structures Geometrical Shape", IASS Working Group I "Pipes and Tanks", Sept. 18-20, 1990, Varna, Bulgaria.

/13/   Mang, F., Knödel, P.: Schweissen und Schweissverbindungen. Chapter 9.3 in /6/.

/14/   Mang, F., Knödel, P., Wolfmüller, F.: New Developments in the Design of Metal Single- and Multi-Chamber-Silos. Proc. Vol. I, 10 years of progress in shell and spatial structures, 30 anniversary of IASS, Madrid, 11-15 Sept. 1989.

/15/   Peil, U., Nölle, H.: Measurement of Wind Load and Response of a Guyed Mast. Eurodyn '90, European Conference on Structural Dynamics, SFB 151, Ruhr-University Bochum, June 5-7, 1990.

/16/   Steidl, G.: Schweißtechnische Qualitätssicherungsmaßnahmen beim Bau der neuen Eisenbahnbrücke Rheinbrücke Maxau. Vortragsmanuskript, Stahlbauforum Baden-Württemberg, 6. April 1990, Versuchsanstalt für Stahl, Holz und Steine, Universität Karlsruhe, (unveröffentlicht).

/17/   Thompson, J.M.T.: Optimization as a generator of structural instability. Int. J. Mech. Sci. 14 (1972), pp 627-629. (comp. /5/)

APPLICATION OF THE CAO-METHOD TO AXISYMMETRICAL STRUCTURES UNDER NON-AXISYMMETRICAL LOADING

H. Moldenhauer[1], K. Bethge[2]

[1] Moldenhauer GmbH & Co. KG
Im Brückengarten 9a, 6074 Rödermark 1
[2] Kernforschungszentrum Karlsruhe GmbH
Institut für Material- und Festkörperforschung IV
Postfach 3640, 7500 Karlsruhe 1

Abstract

Harmonic elements with Fourier series expansion for the displacement functions are commonly used for axisymmetrical structures under arbitrary loading. Several 'General Purpose FE-Programs' offer this option. The application of these elements can be extended to the CAO-method. When axisymmetrical structures under arbitrary loading are shape-optimized, the designer normally restricts any geometry modifications to the r-z plane, i.e. the axisymmetrical geometry must be maintained. This paper shows the application of this technique without any FE-source code modification. The main advantages are:

1. 3D-FE computations are avoided.
2. The method yields less conservative, i.e. more accurate solutions than plane strain models.

CAM-manufactured 2D-prototypes are fatigue-tested showing a striking increase in fatigue life.

Usage of harmonic elements

Many structures in the engineering world can be classified to be axisymmetric, with non-axisymmetrical loading. The application of axisymmetrical elements with Fourier series expansions for the displacement components (u, v, w) in a cylindrical coordinate system (r, z, φ) is widely seen in finite element idealizations for these kind of problems. The reason for this is the significant saving of computing time compared to 3D-idealizations.

For simplicity, let the general loading be replaced by an even Fourier expansion, i.e. the loading is symmetrical with respect to the r-z plane. This assumption leads to the following displacement functions:

$$u(r, z, \phi) = \sum_{n=0}^{\infty} u_n(r, z) \cos n\phi \qquad (1a)$$

$$v(r, z, \phi) = \sum_{n=0}^{\infty} v_n(r, z) \cos n\phi \qquad (1b)$$

$$w(r, z, \phi) = \sum_{n=0}^{\infty} w_n(r, z) \sin n\phi \qquad (1c)$$

A complete Fourier series expansion for a loading without any restrictions differs from eq. (1) by additional antisymmetrical terms with respect to the r-z plane (φ = 0), see for example [1]. (The terms "symmetrical" and "antisymmetrical" should be understood with respect to the φ = 0 plane.)

A fillet shoulder under a lateral force or moment is an example with an axisymmetrical geometry and non-axismmetrical load, see Fig. 1. The fillet radius is constant and thus not optimal with respect to a homogeneous stress distribution at the free surface. The CAO-Method is an efficient software tool to remodel the fillet surface yielding a stress concentration factor near 1. This method is extensively described in [2] and [3].

To represent accurately the moment loading by Fourier expansion, the sum in eq. (1) needs only two terms from n = 0 to n = 1, whereas the lateral force needs indefinitely many terms theoretically. However, in practice two terms are sufficient for the latter case as well, because the higher order effects at the position of the singular force diminish at remote locations due to the principle of Saint Venant. If the lateral force would act directly in the fillet radius, then two Fourier terms would not be sufficient, of course.

It should be emphasized that almost all practical applications of simple bending of axisymmetrical structures are represented well by the zero and first order harmonics if forces are applied remotely from the region to be stress-optimized. Thus, there is only a small computational penalty to be paid compared to plane strain or plane stress idealizations. This penalty is drastically overcompensated by the simulation of real 3D effects with the Fourier approach.

Demonstration example

The fillet shoulder shown in Fig. 1 is already presented in [2] for the plane case. This problem is reinvestigated to account for the axisymmetrical geometry. The starting point of stress optimization is characterized by Tab. 1:

	Theory [4]	Abaqus [5]
Plane strain	1.46	1.34
Axisymmetrical	1.32	1.24

Table 1: Stress concentration factors for fillet shoulder under pure bending with geometrical data from Fig. 1.

The difference between theory [4] and the Finite Element analysis results [5] is mainly due to the application of bilinear 4-node elements, which are to be used with the CAO-Method. The weak capability of this element type for resolution of stress gradients is not critical. As long as there is a stress concentration of the right order, this "driving force" will effectively lower stress concentration. For the final and homogeneous stress pattern the simple 4-node element is adequate as stress gradients are eliminated or at least drastically reduced.

Fig. 2 shows the tangential stress along a surface coordinate before and after optimization. The elimination of any stress concentration was achieved after 6 iterations for both plane and axisymmetrical idealization. The optimized surface, however, differs for these cases as is shown in Fig. 3a) and b). A plane strain idealization for axisymmetrical structures under bending yields conservative results at least for the fillet shoulder. The additional material requirements in the fillet region are overestimated, however. The Fourier approach predicts less material requirements and is therefore more accurate.

Experimental verification

For the experimental verification of the CAO-method the 2D-problem of a bending bar with narrowing cross-section was chosen, which in [2] is described in more detail. Starting from a non-optimized form with a diameter ratio of $D/d = 4$ and a circular transition with a radius of $R = 0.12$ d between the cross-sections, a shape-optimized form was designed by the CAO method resulting in a much smoother transition region. Thereby, the local peak stress in the transition could be reduced after a few iteration cycles from 1.8 to 1.1 times the magnitude of the externally applied stress.

A series of specimens having the non-optimized or shape-optimized design were manufactured by CAM from the German structural steel St 37-2 with a yield strength of 300 MPa. Fig. 4 shows an example of a non-optimized and a shape-optimized specimen. Both types of specimens were prepared by spark erosion to an average roughness of $R_a \leq 2$ μm. Nevertheless, the shape-optimized specimen revealed a slightly worse surface finish as the smooth transition has been produced discontinuously in the form of small stairs, due to the limited capability of the spark erosion machine.

The experiments were carried out in a servohydraulic testing machine under bending fatigue load. Fig. 5 shows a photo of the testing device. The specimen is clamped at one side and loaded by a transversal force at the other side. The tests were performed under load control with a sinusoidal load shape at a frequency of 30 cycles/s. For the cyclic loading a constant load ratio of 0.1 was chosen with a maximum compressive load of 11 KN acting at a distance of 38 mm from the circular transition. The tests were run until a limiting deflection of the specimen was reached in consequence of a crack propagating into the specimen. Table 2 shows the results of the tests.

Non-optimized specimen	Shape-optimized specimen
1.540.800	10.000.000 (*)
1.877.300	10.000.000 (*)
2.045.400	10.000.000 (*)
2.417.800	10.000.000 (*)
2.608.700	10.000.000 (*)
3.205.900	11.000.000 (*)
3.539.800	90.000.000 (*)

(*) did not fail

Table 2:   Number of cycles to reach a limiting deflection

While the non-optimized specimen failed only after an average value of 2.500.000 cycles (see Fig. 6), the shape-optimized specimen withstood at least 10.000.000 cycles without visible crack initiation. This means that the fatigue life of the bending bar has been prolonged by a factor larger than 4 by means of the CAO- method. In one case, the specimen was loaded with the same stress until a 36 times longer life was reached without visible crack initiation. In order to initiate a crack in the shape-optimized specimen, the maximum cyclic load had to be increased by 36 percent and the specimens withstood another 2 to $3.5 \cdot 10^6$ cycles until they reached the limiting deflection related to crack presence.

Conclusions

1. So-called harmonic elements may facilitate the shape optimization of axisymmetrical structures under non-axisymmetric loading by drastic reduction of mesh sizes.

2. For reasons of simplicity, plane samples of a shape-optimized shoulder fillet were fatigue tested by swelling bending in the laboratory. Significant increase in fatigue life makes worth the effort of shape optimization especially if the simple and straight forward CAO-method is used.

References

[1] K.E. Buck, Rotationskörper unter beliebiger Belastung, in: K.E. Buck et al., Finite Elemente in der Statik, Berlin, 1973.

[2] C. Mattheck, Engineering Components Grow like Trees, Materialwiss. und Werkstofftechnik 21 (1990) 143 - 168).

[3] C. Mattheck, H. Moldenhauer, Intelligent CAD-Method based on Biological Growth, Fatigue Fract. Engng. Mater. Struct., 13 (1990) 41 - 51.

[4] R.E. Peterson, Stress Concentration Factors, Wiley, New York, 1974.

[5] Hibbitt, Karlsson and Sorensen, Abaqus User's Manual, Version 4.8, Providence, R.I., U.S.A., 1989.

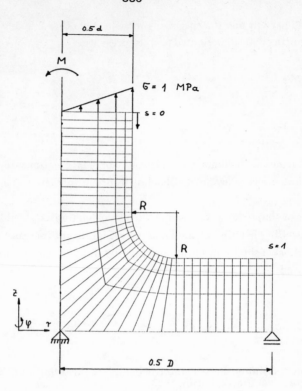

Fig. 1   Fillet shoulder idealization
E = 200000 MPa, ν = 0.3, D/d = 3, R/d = 0.318, s = surface coordinate. Applied
moment produces unit stress at s = 0 for the plane and axisymmetrial model.

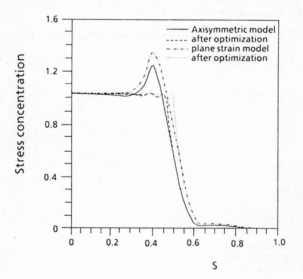

Fig. 2   Stress concentration vs. surface coordinate "s" for plane strain and axisymmetrical
model before and after optimization.

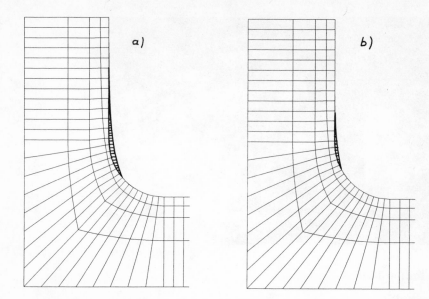

Fig. 3   Optimized fillet shoulder without any stress concentration for the a) plane strain and b) axisymmetrical model after 6 iterations. Shaded region: material requirement for optimization.

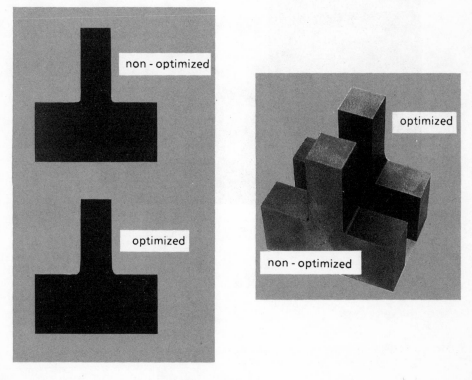

Fig. 4   Non-optimized and shape-optimized specimen

Fig. 5   Testing device

Fig. 6   Crack in the tensile side of the non-optimized specimen.

NUMERICAL SIMULATION OF INTERNAL AND SURFACE BONE REMODELING

T.J. Reiter, F.G. Rammerstorfer and **H.J. Böhm**
Institut für Leichtbau und Flugzeugbau, TU-Wien
Gußhausstr. 27-29/317, A-1040 Wien, Austria

Nomenclature

α, β, γ, k, s, U_{n0}	Modeling constants
C_{i1}, C_{x1}, C_{i2} etc.	Remodeling parameters
$\Delta\rho$	Change of apparent density
ΔX	Growth Increment perpendicular to the surface
E	Young's modulus
ε_{ij}	Strain tensor
σ_{ij}	Stress tensor
ρ_a	Apparent bone density
ρ_c	Density of cortical bone (upper boundary)
U	Strain Energy Density (SED)
U_b	Bulk SED
U_{eff}	Effective SED (stimulus)
U_n	Equilibrium SED

Introduction

Like any living tissue bone undergoes a steady process of material formation and resorption. Special bone cells (Osteoblasts) deposit new material wheras others (Osteoclasts) are removing older bone, creating a balanced state of bone form and architecture as far as mature, healthy and physiologically loaded bone is concerned. It has long been recognized that these so called bone remodeling processes are mainly controlled by the mechanical loading environment to which a bone is exposed (Wolff 1892), giving bone tissue the ability to adapt and even optimize its surface shape as well as its internal structure in respect to its functional requirements. Changes in the actual stress/strain pattern within the bone will stimulate pronounced cell activity resulting in a new equilibrium state. This process of functional adaptation enables bone to endure its mechanical loading with a minimum amount of mass, but on the other hand, as clinical practice shows, it is often detrimental to the long term success of prostheses and implants as used in orthopaedic or dental surgery. Thus the development of methods for predicting the adaptive changes in bone could be of great benefit for clinical applications. Although during the last decades a great deal of theoretical and experimental work has been done to explain the physical and biochemical mechanisms which transmit the mechanical stimulus (local stress and/or strain state) into cell activity (e.g. Kufahl and Subrata 1990; Scott and Korostoff 1990) the process of functional adaptation of bone is not yet fully understood. Thus attention has been focused on the development of phenomenologically based methods, relating mechanically derived stimuli such as stress- or strain tensors, v'Mises stresses etc.

to local bone growth rates *via* simple mathematical equations (Frost 1964; Cowin *et al.* 1985, 1987; Hart 1983; Carter *et al.* 1987, 1989, 1990; Huiskes *et al.* 1987). The optimization strategy on which bone remodeling and functional adaptation is based can advantageously be used in the design of technical structures, too. This offers a good tool especially for structures like composites, which allow density and stiffness variation in a wide range.

Mathematical Model For Numerical Simulation Of Adaptive Bone Remodeling

In accordance with Carter *et al.* 1987 and Huiskes *et al.* 1987 the strain energy density (SED) U, which is given by

$$U = \frac{1}{2} \sigma_{ij} \, \varepsilon_{ij} \tag{1}$$

or the bulk SED

$$U_b = \frac{\rho_c}{\rho_a} \, U \tag{2},$$

which better reflects the strain energy actually stored in the mineralized bone tissue, is taken as an adequate mechanical stimulus for adaptive bone remodeling. To take into account the multiple loading conditions and individual loading time histories experienced by a bone in the course of a typical time period an appropriate superposition of a number of discrete load cases is considered. The individual SED-distributions are weighted according to the corresponding number of load cycles (Carter *et al.* 1987):

$$U_{eff} = \left(\sum_i^c \frac{n_i}{n_t} \, U_{(b)i}^k \right)^{1/k} \tag{3}.$$

Here c stands for the number of load cases, n_i is the number of loading cycles corresponding to load case i, n_t is the total number of loading cycles and k is a modeling parameter weighting the degree of influence of load magnitude and number of loading cycles, respectively.

Following the theory of adaptive elasticity as initially developed by Cowin *et al.* and proposed in a slight modification by Huiskes *et al.*, the difference between the actual effective strain energy density U_{eff}, and a site specific "homeostatic SED" U_n, which is calculated by

$$U_n = U_{n0} + \alpha \rho_a \tag{4},$$

representing the balanced state of no bone remodeling, is utilized as the feedback control variable for determining adaptive changes in bone shape and apparent density. This deviation must exceed a certain threshold level in under- or overloading to cause remodeling activity, so that bone is assumed to react "lazily". This way analogous equations for internal and surface remodeling in a given point can be formulated as illustrated in fig.1:

$$\Delta \rho = \begin{bmatrix} C_{i\,1} \left[U_{eff} - U_{in} \, (1-s_i) \right] & U_{eff} < U_{in} \, (1-s_i) \\ 0 & U_{in} \, (1-s_i) \le U_{eff} \le U_{in} \, (1+s_i) \\ C_{i\,2} \left[U_{eff} - U_{in} \, (1+s_i) \right] & U_{in} \, (1+s_i) < U_{eff} < U_{i\,max} \\ -|C_{i\,3}| & U_{eff} \ge U_{i\,max} \end{bmatrix} \tag{5}$$

$$\Delta X = \begin{bmatrix} C_{x1}\left[U_{eff} - U_{xn}(1-s_x)\right] & U_{eff} < U_{xn}(1-s_x) \\ 0 & U_{xn}(1-s_x) \leq U_{eff} \leq U_{xn}(1+s_x) \\ C_{x2}\left[U_{eff} - U_{xn}(1+s_x)\right] & U_{xn}(1+s_x) < U_{eff} < U_{xmax} \\ -|C_{x3}| & U_{eff} \geq U_{xmax} \end{bmatrix} \tag{6}$$

The index i marks internal remodeling, while the index x is used to identify surface remodeling. $\Delta\rho$ stands for the change in apparent density and ΔX is the surface growth perpendicular to the surface. U_{max} is the maximum SED level to which bone can be exposed without causing actual damage to bone cells yielding overstrain necrosis.

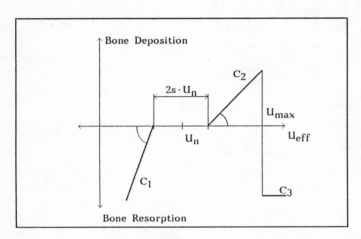

Fig. 1   The relationship between the actual SED, U_{eff}, and the
bone remodeling reaction.

Implementation

An iterative computer procedure has been implemented to allow quantitative predictions of adaptive bone remodeling. The stress and strain fields within the bone tissue and at the surface of the bone, needed for the remodeling scheme formulated above, are derived by linear finite element analysis (DLEARN-Hughes 1987), which has to be performed for each loading case in each timestep. In each timestep the FE-representation of the model is adapted according to the current distribution of U_{eff}.

According to equ.(4) the values of U_{eff} in the surface nodal points are used to adapt the cortical thickness of the bone by relocating these nodel points, thus simulating surface bone remodeling. In order to avoid numerical instabilities resulting from excessive mesh distortion, the nodal points inside the cortex are also moved in a proper manner. To simulate internal adaptive remodeling within the bone tissue the apparent element density ρ_a of every finite element is altered as prescribed by equ.(3). Following an approximation proposed by Carter and Hayes 1977 the Young's modulus of each finite element can be calculated as

$$E = \beta \, \rho_a^{\gamma} \tag{5},$$

treating cortical bone simply as densified trabecular bone differing only in porosity.

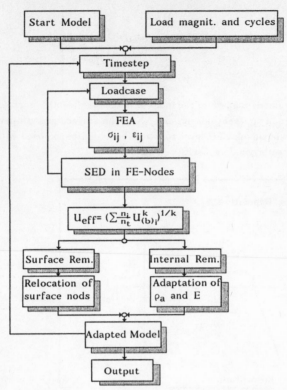

Fig. 2 Flowchart of the adaptive bone remodeling algorithm

Test Problems

Several test problems have been investigated demonstrating the applicability of the numerical approach presented.

Density Distribution in the Proximal Femur

Starting with a uniform apparent density configuration, the distribution in the proximal femur is predicted. Three different typical loadcases, which follow Huiskes *et al.* 1987, are considered, resulting in a bone density distribution (fig. 3) which is in good agreement with those found in real femura.

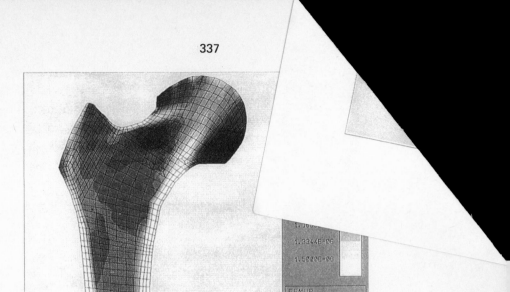

*Fig. 3 Distribution of apparent bone density in the proximal femur, obtained
by computational simulation*

Clamped Beam

To show the principal applicability of the method to structural optimization in engineering
mechanics, in particular, to problems where both shape and material properties are variable
(e.g. composite structures), the optimization of an initially homogeneous, uniformly loaded
beam clamped at both ends (fig. 4) is investigated as a 2D-plane stress-model.

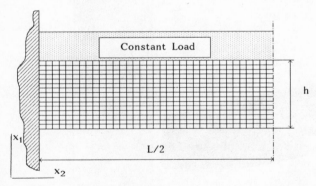

*Fig. 4 Beam under constant load clamped on both sides 2D FE-model
(only the left half is modeled).*

The effects of pure internal remodeling, pure surface remodeling and combined internal and
surface remodeling have been investigated. The use of pure surface remodeling tends to equalize
the SED along the free boundaries only, yielding a shape similar to a beam optimized under
stress constraints (see fig. 5.).

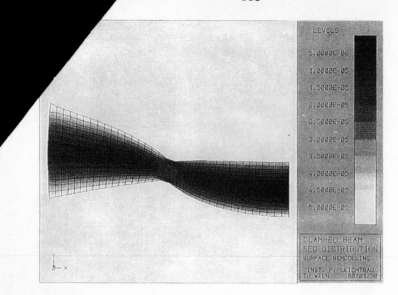

Fig. 5 *Adaptive optimization of a clamped beam under constant load: Distribution of SED after pure surface remodeling.*

In the case of pure internal remodeling bone mass and stiffness are concentrated near the clamped ends and at the surface close to the center leading to a density distribution similar to an I-beam (fig. 6).

Under combined remodeling quite interesting shape and density distribution configurations can be derived, depending on the actual values of the remodeling parameters (see fig. 7).

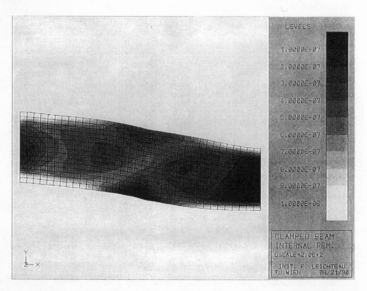

Fig. 6 *Adaptive optimization of a clamped beam under constant load: Distribution of apparent density after pure internal remodeling.*

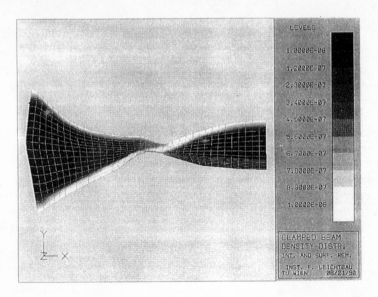

Fig. 7 Adaptive optimization of a clamped beam under constant load: Distribution of apparent density after combined surface and internal remodeling.

Density Distribution around a Single Tooth

A further example belongs to the field of dental surgery. The formation of bone mass around a natural tooth under normal chewing loads (Siegele 1989), supported in the jaw-bone *via* the periodontal membrane is predicted, starting from a configuration of homogeneous density, using a 2D-plane FE-mesh. As shown in fig. 8, bone mass is concentrated around the alveolus, building the well known cortical crest surrounding every single tooth.

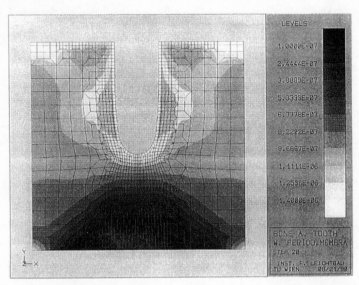

Fig. 8 Distribution of apparent bone density arround a single tooth (Finite Elements representing dental material have not been plotted).

Conclusion

Considering the highly idealized nature of the FE-models used, the results presented in the previous section are in remarkably good, but nevertheless only qualitative agreement with natural bone density distributions. To obtain accurate predictions of *in vivo* bone remodeling responses in a quantitative way as needed to achieve clinical relevance the complexity of bone geometry and architecture (e.g. bone anisotropy, trabecular orientation etc.) has to be taken into consideration. Furthermore the mechanical loading environment as well as the site specific remodeling parameters have to be considered very carefully, which may require further experimental investigations. Numerical approaches to stress induced bone remodeling, combining well established methods of computational mechanics such as the FEM with phenomenological remodeling laws can help to increase the understanding of the stress-remodeling relationship in bone. Despite the fact that the clinical application of algorithms like the one presented here requires further intensive research, the ideas can be used to optimize structures made of materials with varying density and stiffness like composites.

References

Carter, D.R. and Hayes, W.C., " The compressive behaviour of bone as a two-phase porous structure", *J. Bone and Joint Surg.* **59-A**, 1977, pp. 954-962.

Carter, D.R., Fyhrie, D.P. and Whalen, R.T., "Trabecular bone density and loading history: Regulation of connective tissue biology by mechanical energy", *J.Biomech.* **20**, 1987, pp. 785-794.

Carter, D.R., Orr, T.E. and Fyhrie, D.P., "Relationships between loading history and femoral cancellous bone architecture", *J.Biomech.* **22**, 1989, pp. 231-244.

Carter, D.R., Fyhrie, D.P., " Femoral head apparent density distribution predicted from bone stresses", *J.Biomech.* **23**, 1990, pp. 1-10.

Cowin, S.C., Hart, R.T., Balser, J.R. and Kohn, D.H., "Functional adaptation in long bones: Establishing *in vivo* values for surface remodeling rate coefficients", *J.Biomech.* **18**, 1985, pp. 665-684.

Cowin, S.C., "Bone remodeling of diaphyseal surfaces by torsional loads: theoretical predictions", *J.Biomech.* **20**, 1987, pp. 1111-1120.

Frost, H.M., "The laws of bone structure", C.C.Thomas, Springfield, IL., 1964

Hart, R.T., "Quantitative response of bone to mechanical stress", Dissertation, Case Western Reserve University, 1983.

Hughes, T.J.R., "The Finite Element method: Linear, static and dynamic finite element analysis", Prentice Hall, Englewood Cliffs, NJ, 1987

Huiskes, R., Weinans, H., Grootenboer, H.J., Dalstra, M., Fudala, B. and Slooff, T.J., "Adaptive bone-remodeling theory applied to prosthetic-design analysis", *J.Biomech.* **22**, 1987, pp. 1135-1150.

Siegele, D., "Numerische Untersuchungen zur Optimierung der durch Zahnimplantate bedingten Beanspruchungen des Kieferknochens", Dissertation, Universität Kahrlsruhe, 1989.

Wolff, J., "Das Gesetz der Transformation der Knochen", Hirschwald, Berlin 1892.

ENGINEERING ORIENTED APPROACHES OF STRUCTURAL SHAPE OPTIMIZATION[*]

Gu Yuanxian, Cheng Gengdong
Research Institute of Engineering Mechanics
Dalian University of Technology
Dalian 116024, China

ABSTRACT

The present paper describes some engineering oriented approaches employed in structural shape optimization. A geometry-based approach has been used to integrate modelings of three phases, i.e. structural shape modeling, design optimization modeling and finite element modeling. The strategy of user's programming interfaces is proposed to deal with special application problems and make optimization program really flexible. The versatile sensitivity analysis is implemented with semi-analytical scheme, which is particularly efficient to compute the design sensitivity with respect to variations of load, temperature and boundary conditions caused by shape changes. These approaches have been proved to be practically valuable by the development and application of a general purpose package of micro-computer aided optimum design for structures, MCADS. Two application examples of MCADS, dealing with optimum shape design of continuum structures coupling fields of stress and temperature, are presented.

1. Introduction

The optimum shape design of continuum components and structures is a very important subject in the field of engineering optimization. There are many new features in this subject different from the traditional structural optimization [7]. For instance, design variables have changed from the size parameters of element cross-section to the shape parameters of structural boundary. The optimization object is usually to reduce the stress concentration, but not to minimize the structural

---

[*] The project supported by National Science Foundation of China (NSFC)

weight. In addition more types of finite element, such as membrane, plate, shell, brick and axisymmetric brick, etc., are needed for the structural modeling and analysis. These features present difficulties in sensitivity analysis, optimization algorithm and design modeling of structural shape optimization.

To make the shape optimization of continuum structures really valuable for practical design processes of industry, efficient numerical techniques and gineering oriented approaches are still to be developed. First of all, the shape optimization should be based on existing commercial packages of general purpose FEM analysis. As such, complicated practical structures can be modeled with rich element library and various facilities of description of analysis attributes possessed in FEM packages. Sensitivity computation and structural modeling are two critical techniques for bridging optimization and FEM analysis, both of which are dependent upon the description and the control of structural shape.

With the integration of FEM and optimization in mind, a general purpose program for structural shape and size optimization, MCADS (Micro-Computer Aided optimum Design system for Structures), has been developed from a commercial FEM program DDJ-W [6]. The semi-analytic method of sensitivity analysis implemented in MCADS with DDJ-W as a black box is versatile for general cases, and particularly effective for the computation of such special kind of sensitivities with respect to variations of load, temperature and boundary conditions caused by shape changes. The structural modeling of MCADS is a geometry-based approach which makes direct use of engineering parameters and geometric curves/surfaces to describe structural shape, and introduces "natural variable" and "design element" into the shape optimum design.

The user's interface is a well known concept in software programming. Nevertheless, the applicable user's programming interface (rather than data interface) has been seldom provided in engineering softwares, especially in structural optimization. In engineering design processes, there are always some project-related problems can not been foreseen in programming phase and treated with an unified form. Therefore, an programming interface open to user is very useful to deal with such application oriented problems.

These engineering oriented approaches implemented in MCADS and two application examples of shape optimum design of continuum structures with coupling of stress field and temperature field have been presented in this paper.

2. Geometry-based modeling for shape optimization

The method of design modeling used in traditional structural size optimization is to build design model on the base of finite element model and take finite element parameters as design variables. This approach is not suitable for shape optimization of continuum structures. A geometry-based approach is proposed for the modeling of shape optimization with MCADS. Its main ideas can be explained as following: (1) Using engineering parameters and geometric curves/surfaces directly for the design modeling and shape description. (2) Integrating modelings of three different phases and carrying out them in the order of structural geometry modeling, shape optimization modeling and finite element analysis modeling. (3) Introducing "natural variable" and "design element" into shape optimization of continuum structures.

In the geometric modeling, commonly used curves and surfaces, e.g. lines, arcs, splines and quadratic surfaces, have been employed to fit structural boundaries. Geometric parameters interested to engineers, e.g. radius and central position of arcs, coordinates of control nodes of splines, are selected as design parameters to describe and modify structural shapes, of which changeable parameters are chosen as design variables, that are so-called "natural variables". Then, the domain of structure is divided into some basic geometry entities, normally some regular shaped sub-regions. This domain division is aimed at shape optimization as well as finite element mesh generation. Those sub-regions near to changeable boundaries are defined as "design elements", and their edges and shape are controlled by the natural variables. Regular shaped sub-regions are also used as mapping-elements of automatic mesh generation with mapping method. The local part of mesh, load and boundary support within the design element have been updated during the shape optimization.

The basic parameters of this modeling approach are geometric parameters which are most interested to engineering design. All of attribute information, i.e. shape design variables, finite element mesh, loads and boundary conditions, are defined on the geometric entities and related directly to those control parameters of boundary curves and surfaces. Such a geometry-based modeling approach is engineering oriented and makes structural modeling as close to a real design as possible. The design modeling and analysis modeling of this approach have been unified.

This integrated modeling approach has been implemented in MCADS via a mesh generator, MESHG. This program produces not only data of finite element model, and data of shape optimization model as well. For example, the cross-section shape of a train's wheel to be optimized is shown in Fig.1. Boundaries near the hub and the rim are not permitted to change, and the low- and up-boundary of middle part are optimized. We use spline curves to fit changeable boundaries, and chose z-coordinates of ten control nodes of splines as natural variables. The whole domain is divided into three sub-regions, i.e. mapping-elements for mesh generation, and the No.1 sub-region is design element. The mesh shown in Fig.2 and thermo-load caused by temperature field are updated with changes of the boundary shape.

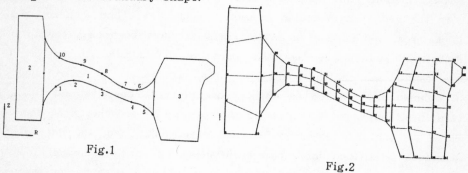

Fig.1

Fig.2

3.Semi-analytic method for sensitivity analysis with load variation

The sensitivity analysis is critical to link shape optimization with FEM analysis. The programming implementation of sensitivity analysis is extremely time consuming for the shape optimization of continuum structures. The difficulty is mainly due to complicated relations between various types of elements and shape design variables. Although there are many papers discussing on various formulas of shape sensitivity analysis, programming implementation of versatile sensitivity analysis is still important for practical application of shape optimization.

The semi-analytic method of sensitivity analysis is an attractive approach for shape optimization based on general purpose FEM analysis packages [1,2]. It has both the advantage of easy programming of finite difference method and the advantage of efficient computing of analytic method. This approach has been implemented in MCADS using an structural analysis package, DDJ-W, as a half-open "black box". The half-open means that the data structure and the subroutine calling of DDJ-W are open to programmer of MCADS, but the program of DDJ-W itself has not been modified. Thus, all of elements and modeling facilities of DDJ-W have been maintained, and any structure that DDJ-W can analyze can also be optimized by MCADS.

The accuracy of semi-analytic method in shape optimization has been deeply studied in [3]. An alternative forward/backward difference scheme has been accepted in MCADS to improve accuracy. The further study on improvement of semi-analytic method by adding second order correction is presented in [5]. According to our studies and experiences, the accuracy of semi-analytic method is usually enough for shape optimization of commonly treated structures modeled with elements such as membrane, brick and axisymmetric brick.

One important feature presented in the paper is that the semi-analytic method is particularly efficient to compute sensitivities with respect to variation of load, temperature, boundary conditions, etc., caused by shape change. This special kind of sensitivities is difficult to treat with analytic methods and is usually ignored.

The structural analysis with finite element method is expressed as

$$K\,U = P \tag{1}$$

and sensitivity of displacement U with respect to design variable x_i is calculated by solving the equations

$$K\frac{\partial U}{\partial X_i} = \frac{\partial P}{\partial x_i} - \frac{\partial K}{\partial x_i}U \qquad (i=1,2,....,n) \tag{2}$$

In the traditional structural optimization, the term $\partial P/\partial x_i$ is usually assumed to be zero, and Eq. (2) is simplified to

$$K\frac{\partial U}{\partial X_i} = -\frac{\partial K}{\partial x_i}U \qquad (i=1,2,....,n) \tag{3}$$

This assumption is acceptable in general cases of size optimization, but it is no longer true in some cases of shape optimization. If load P relates to structural shape, for instance distributed pressure load and gravitational force, thermo-load caused by temperature field, centrifugal force, etc., the term $\partial P/\partial x_i$ can not be ignored. Since it will possess critical influence on sensitivities for such problems as thermo-stressed structural shape optimization. Besides, when boundary conditions are shape dependent, e.g. a boundary support is dependent upon boundary shape, and the $\partial K/\partial x_i$ is also related to the variation of boundary conditions. These shape dependent variations of load, temperature, boundary conditions, etc., can be computed effectively by semi-analytic method without complicated derivation and programming.

Here, we give an example of the train's wheel which design model and mesh of axisymmetric elements have been shown in Fig.1 and 2. Its temperature field caused by braking is shown in Fig.3. We consider maximum thermal stresses arising at nodes 43, 49, 40, 63 and 60, and maximum radial displacement u_{34}. When boundary shape is changed by giving a perturbation $\Delta x_i=1.0$ to design variables in turn, we consider the influence of shape change on thermo-load as well as structural responses. Computed increments of thermal stresses and displacement are given in Table 1. The values on lines corresponding to D_T and D_{NT} are increments of thermal stresses and displacement. They are computed by reanalysis under new thermo-load modified with shape change and old thermo-load, respectively. The values on lines corresponding to S_T and S_{NT} are linear approximations of increments with first order sensitivities calculated by means of semi-analytic method with and without, respectively, consideration of thermo-load variation.

The following conclusions can be drawn from numerical results:

1. The variation of structural response with respect to shape change is caused by two facts, the mesh change and load change.

2. The influence of thermo-load change caused by shape modification is significant for thermo-stressed structures, and therefore can not be ignored in sensitivity analysis and reanalysis.

3. The semi-analytic method is accurate for shape sensitivity analysis of thermo-stressed structures composed of axisymmetric brick elements.

4. The temperature field is less sensitive than thermal stress and displacement to structural shape change. Computational example will be mentioned in section 5.

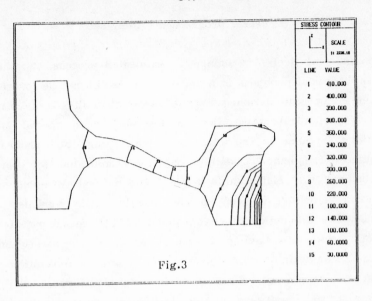

Fig.3

STRESS CONTOUR

SCALE
1: 2236.18

LINE	VALUE
1	410.000
2	400.000
3	390.000
4	380.000
5	360.000
6	340.000
7	320.000
8	300.000
9	260.000
10	220.000
11	180.000
12	140.000
13	100.000
14	60.0000
15	30.0000

Table 1.

		σ_{43}	σ_{46}	σ_{40}	σ_{63}	σ_{60}	u_{34}
Initial value		91.798	77.396	57.937	63.958	56.989	1.6937
$\Delta x_1 = 1.0$	D_T	-0.2632	-2.3835	3.1606	-0.1363	-0.1154	0.00023
	D_{NT}	-0.0656	-3.4452	4.3411	0.0614	-0.2565	0.00059
	S_T	-0.2585	-2.3837	3.2855	-0.1316	-0.1111	0.00011
	S_{NT}	-0.0557	-3.4781	4.4698	0.0734	-0.2497	0.00060
$\Delta x_2 = 1.0$	D_T	1.9073	0.9979	-2.1929	-0.7598	-0.0329	0.00614
	D_{NT}	4.1755	3.2377	-1.8452	-1.5590	0.1503	0.00729
	S_T	1.9502	1.2803	-2.1294	-0.7631	-0.0006	0.00608
	S_{NT}	4.3168	3.5599	-1.7286	-1.5428	0.2053	0.00730
$\Delta x_3 = 1.0$	D_T	-2.2026	0.6814	-1.6457	0.1865	0.7666	0.00556
	D_{NT}	-2.5066	1.7575	-2.0589	2.4451	2.7149	0.02764
	S_T	-2.2272	0.7277	-1.6497	0.2352	0.7005	0.00520
	S_{NT}	-2.4057	1.9791	-1.9676	2.5263	2.5871	0.02785
$\Delta x_4 = 1.0$	D_T	0.6396	-0.2021	0.5147	0.7164	0.6633	-0.00036
	D_{NT}	-0.3343	-1.0171	-0.5444	3.3942	-1.8204	0.04743
	S_T	0.6077	-0.2174	0.4914	0.9285	0.5664	-0.00003
	S_{NT}	-0.2497	-0.9394	-0.0922	4.1579	-2.2200	0.04750
$\Delta x_5 = 1.0$	D_T	-0.1799	0.2709	-0.1843	0.8399	-0.3611	-0.00001
	D_{NT}	-0.1962	0.3844	-0.2079	-3.6641	-2.3054	0.04787
	S_T	-0.2426	0.2409	-0.2265	0.4937	-0.8310	-0.00003
	S_{NT}	-0.2518	0.3763	-0.2476	-3.9478	-2.8664	0.04750
$\Delta x_6 = 1.0$	D_T	0.5201	0.2407	0.3560	-3.1865	1.4938	0.00265
	D_{NT}	0.6530	0.4493	0.4189	-9.6704	-0.2139	0.05068
	S_T	0.5502	0.2620	0.3776	-3.0853	1.6495	0.00035
	S_{NT}	0.7578	0.5093	0.4967	-12.527	0.2757	0.04866
$\Delta x_7 = 1.0$	D_T	-0.1855	-0.0393	-0.0777	1.4369	-3.1492	0.00356
	D_{NT}	-0.6439	-1.0526	-0.2356	0.0336	-10.663	-0.00837
	S_T	-0.1886	-0.0839	-0.0669	1.4164	-3.2214	0.00143
	S_{NT}	-0.6290	-1.1059	-0.1929	0.0504	-10.847	-0.01031
$\Delta x_8 = 1.0$	D_T	0.6504	-0.3003	0.4226	-0.0737	-0.2707	-0.00103
	D_{NT}	-1.2793	-1.4844	-0.7724	1.0069	-0.4058	-0.00348
	S_T	0.4964	-0.2892	0.3659	-0.0588	-0.3985	-0.00135
	S_{NT}	-1.4497	-1.5121	-0.8095	1.0879	-0.4915	-0.00349
$\Delta x_9 = 1.0$	D_T	-0.9210	-1.1121	0.5750	-0.2496	-0.6696	-0.00309
	D_{NT}	-0.9659	-0.6966	0.1555	-0.1486	-0.2070	-0.00210
	S_T	-0.9609	-1.0589	0.5683	-0.2480	-0.6806	-0.00338
	S_{NT}	-1.0569	-0.8364	0.1878	-0.1616	-0.2353	-0.00235
$\Delta x_{10} = 1.0$	D_T	-0.9543	0.7400	-0.7797	0.0422	-0.0731	-0.00081
	D_{NT}	-0.5138	0.5015	-0.3662	0.1038	-0.0492	-0.00065
	S_T	-0.9271	0.7588	-0.7794	0.0386	-0.0775	-0.00087
	S_{NT}	-0.6257	0.5156	-0.3481	0.0948	-0.0477	-0.00046

4. User's programming interface

The Engineering design is usually a complicated process, and there are always some special problems in a design process can not be foreseen and expressed in an unified scheme. Thus, it is difficult for a commercial package of structural optimization to deal with such special problems with definite formulae in the stage of system design and programming. For instance, the shape modeling may be complicated for particular structures, and the relationships between the natural variables and the interpolation curves/surfaces varies greatly with various shaped boundaries. The calculation of changeable load, stiffness parameters and stress of components having a special shape of cross-section, and particular constraint functions, etc., are application-related requirements and can not be expressed in a definite forms.

In order to make software MCADS flexible in practical application of engineering, a programming interface has been developed and opened to users. This interface is composed of some subroutines dealing with some specific application-related problems and a mechanism storing and accessing user defined data. These user opened subroutines are located in one module and designed weakly related to other subroutines. They are called in optimization processes to finish a definite job, but their detailed contents are independent to program. The user is permitted to rewrite the internal contents of these subroutines for particular application problems. The storage/access mechanism of user data is provided to process a group data defined and used for modification of these subroutines. Such an interface is application oriented and supports user's programming on the base of MCADS.

By means of this interface technique, the design model can be extended easily by user himself via modifying user opened subroutines to meet specific requirements of practical applications. For example, shape optimization of turbine engine disk shown in section 5 is completed by programming only two subroutines of user's interface. One is used to determine the relationship between natural variables and control nodes of boundary curves, another is used to compute the vector of centrifugal forces changed with respect to structural shape. While all of other parts of MCADS need not be modified. There are some other subroutines user opened to deal with special requirements of structural optimization.

5. Application examples of shape optimization using MCADS

Two practical application examples of MCADS for the shape optimization of thermo-stressed continuum structures, dealing with a coupling of stress field and temperature field, are presented in this section.

Example 1. The shape optimization of train wheels.

The problem has been mentioned in section 2 and 3. The load-bearing of each wheel of a new heavy train is designed to increase from 21 tone to 25 tone and the radius of wheel is keep, then temperature caused by braking increases with load-bearing. This thermo-load is main load case and produces concentration of thermal stress. The optimization objective is to reduce this concentration of thermal stress by means of shape modification of wheel cross-section. The design model, analysis model and temperature distribution of initial design are shown in Fig.1-3. The thermo-load variation with respect to shape changes has been considered in sensitivity analysis via semi-analytic method. The optimum design is obtained after 5 iterations of optimization, and the maximum thermal stress of wheel has been reduced from 91.8 to 67.9 under 25 tone load-bearing. This value is even lower than the maximum thermal stress 77.6 of old design with 21 tone load-bearing. The distribution of thermal stress (Mises stress) of old design and optimal design under 25 tone load-bearing are shown in Fig.4 and 5, respectively.

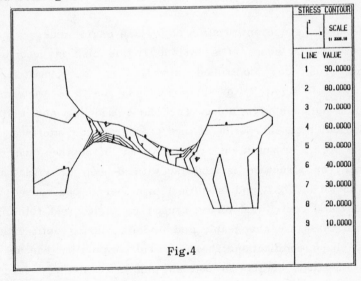

STRESS CONTOUR	
SCALE	1: 2335.10

LINE	VALUE
1	90.0000
2	80.0000
3	70.0000
4	60.0000
5	50.0000
6	40.0000
7	30.0000
8	20.0000
9	10.0000

Fig.4

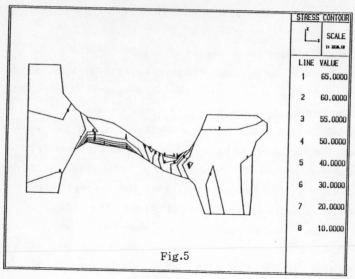

STRESS CONTOUR	
SCALE	
LINE	VALUE
1	65.0000
2	60.0000
3	55.0000
4	50.0000
5	40.0000
6	30.0000
7	20.0000
8	10.0000

Fig.5

The temperature field is assumed not change during shape optimization. Reanalysis of temperature field and thermal stress field have been carried out for optimal design to check its safety. The results show that the change of temperature field is much smaller than the change of thermal stress, and the maximum thermal stress induced by new temperature field is smaller than that induced by old temperature field. This means that the shape optimization of thermo-stressed structures with temperature unchanged field gives a safety design.

Example 2. The shape optimization of turbine engine disk.

The cross-section shape of a turbine engine disk is described with lines and arcs, and controlled directly by 25 geometric design parameters $(L_i, H_i, R_i, A_i, B_i)$ as shown in Fig.6. The 18 parameters of them are selected as design variables. The shape modeling and modification are carried out by a subroutine of user's programming interface. And the centrifugal forces computation are calculated in another user defined subroutine. The structure is modeled with 8-node axisymmetric brick elements shown in Fig.7, and is subject to a thermo-load caused by high temperature and centrifugal forces caused by high-speed rotating. Both of two load cases are changeable and updated during optimization with structural shape modification. The numerical computation has also shown that the variations of thermo-load as well as centrifugal forces with respect to changes of boundary shape play an important role in the sensitivity analysis

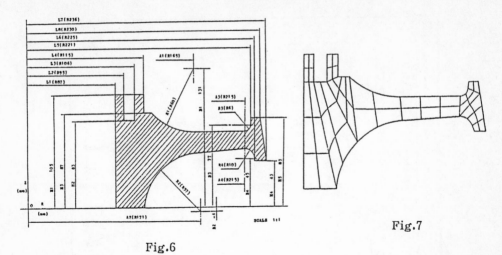

Fig.6

Fig.7

A bi-objective of the shape optimization is to reduce both weight and stress level of disk structure. Various design criteria on structural strength and usage life of disk with consideration of all facts of fatigue, fracture, creep and manufacture have been studied, and expressed in forms of stress constraints. Thus the structural design model is very close to the real state of turbine engine disk and meets requirements of the ENSIP (Engine Structural Integrity Program) of USA.

A new decomposition algorithm of optimization has been proposed in solving this problem. The algorithm makes a decomposition in the space of design variable, i.e. divides design variables into several groups. The original optimization problem is reduced to some sub-problems with fewer variables. One sub-problem with only one group of chargeable variables is solved in each iteration. The group division of variables considers different influence of variables to the design and convenience of design and manufacture. The optimization procedure shows that this decomposition algorithm is useful to large-scale optimization problems.

The result of optimum shape design is that the structural weight and maximum Mises stress of the turbine engine disk has been reduced 18.52% and 14.91% respectively. The boundary shapes and the distributions of Mises stress of original design and optimal design have been shown in Fig.8-9 and color photographs.

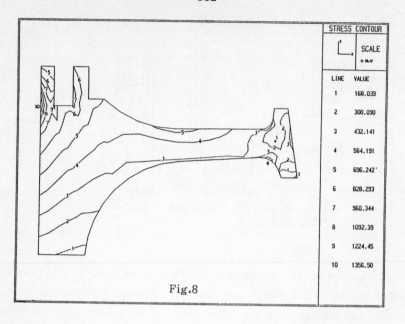

Fig.8

STRESS CONTOUR	
LINE	VALUE
1	168.039
2	300.090
3	432.141
4	564.191
5	696.242
6	828.293
7	960.344
8	1092.39
9	1224.45
10	1356.50

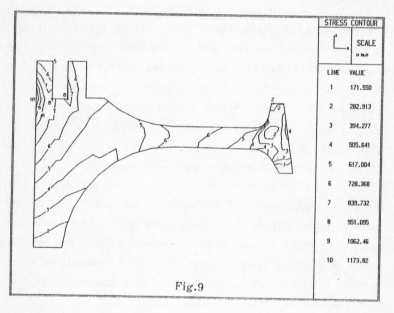

Fig.9

STRESS CONTOUR	
LINE	VALUE
1	171.550
2	282.913
3	394.277
4	505.641
5	617.004
6	728.368
7	839.732
8	951.095
9	1062.46
10	1173.82

REFERENCE

[1] Cheng, G.D. & Liu, Y.W. 1987: A new computational scheme for sensitivity analysis, J. Eng. Opt. 12, 219–235

[2] Haftka, R.T. & Adelman, H.M. 1989: Recent developments in structural sensitivity analysis, J. Stru. Opt. 1, 137–151

[3] Cheng, G.D.; Gu, Y.X. & Zhou, Y.Y. 1989: Accuracy of semi-analytic sensitivity analysis, J. Finite Elements in Analysis and Design, 6, 113–128

[4] Gu, Y.X. & Cheng, G.D. 1990: Structural shape optimization integrated with CAD environment, J. Stru. Opt. 2, 23–28

[5] Cheng, G.D.; Gu, Y.X. & Wang, X.C. 1990: Improvement of semi-analytic sensitivity analysis and MCADS, to appear in Conf. Engineering Optimization in Design Processing.

[6] Gu, Y.X. & Cheng, G.D. 1990: Microcomputer-based system MCADS integrated with FEM, optimization and CAD, J. Computational Mechanics and Applications, 1, 71–81

[7] Bennett, J.A.; Botkin, M.E. (eds) 1986: The optimum shape, automated structural design, New York: Plenum Press

Author Index

H. Baier	79, 135	P. Pedersen	91
M.P. Bendsøe	27	P.U. Post	99
K. Bethge	325		
J. Blachut	155	D. Radaj	181
B. Bochenek	259	F.G. Rammerstorfer	333
H.J. Böhm	333	J. Rasmussen	27, 193
A. Borkowski	301	T.J. Reiter	333
J. Bühlmeier	107	H.C. Rodrigues	27
S. Burkhardt	15	M. Rovati	127
		G. Rozvany	59
G. Cheng	211, 341	P. Sas	249
		R. Schirrmacher	35
M. Danicka	301	Ch. Seeßelberg	135
		B. Specht	79
D. Erb	15	W. Sprangers	71
H. Eschenauer	1, 145		
		A. Taliercio	127
A. Gajewski	275	P. Tiefenthaler	201
J.T. Gierlinski	309	V.V. Toropov	225
Y. Gu	211, 341		
		Th. Vietor	145
L. Harzheim	173		
G. Helwig	135	M. Weck	71
W. Heylen	249		
J. Holnicki-Szulc	309	W. Xicheng	211
H. Huber-Betzer	293		
		Sh. Zhang	181
U. Jehle	35	M. Zhou	59
S. Jozviak	301	M. Zyczkowski	259
S. Kibsgaard	233		
P. Knödel	317		
V.V. Kobelev	165		
B.W. Kooi	283		
G. Krzesinski	51		
S. Lammens	249		
V.P. Malkow	225		
C. Mattheck	15, 173, 293		
H.P. Mlejnek	35		
H. Moldenhauer	325		
G. Müller	201		
H.H. Müller-Slany	267		
N. Olhoff	193		
A. Osyczka	241		

Lecture Notes in Engineering

Edited by C.A. Brebbia and S.A. Orszag

Vol. 40: R. Borghi, S. N. B. Murhty (Eds.)
Turbulent Reactive Flows
VIII, 950 pages. 1989

Vol. 41: W. J. Lick
Difference Equations
from Differential Equations
X, 282 pages. 1989

Vol. 42: H. A. Eschenauer, G. Thierauf (Eds.)
Discretization Methods
and Structural Optimization –
Procedures and Applications
Proceedings of a GAMM-Seminar
October 5-7, 1988, Siegen, FRG
XV, 360 pages. 1989

Vol. 43: C. C. Chao, S. A. Orszag, W. Shyy (Eds.)
Recent Advances in Computational
Fluid Dynamics
Proceedings of the US/ROC (Taiwan) Joint
Workshop in Recent Advances in
Computational Fluid Dynamics
V, 529 pages. 1989

Vol. 44: R. S. Edgar
Field Analysis and
Potential Theory
XII, 696 pages. 1989

Vol. 45: M. Gad-el-Hak (Ed.)
Advances in Fluid Mechanics
Measurements
VII, 606 pages. 1989

Vol. 46: M. Gad-el-Hak (Ed.)
Frontiers in Experimental
Fluid Mechanics
VI, 532 pages. 1989

Vol. 47: H. W. Bergmann (Ed.)
Optimization: Methods and Applications,
Possibilities and Limitations
Proceedings of an International Seminar
Organized by Deutsche Forschungsanstalt für
Luft- und Raumfahrt (DLR), Bonn, June 1989
IV, 155 pages. 1989

Vol. 48: P. Thoft-Christensen (Ed.)
Reliability and Optimization
of Structural Systems '88
Proceedings of the 2nd IFIP WG 7.5 Conference
London, UK, September 26-28, 1988
VII, 434 pages. 1989

Vol. 49: J. P. Boyd
Chebyshev & Fourier Spectral Methods
XVI, 798 pages. 1989

Vol. 50: L. Chibani
Optimum Design of Structures
VIII, 154 pages. 1989

Vol. 51: G. Karami
A Boundary Element Method for
Two-Dimensional Contact Problems
VII, 243 pages. 1989

Vol. 52: Y. S. Jiang
Slope Analysis Using
Boundary Elements
IV, 176 pages. 1989

Vol. 53: A. S. Jovanovic,
K. F. Kussmaul, A. C. Lucia,
P. P. Bonissone (Eds.)
Expert Systems in Structural
Safety Assessment
X, 493 pages. 1989

Vol. 54: T. J. Mueller (Ed.)
Low Reynolds Number
Aerodynamics
V, 446 pages. 1989

Vol. 55: K. Kitagawa
Boundary Element Analysis
of Viscous Flow
VII, 136 pages. 1990

Vol. 56: A. A. Aldama
Filtering Techniques for
Turbulent Flow Simulation
VIII, 397 pages. 1990

Vol. 57: M. G. Donley, P. D. Spanos
Dynamic Analysis of Non-Linear
Structures by the Method of
Statistical Quadratization
VII, 186 pages. 1990

Vol. 58: S. Naomis, P. C. M. Lau
Computational Tensor Analysis
of Shell Structures
XII, 304 pages. 1990

For information about Vols. 1-39 please contact your bookseller or Springer-Verlag.